Geology of the country around King's Lynn and The Wash

The King's Lynn district is one of great scenic contrast, including as it does the wild and remote tidal flats around The Wash, the almost treeless plain of the Lincolnshire and Norfolk marshes, and the rolling heathland, arable clay lands and chalk downland of north-west Norfolk. Each of these areas, although much affected by man's activities during the past 2000 years, owes its essential character to the underlying geology.

This memoir traces the geological history of the district from the muddy seas and volcanicity of the early Palaeozoic of 450 million years ago, via the deserts of the Trias and the shallow tropical and subtropical seas of the Jurassic and Cretaceous, to the last glaciation 18 000 years ago, when much of the district was covered by ice. Finally, it records the history of the period after the retreat of the ice when temperate climates returned and the broad hollow that was to become Fenland and The Wash became progressively filled with freshwater peats, estuarine muds and marine sands as sea level rose in response to the melting of the ice.

The greater part of the district is taken up by The Wash and the marshlands reclaimed from it in the last 1000 years. Both areas are important to the local and national economy, but are in potential conflict, the farmlands seemingly being able to expand only at the expense of the broad intertidal flats and sand banks that are important feeding and breeding grounds for fish, birds and seals.

The oldest rocks at surface in the district are the richly fossiliferous mudstones of the Jurassic Kimmeridge Clay which come to the surface along the eastern margin of Fenland between King's Lynn and Setchey. These readily erodible rocks, together with similar soft mudstones which underlie them, have given rise to the broad tract of low ground now occupied by Fenland and The Wash. The Jurassic clays are overlain by a complex sequence of late Jurassic and Cretaceous sands and clays, including the Sandringham Sands, Dersingham Beds,

Roach and Carstone, which were deposited close to land in a shallow, shelf sea. Elsewhere in Britain, this time interval is represented by either freshwater deposits or as yet poorly described marine deposits, and the west Norfolk sequence is of international scientific interest and importance. The return of widespread marine conditions in the mid Cretaceous gave rise to the Gault, Red Chalk and Lower Chalk: all three formations are unusually thin and laterally variable in the district because of the continuing influence of the underlying geological structure.

The present-day landscape attained much of its form during the Quaternary ice ages when a thick ice sheet covered the whole of the district on at least one occasion; The Wash and the Hunstanton area were covered by a second ice sheet that retreated only 18 000 years ago. The sequence of deposits and erosional features formed by the ice and associated meltwater torrents, wind action and permafrost, together with the marine deposits that accompanied the complementary rise and fall of sea levels during this period, are fully described. Finally, the distribution of the modern sediments and the processes that formed them within The Wash and beneath Fenland are described, and their importance to the progressive reclamation of the marshland area is discussed.

The district, although largely agricultural, remains an important source of silica sand. Brick clay, chalk and building stone were extensively worked in the past, and the Romans worked small amounts of iron ore. The King's Lynn area was explored as a possible source of oil between 1918 and 1921.

This memoir and the geological maps that accompany it provide the geological data necessary to understand the long-term processes that have shaped the present-day landscape, and which are an essential prerequisite to the prediction of possible future changes, whether naturally induced by global changes of climate or man-made in response to local works.

Cover photograph
The rusty brown Carstone, Red Chalk and white Chalk form spectacular cliffs at Hunstanton. The full thickness of the Carstone is exposed in the cliffs and foreshore, and the full thickness of the Red Chalk in the cliffs: this is the type section for both formations. Photo: Alan Hutchinson, King's Lynn.

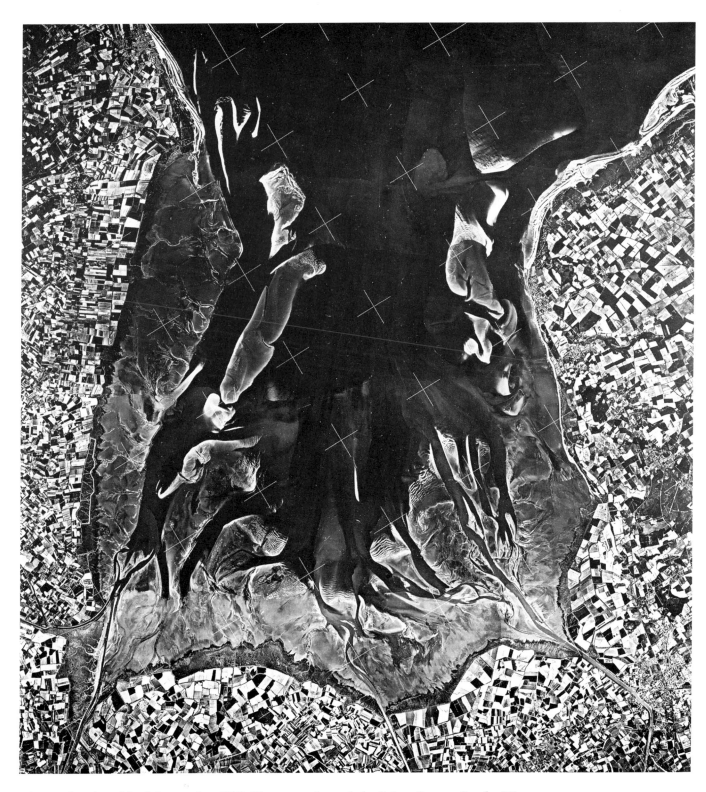

The Wash at low tide, 5 September 1971. Photomosaic made by Fairey Surveys for the Water Resources Board as part of a detailed photogrammetric survey that was used as the base map for the intertidal and adjacent areas during the Wash Water Storage Feasibility Study. The drying banks are close to their maximum exposure.

R W GALLOIS

Geology of the country around King's Lynn and The Wash

CONTRIBUTORS

Biotratigraphy
B M Cox
A A Morter
C J Wood

Geophysics
J D Cornwell
S F Kimbell

Memoir for 1:50 000 geological sheet 145 and part of 129 (England and Wales)

LONDON: HMSO   1994

© *NERC copyright 1994*
First published 1994

ISBN 011 884495 4

Bibliographical reference
GALLOIS, R W. 1994.   Geology of the country around King's Lynn and The Wash.   *Memoir of the British Geological Survey*, sheet 145 and part of 129. (England and Wales).

Author
R W Gallois, BSc, DIC, PhD, CEng, CGeol, FIMM
British Geological Survey, Exeter

Contributors
J D Cornwell, BSc, PhD
Beris M Cox, BSc, PhD
Sarah F Kimbell, BSc
British Geological Survey, Keyworth

A A Morter, BSc
C J Wood, BSc
formerly British Geological Survey

Other publications of the Survey dealing with this and adjoining districts

BOOKS

British Regional Geology
East Anglia and adjoining areas (4th edition), 1961
Eastern England from the Tees to The Wash (2nd edition), 1980

Memoirs
Geology of the country around Grimsby and Patrington, Sheets 90, 91 and 81, 82, 1994
Geology of the Peterborough district, Sheet 158, 1989
Geology of the country around Norwich, Sheet 161, 1989
Geology of the country around Ely, Sheet 173, 1988
Geology of the country around Huntingdon and Biggleswade, Sheets 187 and 204, 1965
Geology of the country around Cambridge, Sheet 188, 1969
Geology of the country around Bury St Edmunds, Sheet 189, 1990

Offshore Reports
The geology of the southern North Sea, 1992

MAPS

1:625 000
Solid geology (south sheet)
Quaternary geology (south sheet)
Aeromagnetic anomaly (south sheet)
Bouguer gravity anomaly (south sheet)

1:125 000
Hydrogeological map of northern East Anglia (two sheets), 1976

1:50 000 (Solid and Drift)
Sheet 81   (Patrington), 1991
Sheet 90   (Grimsby), 1990
Sheet 129/145   (King's Lynn and The Wash), 1978
Sheet 144   (Spalding), 1992
Sheet 158   (Peterborough), 1985
Sheet 161   (Norwich), 1975
Sheet 173   (Ely), 1980
Sheet 187   (Huntingdon), 1975
Sheet 188   (Cambridge), 1981
Sheet 189   (Bury St Edmunds), 1982

1:250 000
Sheet 52N 00   East Anglia, solid geology, 1986
Sheet 52N 00   East Anglia, Quaternary 1992
Sheet 52N 00   East Anglia, aeromagnetic anomaly, 1982
Sheet 52N 00   East Anglia, Bouguer gravity anomaly, 1981

1:253 440 (four miles to one inch) (Solid and Drift)
Sheet 12   Louth, Peterborough, Norwich and Yarmouth, 1986

Printed in the UK for HMSO
Dd 292038   C8   01/95

CONTENTS

FIGURES

PLATES

TABLES

NOTES

The word 'district' is used in this memoir to mean the area represented by the 1:50 000 Geological Sheet 145 and part of 129 (King's Lynn and The Wash). The memoir describes the geology shown on the published (1978) map, but also includes the results of more recent stratigraphical work. Where more recent work has resulted in a revised classification, the memoir follows that shown on the map and the new classification is describes in a footnote.

National Grid references are given in square brackets throughout. They all lie within the 100 km square TF (or 53) except where otherwise indicated.

Numbers preceded by A refer to photographs in the Survey's collection.

Numbers preceded by E refer to thin sections in the BGS Sliced Rock Collection (England and Wales).

SIX-INCH MAPS

The following is a list of the six-inch National Grid geological maps included, wholly or in part in 1 to 50 000 Sheets 145, 129 (southern part) and 146 (Lower Cretaceous outcrop), with the initials of their surveyors and dates of survey. The names of the officers are as follows: F C Cox, R W Gallois and B Young.

Dyeline copies of these six-inch maps may be purchased from the Keyworth Office of BGS.

TF 41 NW	Tydd St Mary	FCC	1970
TF 41 NE	Tydd Gote	FCC	1970
TF 42 SW	Long Sutton	FCC	1970
TF 42 SE	Sutton Bridge	FCC	1970
TF 42 NW	Lutton	RWG	1971
TF 42 NE	Gedney Drove End	FCC	1970
TF 43 SW	Holbeach St Matthew	RWG	1970
TF 43 SE	Dawsmere	RWG	1971
TF 51 SE	Wiggenhall	BY	1970
TF 51 NW	Walpole-Terrington	BY	1970
TF 51 NE	Tilney-Eau Brink	BY	1970
TF 52 SW	Terrington-Wingland	BY	1970
TF 52 SE	Terrington-Clenchwarton	BY	1970
TF 52 NW	Terrington Marsh	BY	1970
TF 52 NE	Admiralty Point	BY	1970
TF 61 SW	Setchey	BY	1970
TF 61 SE	Blackborough End	BY	1971
TF 61 NW	West Winch	RWG; BY	1970
TF 61 NE	Middleton-Bawsey	RWG; BY	1970–71
TF 62 SW	King's Lynn	RWG	1970
TF 62 SE	Roydon Common	RWG	1970–71
TF 62 NW	Wootton Marsh	RWG	1970
TF 62 NE	Sandringham	RWG	1965 & 1970
TF 63 SW	Snettisham Scalp	RWG	1970
TF 63 SE	Dersingham-Ingoldisthorpe	RWG	1965
TF 63 NE	Heacham	RWG	1965
TF 64 SE	Hunstanton	RWG	1965
TF 71 NW	Gayton	RWG; BY	1971
TF 72 SW	Grimston-Congham	RWG	1966
TF 72 NW	Hillington	RWG	1965–66
TF 73 SW	Shernborne	RWG	1965
TF 73 NW	Sedgeford	RWG	1965
TF 74 SW	Ringstead	RWG	1965

PREFACE

This memoir describes an area which at first sight might seem geologically unrewarding because of its low relief and the paucity of rock outcrop, but which has yielded a remarkable amount of geological information in recent years. The initial impetus to survey the district in detail came from a single drainage channel which yielded fossils that were known only from strata close to the Jurassic–Cretaceous boundary in Russia. Later finds confirmed the importance of the west Norfolk sequence which now, following the geological survey and related stratigraphical studies described here, is the best documented late Jurassic to early Cretaceous succession in north-west Europe, and hence is of international importance for correlation purposes.

The decision to explore the possibility of building a water-storage barrage in The Wash, either across its mouth or as a series of large bunds (large banks for retaining fresh water) adjacent to its southern shore, provided a particular economic relevance to the survey and, additionally, yielded much new geological data. In the onshore area, extensive drilling to explore the Upper Jurassic clays beneath the Recent deposits of a large tract of Fenland enabled complex stratigraphical sequences to be identified for the West Walton Beds, Ampthill Clay and Kimmeridge Clay. These have subsequently become the accepted standards for Upper Jurassic sequences throughout Britain. The identification of organic-rich horizons (oil shales) in the Kimmeridge Clay during the study was important in that such horizons might give rise to explosive atmospheres in the aqueduct tunnels. Subsequently this study led to a more comprehensive study of the oil shales as a possible fuel; it also further confirmed the widespread validity of the stratigraphical scheme devised for the Jurassic rocks.

In the offshore area, detailed geological and geophysical surveys to determine the 3-dimensional distribution of the 'mobile' Recent sediments, and their relationship to the underlying Jurassic and Pleistocene bedrock, constitute the most detailed study of such a large area yet undertaken in UK waters. Much of the King's Lynn district has been reclaimed from the sea in historical times and, with the possibility of a greater frequency of storms and rising sea levels, it remains vulnerable to the sea. An understanding of the history and sedimentary processes which have built up the complex Recent sequence that underlies Fenland, and the processes that continue daily to redistribute the sediments within The Wash, is therefore essential if the present balance between the reclaimed lands and the important offshore environments are to be maintained.

The district is predominantly agricultural, but it is an important source of silica sand; brick clay, building stone and chalk have been extensively worked in the past. Iron ore was worked by the Romans, and the King's Lynn area was extensively explored for oil between 1918 and 1921.

This memoir and the geological maps that accompany it set out the geological data necessary to understand the long-term processes that have shaped the present-day landscape and its use. In doing so they provide an important contribution to the continuing debate as to how to maintain the best balance between urban, industrial, agricultural and leisure uses within the district, and how best to respond to both local man-made and global climatic changes that may have a profound effect on the delicate balance of environments in the region.

Peter J Cook DSc
Director

British Geological Survey
Keyworth
Nottingham NG12 5GG

17 May 1994

ACKNOWLEDGEMENTS

In poorly exposed areas of soft rock such as the King's Lynn district, the information obtained from boreholes and the few good exposures is especially valuable. A large number of percussion and cored boreholes were drilled for the Wash Water Storage Scheme, and the assistance is gratefully acknowledged of staff of the Water Resources Board and of their consulting engineers, Messrs Binnie and Partners, who supervised the drilling programme. Especial thanks are due to Mr L E Taylor (WRB) and Mr D H Cowie and Mr T D Ruxton of Binnie and Partners. In addition, four cored boreholes were drilled as part of the geological survey of the district by the Survey's rig under the supervision of Mr S P Thorley. Much has been published about the Red Chalk at Hunstanton. However, the outstanding contribution to our knowledge of its stratigraphy has been that of the late Mr Hamon Le Strange of Hunstanton, whose dedicated fossil collecting over a period of more than 50 years has provided much of the data on which the modern classification is based.

The thanks of the British Geological Survey are due to the many quarry owners, landowners and tenants in the district for their ready co-operation in facilitating access to their properties. The survey of the solid geology was commenced as a contribution to detailed stratigraphical studies being carried out by Dr R Casey, and the author is grateful for many fruitful discussions with him. Many of the sedimentological results described in this memoir are derived from a PhD study jointly supervised by Dr G Evans (Imperial College) and Dr P A Sabine (BGS), and their expert help and encouragement is gratefully acknowledged.

The memoir has been written mainly by Dr R W Gallois. It incorporates information from Mr B Young and specialist contributions on the concealed rocks and their structure by Dr J D Cornwell and Mrs S F Kimbell, and on the stratigraphical palaeontology of the Jurassic by Dr B M Cox and of the Cretaceous by Mr A A Morter and Mr C J Wood. The memoir has been compiled by Dr Gallois and edited by Dr R G Thurrell.

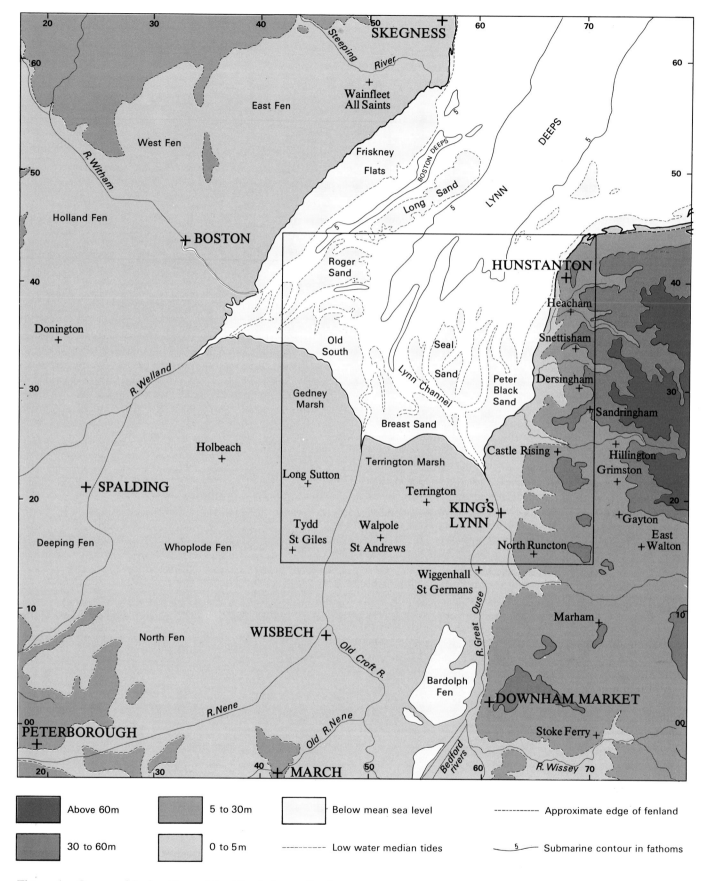

Figure 1 Geographical setting of the King's Lynn district.

ONE

Introduction

GEOGRAPHICAL SETTING

The King's Lynn district contains some of the most attractive and varied scenery in eastern England, ranging from broad intertidal flats and extensive reclaimed marshes to sandy heathlands, arable clay lands and chalk downlands. The district includes most of The Wash, much of northern Fenland (parts of the marshlands of Lincolnshire and Norfolk), and a strip of Norfolk upland that runs from Hunstanton to the River Nar and which forms the eastern limit of Fenland and The Wash (Figure 1).

The scenery of the district can be divided into two distinct types, both related to the geology (Figure 2). The upland area is underlain by Upper Jurassic and Cretaceous rocks which give rise to sandy heathlands and woodlands interspersed with more clayey, mostly arable land. The Chalk forms a low but prominent escarpment which marks the eastern limit of the district. In contrast, the Recent sediments of Fenland produce an almost flat plain, intersected by a complex pattern of natural creeks and artificial drains and embanked rivers, which, in the more recently reclaimed areas, is treeless, bleak and very sparsely populated.

The maximum topographical relief within the upland area is about 50 m, with the lowest points at about 3 m above Ordnance Datum (OD) in the valleys of the Babingley and Gaywood rivers, and the higher points, over 50 m above OD, on the Lower Cretaceous at Rising Lodge near King's Lynn and on the Chalk escarpment near Snettisham (Figure 1).

The eastern part of the district is drained by the River Great Ouse and its tributaries the Gaywood River, the Middleton Stop Drain and the River Nar, together with the Babingley River, the Ingol and the Heacham River. These last three drain directly into The Wash but were also tributaries of the Great Ouse in Pleistocene times. The western part of the district is drained by the River Nene and a network of artifical drains.

All the major population centres in the district except Hunstanton and King's Lynn were well established by the time of the Domesday Survey. Some of those in the upland area are of much greater antiquity and their foundations were probably laid before the Roman conquest. In the Fenland area, the older villages date from Saxon times, having grown up in the rich grazing lands reclaimed from the salt marshes by the construction of the 'Roman Bank' (see Chapter Seven for details). Romano-British remains are present, but they probably derive from small, itinerant communities. A large part of the Fenland area has been reclaimed from The Wash since the 17th century.

The largest town in the district, King's Lynn (population c. 38 000) lies on the eastern edge of Fenland where it occupies a strategic position at the mouth of the River Great Ouse (Plate 1). It was founded in the 11th century as a 'new town' at a point where a bend in the river brings the east bank closest to the upland and near to the much older village and bishopric of Gaywood. The second largest settlement, Hunstanton (population c. 4000), is in this upland area, as are the villages of Heacham, Snettisham, Dersingham, Hillington, Castle Rising, Gayton, Middleton, North Runcton and West Winch. The major settlements within the Fenland part of the district are the small town of Long Sutton and the villages of Lutton, Sutton Bridge, Tydd Gote, West Walton, the Walpoles, the Terringtons, the Tilneys, Clenchwarton and West (or Old) Lynn.

The district is largely agricultural, but with the port, food processing plants and light engineering in the King's Lynn area, and tourism in the Hunstanton area, making important contributions to the local economy. In Victorian and earlier times it was almost entirely agricultural. A variety of minerals have been worked in the past, and one of these, the glass and moulding sands in the Sandringham Sands at Leziate near King's Lynn, continues to be the basis for an important local industry. Gravel is worked for ballast near Hillington, chalk is worked for agricultural purposes at Heacham, Gayton and Hillington, and Carstone for building purposes at Snettisham, all on a small scale. All these materials have been worked at other localities in the district in the past, also on a small scale, together with brick clays dug from the Kimmeridge Clay, Snettisham Clay, Gault and Terrington Beds. Sandstones in the Dersingham Beds, River Terrace Gravels, Boulder Clay and ferruginous pan formed from the Sandringham Sands, have been used locally from time to time for building and construction purposes. There is evidence that the Romans smelted iron in the district from ore dug from the Sandringham Sands. In the 1920s oil shales in the Kimmeridge Clay were extensively explored in the district as a possible oil source, but no commercial working was undertaken.

GEOLOGICAL SEQUENCE

The succession proved in the district is summarised on the inside front cover. The formations below the Kimmeridge Clay do not crop out (Figure 2), but have been proved in deep boreholes at Hunstanton and Wiggenhall. Large parts of the concealed Upper Jurassic sequence have also been proved in boreholes at Gedney Drove End and North Wootton.

Little is known about the structure or stratigraphy of the pre-Mesozoic rocks, but in the Hunstanton and Wiggenhall boreholes metasedimentary rocks of presumed Precambrian age and of probable early Palaeozoic age, respectively, have been proved to underlie the Trias. Geo-

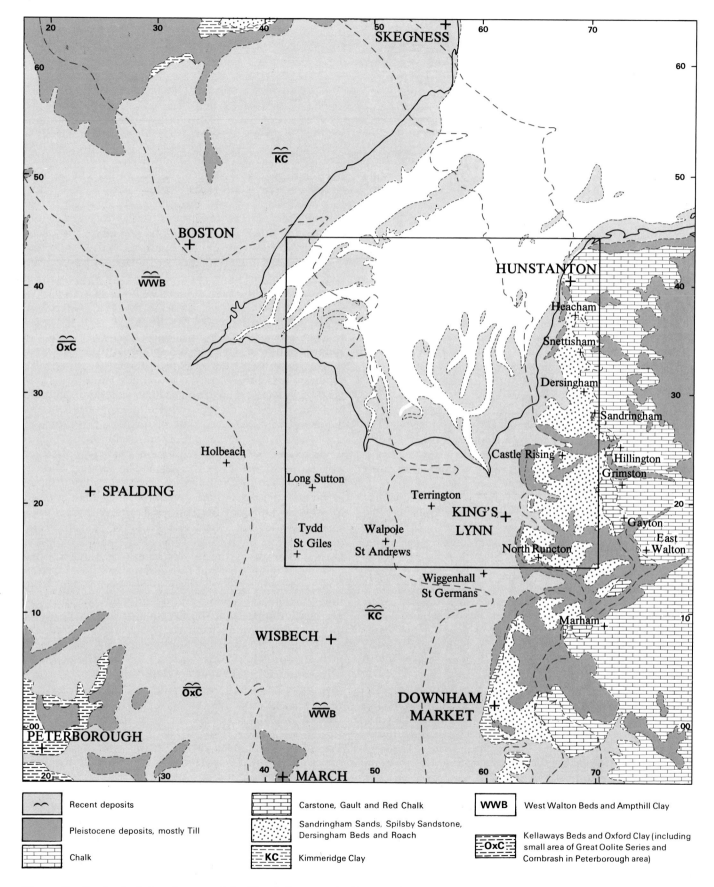

Figure 2   Geological sketch map of the King's Lynn and adjacent districts.

physical evidence suggests that the deep structure of the district is complex and that, in addition to Precambrian and Lower Palaeozoic rocks, Upper Palaeozoic sedimentary rocks and Palaeozoic intrusive igneous rocks may also be present locally. The district is everywhere underlain by Triassic rocks and by a relatively complete Jurassic sequence. The thicknesses of the Mesozoic rocks (inside front cover) are the predicted ranges for the district; they are based on the thicknesses proved in boreholes in the district combined with estimates based on regional trends.

GEOLOGICAL HISTORY

The early geological history of East Anglia is poorly known because rocks older than the late Jurassic are everywhere concealed by younger formations which have rarely been penetrated by boreholes. It seems reasonable, however, to assume that the sequence of sedimentation and earth movements is similar to that of the East Midlands region where these older rocks are exposed. Folded Precambrian and/or early Palaeozoic rocks underlie the Mesozoic rocks beneath most of East Anglia, including the present district, at relatively shallow depths. Metamorphic rocks of presumed Precambrian age have been proved at Hunstanton and in the adjacent districts at North Creake in Norfolk and Glinton in Northants. Indurated mudstones of probable early Palaeozoic age have been proved in the district at Wiggenhall and quartzites of presumed similar age nearby at Spalding and Wisbech.

The early history of the Precambrian rocks beneath the district is presumed to be similar to that surmised for the Precambrian of the Midlands. The Palaeozoic rocks of East Anglia appear to have been folded and lithified during the later stages of the Caledonian earth-movements to form a stable massif, the London Platform, which strongly influenced the pattern and type of sedimentation in the region throughout much of the Mesozoic. This massif probably gave rise to a land area throughout much of the Trias, Jurassic and early Cretaceous. It was finally overstepped in mid Cretaceous times by the successive transgressions of the Gault and Chalk seas.

The oldest Mesozoic rocks proved in the district are Triassic mudstones, sandstones and conglomerates that rest with angular unconformity on the Precambrian and Palaeozoic. They were probably deposited in ephemeral streams and playa lakes that were formed by flash floods in a hot desert. These floods reduced a rugged topography of older rocks to one of low relief; this was further eroded and finally transgressed by the warm, relatively shallow seas of the Jurassic. The earliest Jurassic sea moved southwards across the district and converted it into a relatively shallow, broad marine shelf in which the soft muds of the Lias, with their rich marine fauna of ammonites, bivalves, brachiopods, crinoids, foraminifera and the remains of marine reptiles, were deposited. To the south, the London Platform probably formed a heavily vegetated, low-relief island with a humid subtropical or tropical climate, that covered much of what is now central East Anglia.

Minor earth movements occurred towards the end of the early Jurassic and caused uplift and erosion that removed some of the younger Liassic sediments. The London Platform and other sediment source areas became rejuvenated, with the result that the argillaceous Lias was replaced by predominantly arenaceous rocks of the mid-Jurassic Estuarine 'Series'. These latter sediments are thought to have been laid down in shallow, fresh- and brackish-marine environments on a broad coastal plain traversed by numerous small rivers and tidal creeks that drained through lagoons and swamps into a shallow shelf sea. An embayment of the sea reached the southern part of the district on one occasion and calcareous oolites, part of the Lincolnshire Limestone, were deposited in that area. This coastal plain was subsequently inundated by a shallow, clear sea in which the shelly, sandy limestones of the Cornbrash were laid down.

Progressive deepening of the sea, combined with denudation of the land areas that supplied the sediment, gave rise to the sandy clays of the late Jurassic Kellaways Beds and then to the almost wholly argillaceous Oxford Clay, West Walton Beds, Ampthill Clay and Kimmeridge Clay. At times, the sea adjacent to the London Platform was sufficiently warm, clear and shallow to support coral reefs, and the currents sufficiently strong to produce shoals of calcareous oolite. The nearest of these shoals lay about 40 km south of the present district. Away from the London Platform the late Jurassic formations contain, at some stratigraphical levels, sufficient debris derived from pelagic marine animals and plants to suggest that the sea was at times clear and relatively free from land-derived mud. The late Jurassic formations contain a rich fauna dominated by ammonites and bivalves, with subordinate numbers of brachiopods, echinoids, crinoids, serpulids, foraminifera, ostracods, crustaceans, fish and aquatic reptiles. The nearest land, the London Platform, was probably of low relief and covered by tropical forests of ferns, horsetails, cycads, conifers and ginkgos, and drained by sluggish rivers.

The district was again affected by earth movements in the late Jurassic. The land areas were rejuvenated, with some consequential erosion of the earlier Jurassic rocks, and the district was transformed into an area of predominantly sandy sedimentation. These earth movements were the first of several discrete phases that began shortly after the deposition of the Kimmeridge Clay and continued into the Cretaceous. They were part of widespread tectonic activity associated with the separation of the European and North American continental plates and the opening of the North Atlantic.

The history of the latest Jurassic and early Cretaceous in the district is one of repeated transgression and regression which laid down alternations of sand and clay. Although the sequence Sandringham Sands, Dersingham Beds, Roach and Carstone is entirely marine, and although there is no direct evidence of a shoreline within the district at that time, the lithologies and faunas at a number of levels suggest deposition in very shallow water close to land. The youngest Jurassic rocks in the district, the basal beds of the Sandringham Sands, are marine sands that were probably deposited in shallow water. The unconformity at their base is not obviously angular, but they overstep the Kimmeridge Clay in a south-easterly direction and their basal pebble bed contains phosphatised fragments of ammonites derived from the higher parts of the Kimmeridge Clay. The Sandringham Sands were themselves overstepped in the same southerly direction, towards the London Platform, by the Dersingham Beds.

In the early Cretaceous, the London Platform was probably fringed by a broad coastal plane composed of Cretaceous rocks that were at times submerged and at times emergent and subject to erosion. Renewed earth movements in the mid Cretaceous resulted in considerable erosion of the pre-Jurassic rocks of the London Platform and the surrounding Jurassic and Cretaceous sediments, with the result that the marine pebbly sands of the Carstone rest unconformably on strata ranging in age from mid-Cretaceous in the Hunstanton area to early Cretaceous in the Blackborough area, Jurassic in the West Dereham area and Palaeozoic in the Bury St Edmunds area.

The London Platform continued to be eroded during the period of deposition of the Carstone and became so degraded that it was subsequently overstepped by the clays of the Gault. These clays extend northwards into the southern part of the district where they pass, in the Sandringham area, into the pink and red limestones of the Red Chalk. The Gault contains a rich fauna of mostly thin- and smooth-shelled bivalves and ammonites indicative of deposition in quiet water. Its benthonic faunas became less abundant and diverse with time so that the youngest part of the formation and the succeeding Chalk are composed largely of the calcareous parts of coccoliths and other pelagic animals and algae. The Red Chalk contains an abundant and diverse fauna that includes common echinoderms, polyzoans, brachiopods, bivalves, ammonites and belemnites that indicate condensed deposition in a warm shallow sea.

The purity of the Chalk limestones suggests that the nearest source of silicilastic material was far distant. The lithology and fauna of the Chalk indicate deposition in warm, clear water at depths which were probably greater than at any other time during the Jurassic or Cretaceous in the region. Although only the lowest part of the Chalk is now exposed in the district, it is likely that the formation was once fully represented.

Earth movements in the late Cretaceous and early Eocene caused widespread uplift and erosion in southern England with the result that the Chalk sea was converted to land with lakes and estuaries and fringed by shallow marine shelves in which muddy and sandy sediments were deposited. The gentle eastward dip of the Mesozoic rocks of the district was initiated at that time and was accentuated by later tectonic activity in the Tertiary. No Tertiary sedimentary rocks have been recorded in the district and it seems likely that for much of that period the area was being eroded. Deposition may have recommenced during the Pliocene, but any sediments are likely to have been loose pebbly sands and have been removed by erosion during the Quaternary.

The Quaternary history of the district is complex, and is represented by only a fragmentary record. There is evidence in the Pleistocene deposits for at least two glacial phases, when ice sheets covered all or part of the district, and for two or more phases of temperate climate. During the glacial periods, large volumes of the Jurassic and Cretaceous rocks were eroded and incorporated in boulder clay and gravelly sand, some of which was deposited locally on the lower ground and in valleys and hollows. This erosion, and subsequent erosion during periglacial climates in the late Pleistocene, were largely responsible for the low relief topography that characterises the upland area of the district at the present time.

Sea level has risen by more than 100 m since the time of the maximum extent of the last ice sheet in the region, about 18 000 years ago, and has caused the inundation that produced The Wash and Fenland. Sands washed in by the sea and clay and silt supplied by the rivers have accumulated in Fenland, mostly during the last 10 000 years, as a complex sequence of marine, brackish and freshwater deposits. The more landward areas have been reclaimed by man, mostly during the last 1000 years, and now form the rich farmland of Fenland.

PREVIOUS WORK

Early observations on the geology and physical geography of the district occur in works relating to the drainage of Fenland and those describing the agriculture of Norfolk. Skertchly (1877, pp.307–314) has provided a comprehensive bibliography of these early publications together with a list of works published between 1702 and 1876 relating to the geology of the Fenland part of the district.

The geology of west Norfolk, particularly that of the Sandringham Sands, attracted the attention of several early geologists. Notable among these were William Smith, who delineated the Kimmeridge Clay and several of the Lower Cretaceous formations on his map of Norfolk (1819), and Woodward (1833) and Rose (1835–36), who produced the first systematic accounts of the geology of Norfolk. Additional contributions to the stratigraphy of particular parts of the Lower Cretaceous formations were made by Taylor (1823), Fitton (1836), Trimmer (1846), Wiltshire (1859, 1869), Seeley (1861a and b; 1864a and b), Rose (1862), Gunn (1866), Lowe (1869), Teall (1875), Harmer (1877) and Keeping (1883), mostly within the framework of Rose's (1835–36) more general account.

The first comprehensive account of the geology of the district was provided by the 19th century Geological Survey memoirs and maps. The original geological survey of the southern part of the district was at the one-inch scale by Whitaker et al. (Sheet 65, published 1886) and the northern part by Whitaker et al. (Sheet 69, published 1886). The accompanying memoirs to these sheets, *The geology of Norfolk, south-western and northern Cambridgeshire* (Whitaker et al., 1893) and *The geology of the borders of the Wash* (Whitaker and Jukes-Browne, 1899) include descriptions of the geology of the eastern part of the King's Lynn district: the western part is included in the regional memoir *The geology of Fenland* (Skertchley, 1877). The stratigraphy of the Jurassic

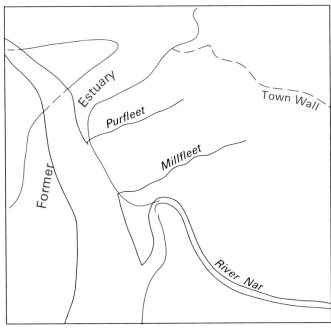

Plate 1 King's Lynn and the former estuary of the River Great Ouse.

Prior to the opening of the new outfall (top left) in 1852, the Great Ouse entered the sea via a broad estuary beneath what is now North Lynn and Wootton Marshes. King's Lynn was founded in the 11th century as a new port at the head of this estuary. (Aerofilms).

and Cretaceous rocks of the district is referred to in the stratigraphical memoirs *The Jurassic rocks of Britain* (Woodward, Vol. V, 1895) and *The Cretaceous rocks of Britain* (Jukes-Browne, Vol. I, 1900; Vol. II, 1903). Details of the water supply from underground sources are given in the Water supply memoir for Norfolk (Whitaker, 1921).

The six-inch geological survey of the district was commenced in 1965 by Dr R W Gallois as part of a combined study with Dr R Casey of the stratigraphy of the late Jurassic and early Cretaceous rocks of north-west Norfolk. This survey was extended in 1970 to cover the whole of the King's Lynn district by Dr F C Cox, Dr Gallois and Mr B Young. The areas surveyed by the respective officers are listed on p.viii. In 1972 the Survey was commissioned by the Water Resources Board to carry out geological and geophysical investigations of The Wash and part of the adjacent land area as part of the Feasibility Study for the Wash Water Storage Scheme. Geological surveys were carried out in the offshore area by Drs C D R Evans and R T R Wingfield with the assistance of Messrs G C Bradley, G K Lott, M Reed and A R Turner, and a geophysical survey was made of the same area by Mrs S E Deegan and Mr R A Floyd with the assistance of Miss R S Brown (reported in Wingfield et al., 1979). Dr Gallois was responsible for the onshore survey (reported in Gallois, 1979a) and the stratigraphy of the solid geology in the offshore area. The results of both surveys have been incorporated in the 1 to 50 000 map.

Because of the poorly exposed nature of the ground, little systematic work was done on the Upper Jurassic and Lower Cretaceous rocks between the time of the old one-inch survey and the present six-inch survey. Particular aspects were discussed by a number of authors, notably the sedimentology of the Sandringham Sands by Schwarzacher (1953), the heavy minerals of the Red Chalk by Rastall (1925), aspects of the Pleistocene by West and Donner (1956), Straw (1960; 1979; 1983) and Ventris (1985), and the age of the Holocene transgressions by Willis (1961). These works have enabled the geology of the district to be placed within its regional setting.

Stratigraphical works that have stemmed from projects carried out in conjunction with the present survey include classifications of the West Walton Beds and Ampthill Clay (Gallois and Cox, 1977), and the Gault (Gallois and Morter, 1982), the first zonation of the beds (Boreal Volgian to Valanginian) close to the Jurassic–Cretaceous bounday (Casey, 1961a; 1963; 1973) and the first description of the geology of the submarine part of The Wash (Wingfield et al., 1979).

There is too little mineral working in the district for it to have been systematically described. The Kimmeridge Clay was the subject of an extensive drilling programme in west Norfolk between 1916 and 1923 to explore its possible potential as a source of oil, but there is little geological record of this work other than two accounts by Forbes-Leslie (1917a and b).

TWO
Precambrian and Palaeozoic

Indurated Precambrian and Palaeozoic rocks have been proved by boreholes to lie at relatively shallow depths beneath an unconformable Mesozoic cover throughout much of the East Midlands and the north-western part of East Anglia. This area is referred to as The Wash region in this memoir (Figure 3). No economically interesting mineral has yet been proved in these basement rocks, but the presence of a moderately thick Mesozoic sequence in the north-eastern part of the region, close to the North Sea gas- and oil-fields, has led to the drilling of deep hydrocarbon-exploration boreholes within the King's Lynn district at Hunstanton and Wiggenhall, and in most of the adjacent districts (Table 1).

The King's Lynn district is situated between the London Platform massif and the North Sea Mesozoic basin: as a result, the depth to the pre-Permian surface increases northwards across the district from about 400 to 800 m below OD (Figure 3).

Rocks of presumed Precambrian or early Palaeozoic age proved in boreholes in the region include rhyolite tuffs at Glinton (Kent, 1962) and North Creake (Kent, 1947), and schistose and gneissic rocks at Hunstanton.

The first two of these occurrences have been likened to Precambrian rocks that crop out in Charnwood Forest in Leicestershire, and the Glinton occurrence probably extends westwards at depth to that area (Horton, 1989, fig. 2). Sedimentary rocks that have tentatively been assigned to the Precambrian, but which are probably early Palaeozoic in age, include quartzites at Spalding and Wisbech, and phyllitic slates with volcaniclastic layers at South Creake and Wiggenhall. These rocks have been sufficiently thermally metamorphosed to suggest that they are older than fossiliferous Palaeozoic rocks proved elsewhere in the East Midlands and East Anglia. Pharaoh et al. (1987) have suggested, on the basis of geochemical studies, that the volcanic rocks at Glinton and North Creake are petrologically unrelated to the Charnian igneous suites and are probably of Ordovician age.

The basement rocks proved in boreholes in the area immediately south and east of the King's Lynn district consist almost wholly of steeply dipping, indurated, poorly cleaved, marine mudstones of late Silurian or early Devonian age that probably formed part of a thick basi-

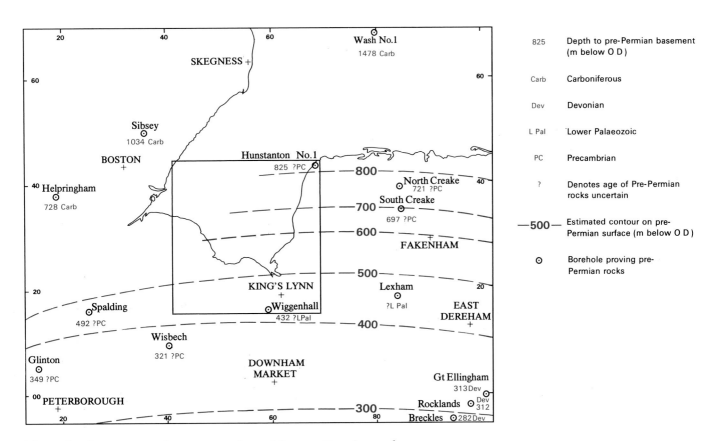

Figure 3   Contours on the upper surface of the pre-Permian rocks.

Table 1 Sources of data used in Figures 3, 8 and 10.

Borehole	National Grid reference*	Drilled for	Reference
Bardney No. 1	1192 6862	British Petroleum	Company report
Breckles	TL 9551 9469	Norris Oil	Cox et al., 1989
Denver Sluice	5911 0106	Water Resources Board	Gallois, 1979a
Glinton	1502 0526	British Petroleum	Kent, 1962
Great Ellingham	TM 0262 9847	Superior Oil	Cox et al., 1989
Helpringham	1753 3884	British Petroleum	Company report
Hunstanton No. 1	6923 4270	Place Oil	Company report
Lexham	850 180	Norris Oil	Company report
North Creake	8568 3864	British Petroleum	Kent, 1947
Parson Drove	3793 1052	BGS	Horton, 1989
Rocklands	TL 995 966	Norris Oil	Cox et al., 1989
Severals House	6921 9639	English Oilfields	Gallois, 1988
Sibsey	361 502	Ball and Collins	Company report
South Creake	8573 3405	British Petroleum	Company report, 1969
Spalding	2434 1478	Texaco	Company report
Tydd St Mary	4307 1737	BGS	Shephard-Thorn, 1985
Wash No. 1	7457 6474	Burmah Oil	Company report
Wiggenhall	5941 1537	Texaco	Company report
Wisbech	4066 0843	Texaco	Company report

* Grid square TF unless otherwise stated.

nal sequence which was extensively folded during the Caledonian orogeny (Gallois, 1988). Their interface with the Precambrian and any older Palaeozoic rocks must have been complexly deformed and it is, therefore, possible that the weakly metamorphosed basement sediments of the southern part of the King's Lynn district are part of the same Palaeozoic suite.

The structure of the basement rocks of the region is not known. Wills (1978) suggested that the high-grade metamorphic rocks of the Hunstanton and North Creake boreholes form a concealed ridge of Precambrian rocks that runs westwards beneath the King's Lynn district to Glinton and Charnwood Forest. However, the presence of weakly metamorphosed sediments in the intervening area, and the complexity of the gravity and magnetic data of the region, suggest that if this ridge exists, it is unlikely to be continuous.

STRUCTURE OF THE BASEMENT ROCKS

In the absence of sufficient borehole data, the lithologies and structure of the basement rocks can only be based on the interpretation of the geophysical data. Geophysical surveys that have been carried out in the King's Lynn and adjacent districts include regional gravity surveys, aeromagnetic surveys and seismic surveys. In addition, geophysical logs recorded in the deep boreholes have enabled some of the other physical properties to be more accurately quantified. An example of such logs, for the Wiggenhall Borehole, is given by Chroston et al. (1987, fig. 4).

The land (including intertidal) areas in The Wash region have been included in regional gravity surveys by the Geological Survey (Institute of Geological Sciences, 1981). The map shown in Figure 4 is based on digital data reduced to Ordnance Datum using a density of 2.1 Mg/m³ for land areas and 1.8 Mg/m³ for the intertidal areas. The station density of gravity readings is sufficient

to define accurately the larger anomalies, but there may be minor local features, defined by only one or two stations, due to rapid local variations in the thickness of the drift deposits.

The Wash region was included in an aeromagnetic survey flown for the Geological Survey in 1956 (Institute of Geological Sciences, 1982). The total magnetic field intensity measurements were made at a barometric altitude of 460 m along north-south flight lines 2 km apart and east–west tie lines 10 km apart. A linear geomagnetic field was removed from the observed data. The results shown in Figure 5 are based on digital aeromagnetic data in the British Geological Survey's database.

Several attempts have been made to use these and additional geophysical data to determine the lithologies and structure of the basement rocks, and these are summarised below.

Previous interpretations

The Wash region was included in a study of the Bouguer gravity anomaly and aeromagnetic maps of south-central England by Linsser (1968), who concluded that the magnetic and Bouguer anomalies, despite having similar regional trends, were not caused by the same rock masses. Linsser reports that where the interpreted boundaries of the anomalous bodies coincide, the lower density units seem to have the higher magnetisation; he suggested that the explanation for this was that dense basic rocks responsible for the magnetic anomalies had sunk to produce fault-bounded basins that subsequently became filled with thick, low-density sediments.

Chroston and Sola (1975) also examined the gravity and aeromagnetic evidence; they suggested that the gravity highs could be correlated with Precambrian rocks and the lows with thick sequences of Palaeozoic sedimentary rocks. They interpreted the aeromagnetic anomalies as due to lateral variations at depth within the Precambrian and Palaeozoic rocks. In a later study, Sola and Chroston

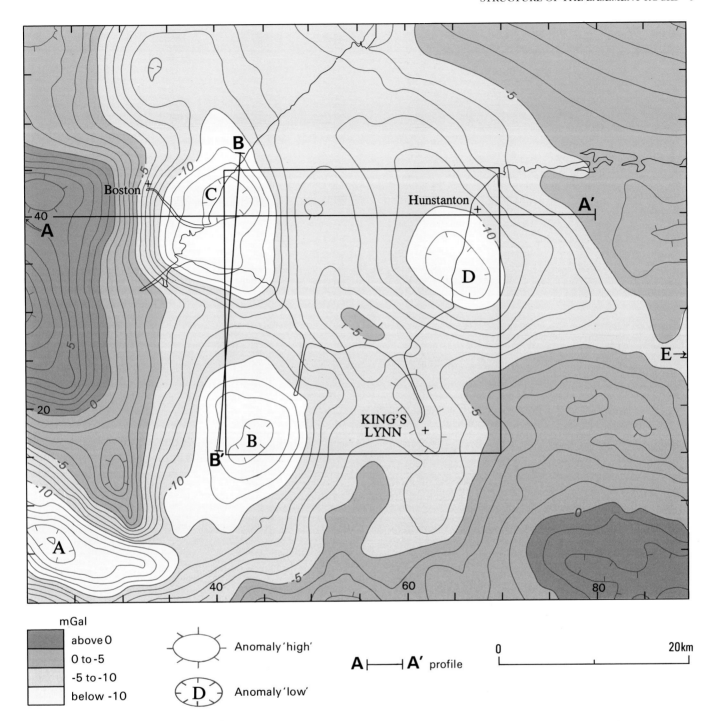

above 0

0 to -5

-5 to -10

below -10

Anomaly 'high'

D   Anomaly 'low'

A ⊢——⊣ A' profile

0   20km

Figure 4 Bouguer gravity anomaly map.

(1978) concluded, from the results of a gravity survey in The Wash, that the relief of the basement surface could not explain the observed anomalies, and also that the pronounced aeromagnetic anomaly in that area is associated with a Bouguer gravity anomaly high. The presence of numerous magnetic anomalies around The Wash suggested to Allsop and Jones (1981) that the basement probably consisted largely of Precambrian rocks. On the pre-Permian geological map of the UK. (Smith, 1985), this area is shown as one of Precambrian to Silurian

rocks flanked to the south and east by various Lower Palaeozoic rock units and to the north and north-west by Carboniferous rocks.

Seismic refraction surveys have been carried out in the region (Figure 3), including short seismic lines designed to provide information at individual sites on the velocities of the basement rocks and the depths to their upper surface (Chroston and Sola, 1982), and a 115 km-long seismic line from The Wash to Corton Sands, near Lowestoft (Chroston, 1985). The short refraction lines indi-

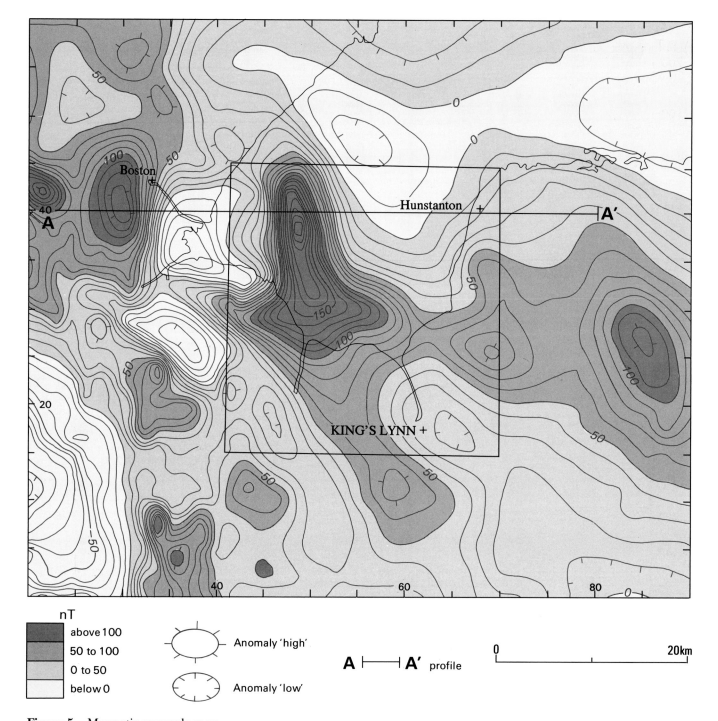

nT
- above 100
- 50 to 100
- 0 to 50
- below 0

⬭ Anomaly 'high'

⬭ Anomaly 'low'

A ⊢———⊣ A'   profile

0   20km

Figure 5   Magnetic anomaly map.

cate basement velocities in the range 5.6 to 6.0 km/s; these are considered to be indicative of metamorphic or igneous rocks and thus provide evidence consistent with that of the aeromagnetic and gravity surveys. Lower velocities were obtained for the Palaeozoic rocks that form the pre-Permian basement in the south-eastern part of The Wash region (Chroston and Sola, 1982). The long refraction line indicated that the thickness of the Palaeozoic sediments is probably not great in the northern part of East Anglia, and that much of Norfolk is underlain by

a metamorphic basement with a velocity of 6.0 km/s (Chroston, 1985). Whitcombe and Maguire (1980) have reported velocities of between 5.40 and 5.65 km/s for the Precambrian rocks at outcrop in Charnwood Forest and 6.40 km/s for those at depth. In the eastern part of The Wash region, basement velocities range between 5.7 and 6.0 km/s, and on this evidence it seems unlikely that the rocks in that area are directly comparable with those of Charnwood Forest. More data are required, particularly to resolve the problem of the effect of anisotropy on

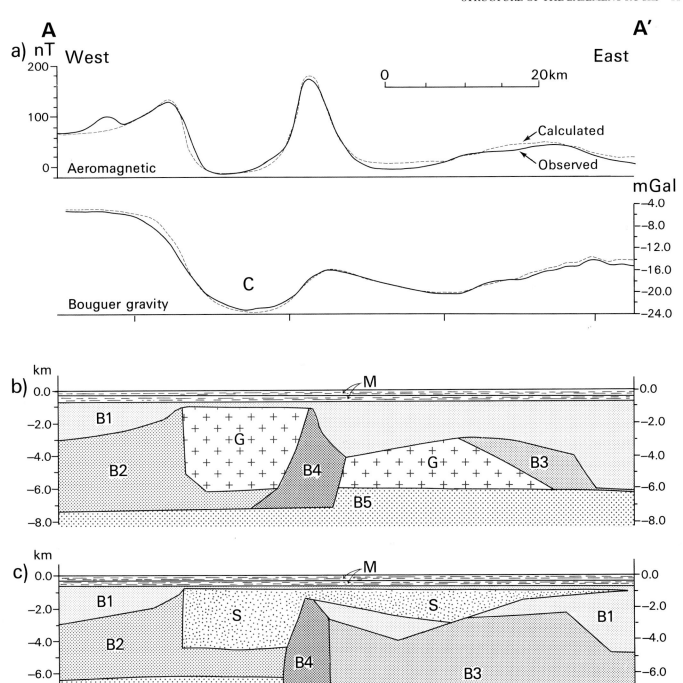

Figure 6   Interpretation of possible basement structures from Bouguer gravity anomaly and aeromagnetic profiles.

Aeromagnetic and Bouguer gravity anomaly profiles (a) along profile AA′ (see Figures 4 and 5). Two models represent alternatives for the pre-Mesozoic basement rocks: (b) granite intrusions and (c) sedimentary basins. The calculated profiles for the granite model are shown in (a). The models are based on the following estimated parameters.

Background fields: aeromagnetic 0 nT, gravity 10 mGal.

Densities and magnetic susceptibilities of the model components: M Mesozoic rocks (density 2.20 and 2.40 Mgm/m³, susceptibility 0 SI units). Basement rocks: B1 (2.73 Mgm/m³, 0 SI), B2 (2.78 Mgm/m³, 0.02 SI), B3 (2.73 Mgm/m³, 0.01 SI), B4 (2.73 Mgm/m³, 0.025 SI), B5 (2.73 Mgm/m³, 0.02 SI). G (granite model) (2.63 Mgm/m³, 0.002 SI), S (sedimentary basin model) (2.60 Mgm/m³, 0 SI).

the velocities of the basement rocks (Evans and Allsop, 1987).

Bouguer gravity anomaly lows occur on either side of The Wash, near Hunstanton and near Boston (Figure 4), but these could indicate either granitic bodies in the basement or low-density sedimentary basins at, or near, the basement surface (Figure 6). Evidence from a seismic reflection survey indicated that the top of the basement south of Hunstanton might be displaced by a fault with a throw of several hundred metres. This and the other available geophysical evidence suggested to Allsop (1983) that the Hunstanton gravity feature was due to a sedimentary basin.

Subsequent seismic refraction surveys (Chroston et al., 1987) on the east side of The Wash (Figure 3), however, provided data which favours the interpretation of the Hunstanton gravity low (Low D in Figure 4) as a granite. The observed basement velocities in the area of the gravity low are too high (5.1 to 5.8 km/s) to be low density sedimentary rocks. The same authors also suggested that the five gravity lows in The Wash region (lows A to E in Figure 4) are the cupolas of a concealed granite batholith (Chroston et al., 1987, fig. 5). Allsop (1987) considered these granites to form part of a belt of Caledonian intrusions that extends north-westwards towards the Lake District.

Present interpretation

The structure of the basement rocks of The Wash region is complex and probably involves a great variety of rock types. Most of these have not been sufficiently well sampled by the small number of deep boreholes to provide correlations with the geophysical anomalies. Correlation is complicated in some areas by the fact that the gravity and aeromagnetic data indicate distinct bodies at different depths. An attempt has therefore been made to simplify the geophysical data by outlining those areas where large bodies with low densities and those with high magnetic susceptibilities occur either at, or close to, the basement surface (Figure 7). Estimated depths to magnetic basement, and lineaments (linear features along possible fault lines) interpreted from the geophysical data, are also shown in Figure 7.

In the eastern half of The Wash region the magnetic anomalies are characterised by low gradients and small amplitudes (Figure 5), features indicative of more-deeply seated magnetic bodies than occur in the west. The magnetic bodies have two main trends: one east-south-east, which is characteristic of both the main aeromagnetic and Bouguer gravity anomalies in East Anglia, and a second south-east or south-south-east, which is also indicated by some of the anomalies in the west of the region.

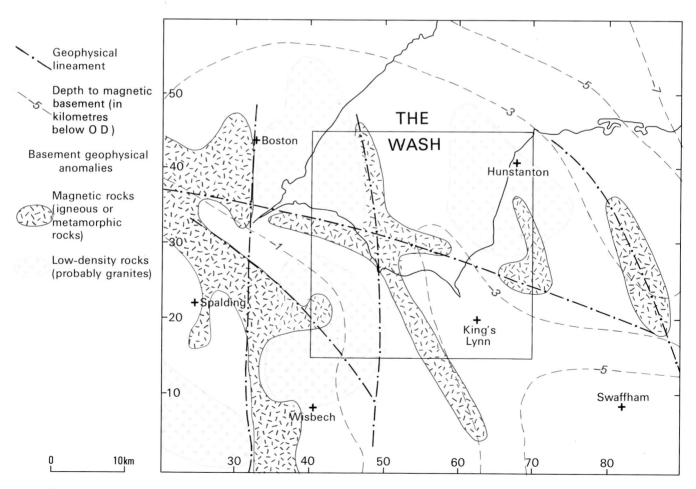

Figure 7   Sketch map showing the postulated structure of the concealed basement rocks.

The south-south-east trend is shown by the discrete magnetic highs [69 26 and 85 27] near King's Lynn and near Fakenham respectively. Both anomalies occur close to the margins of the Bouguer gravity anomaly lows [64 35 and 00 26] centered near Hunstanton and near Fakenham respectively. The magnetic anomalies appear to be due to bodies at a higher level than those responsible for the east-south-east trends.

The Fakenham gravity low (Low E, to the east of Figure 4), is the most easterly of the anomalies suggested as being due to a concealed granite body. Magnetic anomalies at the eastern margin of Figure 5 are associated with this low and may be due to host rocks which flank the western margin of the proposed intrusion as they do the other flanks (Cox et al., 1989, fig. 6c). The combination of a weakly magnetic granite and more strongly magnetised host rocks appears elsewhere in the King's Lynn district. It is also comparable with geophysical signatures of the Wensleydale Granite beneath the Askrigg Block (Wilson and Cornwell, 1982) and those of a presumed granite beneath Market Weighton (Bott et al., 1978). That all three areas might be geologically connected at depth, is suggested by a north-west–south-east-trending belt of pronounced aeromagnetic anomalies that extends from The Wash to the Lake District (Allsop, 1987).

The depth to the magnetic bodies decreases westwards across The Wash region, and it is likely that many of those west of Grid Line 50E have subcrops beneath the Mesozoic. The individual subcrops are probably small, but the magnetic bodies increase in size with depth. Volcanic rocks have been proved in the Glinton Borehole, and the anomalies could represent intrusive or extrusive intermediate igneous rocks, or metamorphic rocks. Potassium-argon dates from two specimens of rhyolitic tuff from Glinton indicate an early Devonian age, but this may be a Caledonian orogeny imprint on early Palaeozoic or Precambrian rocks (Horton, 1989). South of the King's Lynn district, a borehole [TL 2903 7839] sited on coincident magnetic and gravity highs at Warboys, Cambridgeshire, proved diorite of ?Caledonian age with a possible Hercynian overprint (Bullerwell in Stubblefield, 1967; Allsop, 1985). Within the district, however, the absence of clearly defined coincident gravity and magnetic highs suggests that similar intrusions are not present. In the Spalding area the extensive gravity high (Figure 4), associated in places with magnetic anomalies, indicates the existence of an extensive development of basic igneous rocks.

The granite cupola interpreted as being responsible for the gravity low [40 40] near Boston (Low C in Figure 4) appears to be largely surrounded by magnetic host rocks which provide a well-defined guide to the subcrop margins (Figure 7). These magnetic rocks have especially sharp boundaries to the east and west where the linearity of the gradient zones is suggestive of fault-defined contacts. On the assumption that the granites are weakly magnetic, the absence of magnetic anomalies within the area of the Boston gravity low suggests that the postulated granite cupola might there form a subcrop at the basement surface.

The Bouguer gravity anomaly [44 17] near Wisbech (Low B in Figure 4) has approximately the same ampli-tude and area as the Boston low, but differs in that magnetic anomalies occur in its central part [42 22] (Figure 5), perhaps due to magnetic roof rocks. Its western boundary coincides with the edge of pronounced magnetic anomalies [36 20; 33 09 and 38 07] of shallow origin. The Boston and Wisbech gravity lows are separated by conspicuous ridge features on both the Bouguer gravity anomaly (Figure 4) and magnetic anomaly (Figure 5) maps: the ridges are regarded as parts of a major lineament (Figure 7).

The Bouguer gravity and aeromagnetic profiles along one transect across these major features have been interpreted using two-dimensional modelling techniques (Figure 6). Alternative interpretations, for a granite model and sedimentary-basin model, are shown although the first alternative is favoured. The observed gravity data were modified by removing a regional field. A density contrast of 0.1 Mg/m^3 between the basement rocks and the granite was chosen for the granite models, assuming a value of 2.63 Mg/m^3 for the granite. With this contrast the main low-density body in Profile AA′ (Boston–Hunstanton) must lie close to the basement surface if the sharp changes in gradient are to be reproduced. The interpretation of this profile is complicated by the possibility that the postulated igneous complex immediately west of the Boston Bouguer gravity anomaly low probably has a high density, resulting in a density contrast of more than 0.1 Mg/m^3 (B2 in Figure 6). This possibility is suggested not only by high Bouguer anomaly values (+ 8 mGal at [19 30]), but also by the high gradients associated with the south-south-east extension of this anomaly high [at 31 07]. The magnetic host rocks to the granite model can be modelled satisfactorily, but their contact with the Boston intrusion (C) must be vertical or dip inwards. The granites have to be assumed to be slightly magnetic.

The gravity low [22 06] (Low A in Figure 4) near Peterborough has a more elongate form than those near Boston and Hunstanton. It appears to be flanked by magnetic anomalies of shallow origin (such as at [15 06]), and to coincide with a magnetic gradient zone. A seismic refraction survey (line LU1 in Figure 7) over the area of the gravity low near Spalding recorded a basement velocity of 5.7 km/s (Arter, 1982). As with the seismic results for the Hunstanton area (Chroston et al., 1987), this suggested that a granite intrusion or a thick sequence of acid volcanic rocks is responsible for the gravity low.

The gravity and aeromagnetic data for The Wash region show several long linear anomaly alignments or displacements (shown as lineaments in Figure 7). These are thought to indicate major faults in the basement rocks; in some cases these faults may have been reactivated in the Mesozoic. One of the most pronounced lineaments in eastern England extends over a distance of about 200 km from Derbyshire to near Norwich (referred to as the Grantham lineament by Cornwell and Walker 1989; cf. Lee et al., 1990). It crosses the King's Lynn district with an east-south-east–west-north-west trend (Figure 7) and appears from indirect evidence to have had a long geological history, and may have influenced the basement rocks in the district. Other lineaments in the region, trending north-west–south-east and north–south, form the margins of contrasting lithological units in the basement rocks, and may represent splay faults associated with the Grantham lineament. North–south-trending lineaments form the margins of the Boston granite cupola. Evidence from farther south in East Anglia suggests that these could be part of an important fault system which has subsequently influenced structures in the Mesozoic rocks, and which may even have been responsible for the north–south strike of these rocks in the King's Lynn district.

The geology of the basement rocks can be summarised as follows. In the central part of The Wash region the Permo-Triassic rocks are probably underlain by a complex of intrusive and extrusive igneous rocks, metamorphic rocks and less altered sedimentary rocks including slates and quartzites. They are likely to be mostly Precambrian or early Palaeozoic in age and to have been much affected by the Caledonian orogeny. The seismic velocities of the basement rocks are lower than those of the nearest outcrops of Precambrian rocks in Charnwood Forest, Leicestershire. Furthermore, the Charnwood Forest rocks have no distinctive aeromagnetic or gravity anomaly that matches those in The Wash region.

In the eastern part of the region the dominant trend of the aeromagnetic and Bouguer gravity anomalies is east-south-east, a trend shown by many anomalies in East Anglia and by Mesozoic structures in the southern North Sea. In the western part, south-easterly trends are superimposed on anomalies with east-south-easterly trends. The complexity in the distribution of the aeromagnetic anomalies in particular appears to reflect the combined effect of a north-west–south-east (presumed Caledonian) trend, major faults and the granite intrusions. The larger gravity anomalies are probably largely due to igneous intrusions, and it is in these areas that the closest correlation with the aeromagnetic anomaly pattern exists.

DETAILS

The rocks classified as probable Precambrian in the Hunstanton No.1 Borehole [6923 4270] have been described (Place Oil Company completion report, 1969) as a weathered, quartz-rich metamorphic rock with gneissose texture, which includes minor amounts of feldspar and chlorite. At North Creake [8562 3862], about 17 km east-south-east of Hunstanton, rocks of probable similar lithology were described by Phemister (in Kent, 1947, pp.12–13) as agglomerate or tuff of probable Charnian age. Similar tuffs were proved at Glinton [1500 0528], about 65 km south-west of Hunstanton, (Kent, 1962).

In the intervening area, at Wiggenhall [5941 1537], Wisbech [4066 0843] and Spalding [2434 1478], markedly different lithologies have been proved. At Wiggenhall the basement rocks consist of indurated and poorly cleaved, slightly silty and silty mudstones, finely micaceous in part, with thin siltstone bands. The upper 90 m or so of the mudstones are mostly pale greenish grey, and the underlying beds dark purplish grey and dark grey. Vein quartz is abundant at several levels, mainly in the lower part of the sequence, and many of the rock cuttings are sharply angular, suggesting tectonic brecciation. Petrographical studies (Texaco well completion report, 1971) show quartz, chlorite, sericite and plagioclase to be common, with much of the chlorite and sericite showing a preferred cleavage orientation. Although these rocks have been compared with Precambrian lithologies that crop out in Charnwood Forest, they are lithologically similar to, and only slightly more metamorphosed than mudstone sequences elsewhere in the northern half of East Anglia that have been dated as late Silurian to early Devonian in age (for example, the basement rocks in the Soham, Lakenheath and Four Ashes boreholes (Gallois, 1988)). White-mica crystallinity studies of the phyllites from the Wiggenhall Borehole and of similar rocks from elsewhere in the east Midlands (Pharaoh et al., 1987, pp.361–363), have shown those at Wiggenhall to have the highest metamorphic grade in the region.

The basement rocks proved in the Wisbech and Spalding boreholes have both been described (Texaco well completion reports, 1971) as hard, pale grey quartzites containing fragments of volcanic rocks. These too have been likened to Charnian lithologies, but are here presumed to belong to the same sedimentary suite as the basement rocks proved at Wiggenhall. There is, as yet, no reliable evidence to confirm or deny this, although the geophysical and geochemical evidence presented above suggests that the basement rocks penetrated in boreholes in the Wash region are part of a Lower Palaeozoic suite and are not related to the Charnian.

THREE

Permian and Triassic

During the late Carboniferous and throughout the Permian, much of the central part of East Anglia was probably occupied by a rugged topography of Precambrian and Palaeozoic rocks that were undergoing intense weathering in a hot arid climate. In contrast, contemporaneous rapid subsidence and sedimentation occurred in the North Sea, mostly in fault-controlled basins, and some of the more southerly of these reached as far as the north Norfolk coast. As a result, the southern limit of the preserved Carboniferous rocks and the southern depositional limit of the Permian lie close to the northern boundary of the King's Lynn district (Figure 8). Rapid increases in the depositional thicknesses of the late Carboniferous and Permian sediments appear to have occurred across an east–west line close to the present-day north Norfolk coast, and they may have been fault-controlled (Figure 9). The same structure may have subsequently given rise to the abnormally straight Pleistocene coastline between Hunstanton and Cromer.

Triassic rocks are almost certainly present everywhere beneath the King's Lynn district: they are likely to thin steadily southwards across it. They have been proved only

at Hunstanton (312 m thick) and Wiggenhall (90 m), but are known from sufficient other boreholes in the region to indicate a regular pattern of thickness variations in the Wash area, and an irregular pattern farther south (Figure 8). The borehole data also show this southward thinning to be due to the overstep of the pre-Permian rocks of the London Platform by successively younger Triassic formations (Figure 9). Although there is no palaeontological evidence in the district to confirm it, the regional data suggest that the oldest Triassic rocks in the district, those at Hunstanton, are early Triassic in age and that those in the southern part of the district are late Triassic. In the area immediately south of the district, the Triassic rocks proved in boreholes are thin and locally variable in thickness and they appear to infill an irregular topography cut in the basement rocks. It is probable that the area of thin variable thicknesses extends beneath the south-western part of the King's Lynn district.

The youngest Triassic rocks in the region were classified as 'Rhaetic' at the time when the boreholes were drilled: they belong largely with the Penarth Group. They form a thin but laterally persistant deposit which

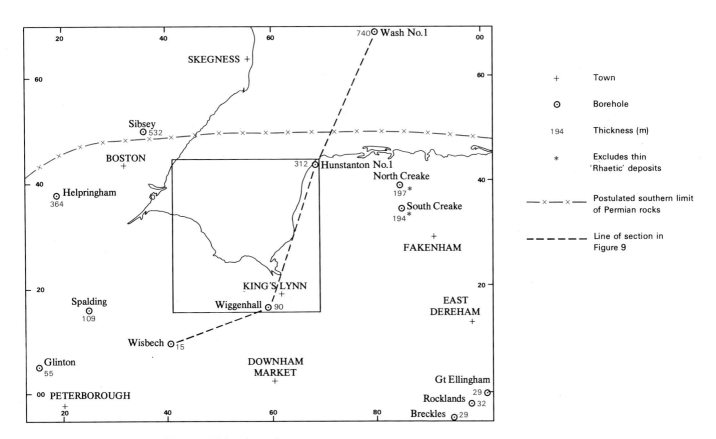

Figure 8 Thicknesses of Permo-Triassic rocks.

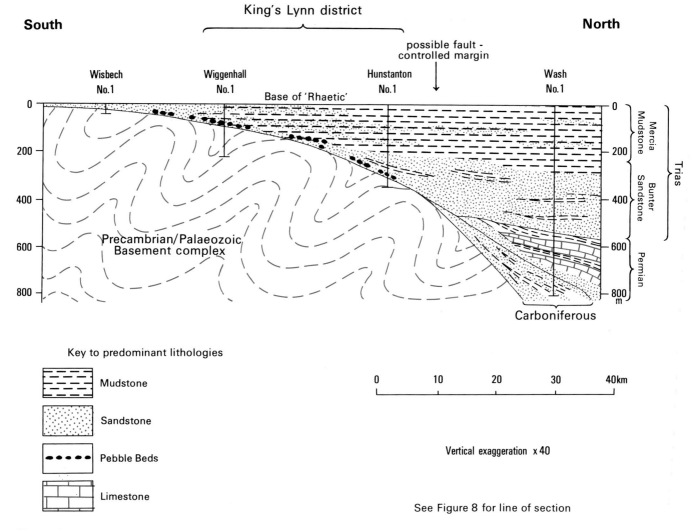

Figure 9 Cross-section through the Permo-Triassic rocks in the region.

marks a widespread marine transgression that appears to have inundated an area of low relief. This transgression heralded the Jurassic transgressions that were to cover much of the London Platform.

DETAILS

A total of 311.5 m of Triassic rocks was proved in the Hunstanton No. 1 Borehole [6923 4270]. The sequence, based on the Place Oil and Gas company's well-completion report (1969), can be summarised as follows:

	Thickness m	Depth m
PENARTH GROUP (formerly 'Rhaetic') Limestone, fine-grained, muddy, pale grey, interbedded with calcareous mudstones	6.7	530.7
MERCIA MUDSTONE GROUP (formerly Keuper Marl) Mudstone, silty, reddish brown, greenish grey and pale grey, with gypsum- anhydrite concretions at several levels; beds of fine- and medium-grained grey sandstone at several levels, especially in middle and lower part	197.5	728.2
SHERWOOD SANDSTONE GROUP (formerly Bunter Sandstone) Sandstone, yellowish grey, medium-grained with scattered small quartzite and vein quartz pebbles; thin interbeds of silty, reddish brown mudstone become common below 719 m and dominant below 810 m; conglomerate at base with pebbles of quartzite and metamorphic rocks	107.3	835.5

In the Wash No.1 Borehole [7457 6474], about 18 km north-north-east of Hunstanton, 739.75 m of Permo-Triassic strata were proved. The sequence between 767.18 and 1506.93m can be summarised as follows (Burmah Oil company's well-completion report, 1969):

	Thickness m	Depth m
TRIASSIC		
PENARTH GROUP (formerly 'Rhaetic')	7.32	774.50
MERCIA MUDSTONE GROUP		
Mudstone, red, silty with a few sandstone interbeds	112.77	887.27
Mudstones with anhydrite in lower part and with sandstone interbeds in upper ('Muschelkalk')	107.60	994.86
SHERWOOD SANDSTONE GROUP		
Sandstones with shale interbeds and some anhydrite	310.64	1308.51
PERMIAN		
UPPER PERMIAN MARLS: mudstone with anhydrite	20.42	1328.93
UPPER MAGNESIAN LIMESTONE	29.87	1358.80
MIDDLE PERMIAN MARLS: mudstone with anhydrite	21.94	1380.74
MIDDLE MAGNESIAN LIMESTONE	62.18	1442.92
LOWER PERMIAN MARLS: mudstone with anhydrite	17.07	1459.99
ROTLIEGENDES: sandstone	46.94	1506.93

The Wiggenhall No. 1 Borehole [5941 1537] proved 90.22 m of Triassic rocks; the following sequence, between 344.42 and 434.64 m is based on the Texaco company's well-completion report (1971):

	Thickness m	Depth m
Sandstone, pale grey, fine- to medium-grained, interbedded with pale grey to greenish grey mudstone and silty mudstone; dolomite present at several levels and pink and white gypsum common at other levels	36.58	381.00
Interbedded sandstone and mudstone, as above but with numerous thick pebble beds containing slate, quartz, granite, volcanic and metamorphic rocks in a sandstone matrix	53.64	434.64

There is no definite palaeontological evidence for the age of the Wiggenhall sequence. Palynomorphs obtained from samples throughout the sequence indicate a 'Rhaeto-Liassic' age, but all are long-ranging forms and may have come from cavings from the overlying Jurassic rocks. The regional distribution of the Triassic rocks suggests that the Wiggenhall sequence is a marginal deposit equivalent to all or part of the Mercia Mudstone of the northern part of the district. A similar marginal sequence was proved in the Wisbech No. 1 Borehole [4066 0843].

FOUR

Jurassic

An almost complete Jurassic sequence, which includes representatives of all the major divisions that crop out in the East Midlands, occurs everywhere beneath the Cretaceous rocks of the King's Lynn/Wash district. The sequence is thinner than that at outcrop because of attenuation in a south-easterly direction as the London Platform is approached (Figure 10). In the Fenland and offshore areas of the district, much of the Upper Jurassic[1] has been removed by erosion in the Quaternary. The full thickness of the Jurassic was penetrated in the Hunstanton No.1 Borehole and all but the highest beds were proved in the Wiggenhall Borehole. Within the district, much of the sequence was continuously cored in the Tydd St Mary Borehole, and most of the Upper Jurassic was continuously cored, albeit in separate boreholes, for the Wash Water Storage Feasibility Study. In the adjacent

districts, large parts of the sequence were cored in the Denver Sluice and Parson Drove boreholes.

The sequence contains evidence of nearshore environments, condensed deposition and erosion at several levels, the more important stratigraphical breaks being at the base of the Lias (late Sinemurian on Rhaetian), the base of the Middle Jurassic (Bajocian on early Toarcian) and the base of the Sandringham Sands (late Volgian/Portlandian on Kimmeridgian). All these erosion surfaces become increasingly marked in the direction of the London Platform. There is evidence in the region of minor folding and faulting, which occurred after the deposition of the Kimmeridge Clay and before that of the Sandringham Sands. This can be correlated with a phase of major tectonic activity, commonly referred to as the 'Cimmerian Orogeny', which affected the whole of north-west Europe and which was related to a major pulse in the opening of the Atlantic Ocean. The effects are slight in the district, but become more pronounced close to the edge of the London Platform (e.g. Gallois, 1988, fig.8).

The thickness of Jurassic rocks proved in boreholes in the Wash region, together with postulated isopachytes

---

1 To enable direct comparisons to be made with earlier geological literature and the published geological maps, the term Upper Jurassic is used in this memoir sensu Arkell (1947), and includes the deposits of the Callovian, Oxfordian, Kimmeridgian and Volgian/Portlandian stages.

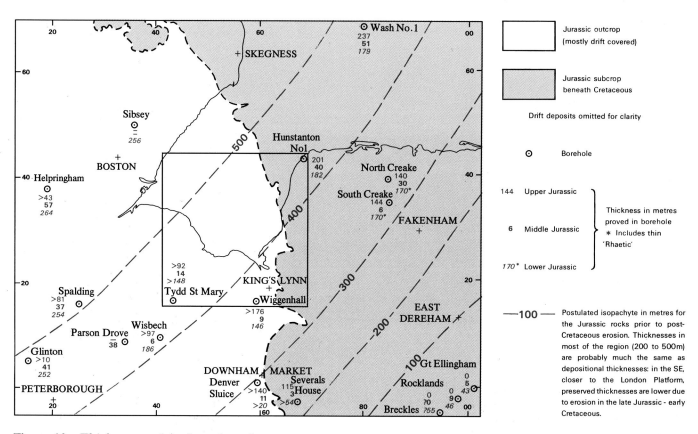

Figure 10 Thicknesses of the Jurassic rocks.

for the depositional thicknesses, are shown in Figure 10. Although only the highest beds crop out, Jurassic rocks underlie about four fifths of the district at shallow depths beneath Quaternary deposits. The Lower and Middle Jurassic sequences proved in boreholes in the region can be matched in lithological and faunal detail with those described at outcrop in the East Midlands, and the Upper Jurassic sequences can be similarly compared with those in boreholes and at outcrop between the River Humber and Oxfordshire.

LOWER JURASSIC

At outcrop in the East Midlands, the Lower Jurassic can be divided into three predominantly argillaceous formations, the Lower Lias (mudstones and calcareous mudstones), Middle Lias (silty, sandy and ferruginous mudstones) and the Upper Lias (bituminous and calcareous mudstones). The geophysical logs of the Hunstanton No.1 and Wiggenhall boreholes show the complete Lias Group to be present in both (Figure 10); the highest part of the Lower Lias and the whole of the Middle and Upper Lias were cored in the Tydd St Mary Borehole. A similar sequence to that at Tydd St Mary was proved in the continuously cored Denver Sluice Borehole in the adjacent Wisbech (Sheet 159) district, and the full thickness of the Lias was proved in hydrocarbon-exploration boreholes in the surrounding region at Wisbech, Spalding, North Creake, South Creake and in the southern North Sea.

Details

Tydd St Mary Borehole

The following sequence, proved between 148.45 and 296.66 m in the Tydd St Mary Borehole [4307 1737] (Shephard-Thorn and Horton, 1985) is based on a provisional log provided by Dr Shephard-Thorn:

	Thickness m	Depth m

LINCOLNSHIRE LIMESTONE (see p.21 for details)

UPPER LIAS
Mudstone, calcareous, medium grey, becoming finely laminated greenish grey in lowest part; small calcareous concretions, phosphatised patches and phosphatic pebbles at several levels; shelly at many levels with pyritised preservation; fauna dominated by bivalves, belemnites and ammonites, including *Harpoceras;* bioturbated at several levels, with common pyritised burrows and trails; passing down into — 18.20, 166.65

Fish Beds
Mudstone, silty, bituminous, brownish grey, finely interlaminated with greenish grey mudstone; thin cementstone beds at several levels; common fish debris throughout; poorly preserved bivalves and ammonites,

including *Harpoceras*, at several levels; sharp, irregular base — 14.50, 181.15

MIDDLE LIAS
Marlstone Rock Bed
Limestone, fine-grained, silty, with common berthierine ('chamosite') and limonitised ooliths scattered throughout and in lenses and burrowfill concentrations; yellowish brown, grey and greenish grey; shelly in part with common belemnites, ammonites, brachiopods, bivalves and shell debris; becoming silty at base and passing down into — 3.29, 184.44

Middle Lias Silts and Clays
Siltstone, micaceous, calcareous in part, greenish grey; bioturbated throughout with large burrows at some levels; finely laminated below 187.74 m; shelly at some levels with common rhychonellid brachiopods, belemnites and bivalves; berthierine ooliths and siderite cement common below 189.14 m; hard phosphatic siderite mudstone with berthierine ooliths at base — 5.43, 189.87

LOWER LIAS
Mudstone, pale and medium grey and olive-grey; slightly silty in part, but mostly smooth-textured; small siderite, phosphatic and cementstone concretions at several levels; some thin interbeds of siltstone, siderite mudstone and berthierine-oolith mudstone; bioturbated at many levels with common *Chondrites*, tracks and trails, some pyritised; shelly at many levels and with shells scattered throughout; fauna dominated by belemnites, ammonites and bivalves, with scaphopods, serpulids, gastropods, pentacrinoids, rhynchonellid brachiopods, ostracods, foraminifera and coalified wood common at some levels — 106.79, 296.66

The lithologies and faunas of the Lower Lias are characteristic of the *Uptonia jamesoni, Tragophylloceras ibex* and *Prodactylioceras davoei* zones; Horton (1989, fig. 5) has implied, on the basis of geophysical marker bands, that the *Echioceras raricostatum* Zone is also represented in the lowest part of the borehole. The Middle and Upper Lias lithologies and faunas can be closely matched with those that crop out over large areas of the East Midlands. The Middle Lias Silts and Clays (*Amaltheus margaritatus* Zone) are characteristically rhythmically interbedded, as at outcrop, and the overlying Marlstone Rock Bed (*Pleuroceras spinatum* Zone, possibly extending up into the *Dactylioceras tenuicostatum* Zone) shows the same range of lithologies as that at outcrop.

The Upper Lias commences with the characteristic bituminous laminated mudstones of the Fish Beds Member (*Harpoceras falciferum* Zone) which here rests with minor sedimentary break on the Middle Lias. The overlying calcareous mudstones (*H. falciferum* and *Hildoceras bifrons* zones) have been suggested by Horton (1989, fig.5), on geophysical grounds, to include in their lower part a correlative of the Cephalopod Limestones Member (mudstones with thin limestones and nodular cementstones locally rich in ammonites) of the East Midlands outcrop.

The highest part of the Lias, probably a few tens of metres, has been removed by erosion at the base of the Lincolnshire Limestone at Tydd St Mary.

MIDDLE JURASSIC

In the southern part of the district, in the area closest to the London Platform, the Middle Jurassic sequence is condensed and composed largely of brackish and fresh-water sediments (Table 2). The sequence thickens north-wards and westwards, with an associated increase in the proportion of marine limestones, mostly Lincolnshire Limestone and Blisworth Limestone. The full range of thickness beneath the district is from 9 to about 50 m, and although the borehole data for the region is sparse, it is clear that there are rapid local variations, particularly in the thinner sequences in the south and east where ir-regular erosion surfaces and local erosional channels are known to be present.

Predominantly fresh- and brackish-water sequences were proved in the Tydd St Mary and Wiggenhall bore-holes, 14 m and 9 m thick respectively. About 40 m of sandstones, mudstones and limestones representing a range of marine to freshwater environments were assigned to the Middle Jurassic in the Hunstanton No.1 Borehole on the basis of geophysical logs and rock cuttings.

Lincolnshire Limestone

The oldest Middle Jurassic formation recognised in the dis-trict is a 1.7 m-thick representative of the Lincolnshire Limestone proved in the Tydd St Mary Borehole. This is its most south-easterly proven occurrence to date; its subcrop in the district probably extends as far east as a line from Tydd Gote [450 180] to the centre of The Wash. The Northampton Sand and the Grantham Formation (former-ly the Lower Estuarine 'Series'), which form extensive out-crops in the East Midlands, die out beneath western Fen-land and are probably overstepped by the Lincolnshire Limestone in the adjacent Spalding (Sheet 144) district.

Upper Estuarine 'Series'

The overlying Upper Estuarine 'Series' can be divided at outcrop in the Peterborough (Sheet 158) district (Hor-ton, 1989, pp.10–11) into a lower freshwater sequence, composed largely of brownish grey seatearth mudstones with abundant rootlets and listric surfaces, and an upper sequence of rhythmically interlaminated brackish-marine and freshwater sediments. In the upper sequence, the more marine parts are commonly represented by shelly limestones and mudstones, and the freshwater parts by seatearth mudstones, siltstones and sandstones, all with common rootlets. In the district and in much of central and eastern Fenland, where the intervening Lincolnshire Limestone is largely absent, the freshwater parts of the Upper Estuarine 'Series' (Bathonian) are lithologically indistinguishable from parts of the Grantham Formation (formerly Lower Estuarine 'Series') of western Fenland. It can generally be assumed, however, because of the suc-cessive south-easterly overstep of the Middle Jurassic for-mations, that the thinner Estuarine 'Series' sequences (including those in the southern part of the present dis-trict) contain only representatives of the younger forma-tion.

Blisworth Limestone and Blisworth Clay

The Blisworth Limestone and overlying Blisworth Clay are thin, but laterally persistent, marine and brackish ma-rine deposits respectively, which probably underlie the whole of the district. The marine transgression at the base of the limestone is everywhere marked by an ero-sion surface which has locally cut down into the underly-ing Estuarine 'Series'.

Details

Tydd St Mary Borehole

The following sequence, based on a provisional log provided by Dr Shephard-Thorn, was continuously cored between 134.75 and 148.45 m in the Tydd St Mary Borehole:

	Thickness m	Depth m
CORNBRASH (see p.22 for details)		
BLISWORTH CLAY		
Mudstone, silty, brownish grey and dark olive-grey; listric surfaces and rootlets common at many levels; shelly in lowest part, with common oysters, some pyritised, including *Liostrea*	1.67	136.42

Table 2 The Middle Jurassic formations of the King's Lynn district.

Stage	Formation	Lithology	Thickness
Bathonian	Lower Cornbrash	not proved	—
	Blisworth Clay	soft silty mudstone with oysters	1 to 2 m
	Blisworth Limestone	shelly oolitic limestone	1 to 2 m
	Upper Estuarine 'Series'	sands, silts and clays with seatearths	4 to 8 m
Bajocian	Lincolnshire Limestone	shelly oolitic limestone	up to 2 m
	Grantham Formation (Lower Estuarine 'Series')	not proved	—
	Northampton Sand	absent	—

	Thickness	*Depth*
	m	m

BLISWORTH LIMESTONE
Limestone, dense, oolitic and shell-fragmental,
with a micritic matrix; bluish grey and grey;
shelly at some levels, with common bivalves;
oyster-rich bed at base   2.11   138.53

UPPER ESTUARINE 'SERIES'
Mudstone, silty, greenish and brownish grey
with common listric surfaces, carbonised
rootlets and rootlet traces; partings of silt and
fine-grained sand at several levels; sparse
bivalve fauna including oysters; minor
erosion surface at base   2.04   140.57
Thinly interbedded, brownish grey, silty
mudstone, very pale grey siltstone and pale
grey, silty, fine-grained sandstone; common
rootlets, some pyritised; sphaerosiderite
pellets in the sandstone; small clay-ironstone
nodules in the basal siltstone; erosion surface
at base with carbonaceous listric clay
extending down irregularly into underlying
limestone   6.20   146.77

LINCOLNSHIRE LIMESTONE
Limestone, pale greenish grey, clayey
(weathered?) at top; passing down into
compact, oolitic limestone with micritic
matrix; shell-fragmental in part, with some
large bivalves including oysters; cross-bedding
picked out by ooliths and shell debris;   1.68   148.45

UPPER LIAS (see p.19 for details)

Wiggenhall Borehole

The beds assigned to the Middle Jurassic in the Wiggenhall
Borehole, from 188.97 to 198.12 m, on the basis of geophysical
logs and rock cuttings, consist of sandy and oolitic, shell-fragmental white limestones with much sparry vein calcite (Blisworth Limestone), dark grey oyster-rich mudstone (Blisworth
Clay and/or Upper Estuarine 'Series') and coalified wood.

Comparison with borehole sections elsewhere in East Anglia
suggests that a thin representative of the Upper Estuarine 'Series' is present, overlain by the Blisworth Limestone and Blisworth Clay. Sandy limestones in the highest part of the sequence probably belong with the Upper Cornbrash.

UPPER JURASSIC

The Upper Jurassic rocks of the region consist almost
wholly of soft mudstones and calcareous mudstones
which give rise to the great embayment that forms Fenland and The Wash. With the exceptions of the arenaceous Kellaways Beds and Roxham Beds, at the bottom
and top respectively, the sequence consists of up to 230m
of fossiliferous mudstones (Table 3) laid down in shelf
and near-shore environments. These mudstones contain
a rich marine fauna in which ammonites and bivalves
predominate: they are zoned on the basis of the former
(Table 4). Minor, but widespread sedimentary breaks
occur at the bases of the West Walton Beds and the Kimmeridge Clay, and lesser breaks occur within the
Ampthill Clay and Kimmeridge Clay, especially in the
south-eastern part of the region close to the London
Platform.

William Smith recognised two Upper Jurassic formations, the Oxford Clay and the Oaktree Clay (subsequently renamed Kimmeridge Clay), on his maps of the
counties of Cambridgeshire, Norfolk and Suffolk. Seeley
(1861a) introduced an intermediate formation, later
named the Ampthill Clay, for mudstones which are lithologically and faunally distinct from the Oxford and Kimmeridge clays at outcrop in Bedfordshire and Cambridgeshire. Continuously cored boreholes drilled in the
western part of the present district for the Wash Feasibility Study showed the Ampthill Clay to be separated from
the Oxford Clay by a distinctive sequence of silty and calcareous mudstones, and these were termed the West Wal-

Table 3 The Upper Jurassic formations of the King's Lynn district.

Stage	Formation	Lithology	Thickness
Volgian/Portlandian	Sandringham Sands (Roxham Beds & Runcton Beds)	Fine-grained glauconitic and clayey sands	5 to 8 m
Kimmeridgian	Kimmeridge Clay	Soft mudstones, calcareous mudstones and oil shales	95 to 120 m
Oxfordian	Ampthill Clay	Soft mudstones and calcareous mudstones	50 to 55 m*
Oxfordian	West Walton Beds	Silty mudstones, calcareous mudstones and muddy limestones	14 to 15 m*
Callovian	Oxford Clay	Soft mudstones, calcareous mudstones and bituminous mudstones	30 to 60 m*
Callovian	Kellaways Sand and Kellaway Clay	Silty and sandy mudstones	3 to 5.5 m*
Callovian	Upper Cornbrash	Limestone and mudstone	up to 2 m*

* Proved in boreholes only.

ton Beds by Gallois and Cox (1977). Along the eastern margin of Fenland, between King's Lynn and Denver, the Kimmeridge Clay is overlain with marked lithological contrast by the Sandringham Sands. This formation has been divided into four members, the lowest two of which, the Roxham Beds and Runcton Beds, are Jurassic in age. For convenience, they are described with the remainder of the formation in Chapter Five (Cretaceous).

Cornbrash

A thin, laterally persistent representative of the Cornbrash is probably present everywhere beneath the district. It has been recorded in the Hunstanton No. 1 and Wiggenhall boreholes, but has only been sampled in the Tydd St Mary Borehole. It was also cored in the Denver Sluice and Parson Drove boreholes in the adjacent districts, and was recorded in the North Creake, Wisbech and Spalding boreholes, but only by means of geophysical logs (Penn et al., 1986). Taken together, the data for the region suggest that the distribution of the Lower Cornbrash (Bathonian) is patchy and probably restricted to western Fenland, but that the Upper Cornbrash (Callovian) is everywhere present.

Details

Tydd St Mary Borehole

The following sequence was recorded by Drs B M Cox, I E Penn and Shephard-Thorn between 131.66 and 134.75 m in the Tydd St Mary Borehole:

	Thickness m	Depth m
KELLAWAYS BEDS (see this page for details)		
UPPER CORNBRASH		
Mudstone, medium grey, very shelly with much shell debris including large pectinaceans, oysters and other bivalves; pyritic in part; burrowed surface at base	0.19	131.85
Limestone, shell-fragmental, fawn to grey; burrowed, with some clay infillings; argillaceous and medium grey in lower part with shell-fragmental lenses; fauna includes *Rhynchonelloidella* sp. and bivalve debris including *Isognomon*	1.57	133.42
Mudstone, silty, calcareous and muddy siltstone; medium grey; much finely comminuted shell debris and large fragments including *Entolium, Isognomon, Modiolus* and *Pinna*; pebbles, some serpulid-encrusted, at base including terebratulid and *Modiolus* resting on burrowed surface	1.33	134.75
BLISWORTH CLAY (see p.20 for details).		

Kellaways Beds

Kellaways Beds have been proved in boreholes to be persistent throughout the Wash region and to consist everywhere of a lower clay member and an upper sand or 'rock' member (Penn et. al., 1986, figs 6–8). In the King's Lynn district, the formation was cored in the Tydd St Mary Borehole (4.6 m thick) and was recorded in the

Hunstanton No.1 (4.3 m) and Wiggenhall (3.4 m) boreholes. It has been proved in boreholes in the adjacent districts at Sibsey (6.1 m), Wisbech (5.3 m), Spalding (5.3m), Parson Drove (4.6 m) and Denver (2.0 m), and is likely to be between 3 and 5.5 m thick everywhere in the King's Lynn district.

Details

Tydd St Mary Borehole

The following section was recorded by Dr Shephard-Thorn between 127.04 and 131.66 m in the Tydd St Mary Borehole:

	Thickness m	Depth m
OXFORD CLAY (see p.24 for details)		
KELLAWAYS SAND		
Sandstone, very fine-grained, silty in part, calcareous in part; greyish brown, pale grey and olive-grey; shelly at many levels, especially in top part, with common belemnites, ammonites and oysters; *Modiolus, Mytilus?, Protocardia* and other bivalves also present; bioturbated at most levels; passing down into siltstone in lowest part	2.53	129.57
KELLAWAYS CLAY		
Mudstone, olive-grey, silty in part; common ammonites, bivalves and gastropods, some pyritised, at many levels, including small oysters and *Modiolus*; pyritised trails and burrowfills; small phosphatic nodules in middle and basal part of bed	2.09	131.66
UPPER CORNBRASH (see this page for details)		

Oxford Clay

The Oxford Clay has an extensive subcrop, some 30 km in width, beneath the Quaternary deposits of western Fenland which extends from Wisbech to Peterborough and from there northwards to Sleaford and northern Lincolnshire. The formation thins southwards and eastwards across the region as it approaches the London Platform, due partly to attenuation and partly due to the overstep of the West Walton Beds. The Oxford Clay dips gently east to north-east beneath the whole of the present district, where its upper surface lies at depths ranging from about 80 to 230 m below OD. It may form rockhead in the floor of a deep drift-filled valley in the Long Sutton area, but there is no borehole evidence to confirm this.

In the outcrop area nearest to the King's Lynn district, the Oxford Clay varies from about 55 to 65 m in thickness. In the Peterborough (Sheet 158) district, Horton (1989) has estimated the full thickness of the formation from borehole and outcrop evidence to be from 63 to 65 m. In central Fenland, the continuously cored Parson Drove Borehole [3793 1052] proved the complete Oxford Clay sequence to be 64.7 m thick (Horton, 1989, p.15). In south-west Norfolk, the Severals House [TL 692 964] and Denver Sluice [5911 0100] boreholes proved about 40 m and 38 m of the formation respectively. Bore-

Table 4 Zonal scheme for the Oxfordian, Kimmeridgian and Volgian/Portlandian sequences of the King's Lynn district.

Stage and substage			Zone[4]	Formation in district
Volgian (pars)[1]	Portlandian (pars)[2]		*Subcraspedites lamplughi*	Runcton Beds
			Subcraspedites preplicomphalus	
			Subcraspedites primitivus	
			Paracraspedites oppressus	Roxham Beds
			Titanites anguiformis	
			Galbanites kerberus	
			Galbanites okusensis	
			Glaucolithites glaucolithus	
			Progalbanites albani	
Kimmeridgian[1]	Kimmeridgian[2]	Upper	*Virgatopavlovia fittoni*	
			Pavlovia rotunda	
			Pavlovia pallasioides	
			Pectinatites pectinatus	
			Pectinatites hudlestoni	
			Pectinatites wheatleyensis	
			Pectinatites scitulus	Kimmeridge Clay
			Pectinatites elegans	
		Lower	*Aulacostephanus autissiodorensis*	
			Aulacostephanus eudoxus	
			Aulacostephanus mutabilis	
			Rasenia cymodoce	
			Pictonia baylei	
Oxfordian		Upper	*Amoeboceras rosenkrantzi*	
			Amoeboceras regulare	Ampthill Clay[3]
			Amoeboceras serratum	
			Amoeboceras glosense	
		Middle	*Cardioceras tenuiserratum*	West Walton Beds[3]
			Cardioceras densiplicatum	
		Lower	*Cardioceras cordatum*	Oxford Clay (pars)
			Quenstedtoceras mariae	

1 Sensu Casey, 1967; 1973
2 Sensu Cope et al., 1980
3 The West Walton Beds and Ampthill Clay comprise the Corallian Group
4 The zones of the Upper Jurassic, originally based on faunal ranges, are now defined on the basis of a combination of faunal, lithological and sedimentary features and are here considered to be chronostratigraphical. They are referred to in the text by the species from which they take their name (e.g. Eudoxus Zone).

〜〜〜 erosion surface

▨ absent because of erosion

hole correlations based on geophysical logs have shown the Oxford Clay to range in thickness in its Fenland subcrop from about 40 m (in the south-east) to about 60 m (in the north-west) (Penn et al., 1986).

Within the present district, the continuously cored borehole at Tydd St Mary [4307 1737] proved the full thickness to be 58.0 m, and the full thickness was also penetrated in the deep hydrocarbon-exploration boreholes at Hunstanton and Wiggenhall, where geophysical evidence shows it to be about 29 m and 40 m thick respectively (Penn et al., 1986).

Throughout the Fenland region, including the present district, the Oxford Clay can be divided into a lower part (Lower Oxford Clay) composed mostly of dark grey mudstones with thin bituminous bands, and an upper part (Middle and Upper Oxford Clay) consisting of paler grey, more calcareous and, locally, more silty mudstones with siltstone and cementstone bands. The Lower Oxford Clay is well exposed in the Peterborough area where the high calorific value of the bituminous bands has been significant in the development of the brickmaking industry. Many of the bituminous horizons contain large numbers of crushed ammonites and bivalves. The Middle and Upper Oxford Clay are characterised by a sparse but varied fauna that includes brachiopods, bivalves, ammonites, belemnites, crustaceans and crinoids. This faunal and lithological assemblage suggests deposition in nearershore environments than those for the Lower Oxford Clay.

The Oxford Clay sequence proved in the Tydd St Mary Borehole can be matched in lithological and faunal detail with that proved in cored boreholes elsewhere in Fenland, and with that at outcrop in the Peterborough area. Analysis of the geophysical-log signatures, especially the total-gamma-ray logs (Penn et al., 1986), confirms that the formation remains relatively constant in lithology throughout the King's Lynn and adjacent districts.

Details

Tydd St Mary Borehole

The most complete Oxford Clay sequence proved to date in the district is that in the Tydd St Mary Borehole between 70.65 and 127.04 m; the following section is based on a provisional log provided by Dr Shephard-Thorn:

	Thickness m	Depth m
WEST WALTON BEDS (see p.25 for details)		
UPPER OXFORD CLAY		
Cordatum and Mariae zones		
Mudstone, soft, uniformly pale grey, slightly silty, calcareous; locally more calcareous and with sub-conchoidal fracture; passing down into medium and pale grey mudstones and silty mudstones; sparsely shelly but with pyritised *Pinna* locally abundant; other bivalves include *Entolium*, *Gryphaea* and *Thracia*: abundant pyritised trails and burrows also present; small pyritised ammonites including *Cardioceras* and *Quenstedtoceras* occur throughout;	19.60	90.25
MIDDLE OXFORD CLAY		
Lamberti Zone		
Mudstone, very silty, bioturbated, medium grey and pale, faintly greenish grey, calcareous; Lamberti Limestone equivalent at top; shelly at some levels with common *Gryphaea*, other bivalves and ammonites including *Hecticoceras*, *Kosmoceras* and *Quenstedtoceras*, common shell debris and abundant pyritised trails and burrows; base of zone taken at a shelly *Gryphaea*-rich band that approximately marks a downward change to less silty mudstone	5.15	95.40
Athleta Zone (pars)		
Mudstone, silty and slightly silty, pale grey and pale, faintly brownish grey, calcareous; passing into soft muddy limestone at several levels; moderately fossiliferous with partially pyritised bivalves abundant at many levels; common *Bositra* and *Nicaniella*, encrusted *Gryphaea*, nuculoids, *Oxytoma*, *Thracia*, *Dicroloma*, common *Genicularia*, *Trochocyathus*, belemnites and rhynchonellid brachiopods; ammonites include *Hecticoceras*, *Kosmoceras* and *Peltoceras*; base of Middle Oxford Clay taken at the upper limit of slightly bituminous mudstones	17.77	113.17
LOWER OXFORD CLAY		
Athleta Zone (pars)		
Mudstone, silty and slightly silty; pale grey and pale, faintly brownish grey, calcareous; passing into soft muddy limestone at several levels; interbedded with fissile, brownish grey, slightly bituminous mudstones crowded with tiny flakes of bivalve shell and spat; moderately shelly (calcareous beds) to very shelly (bituminous beds) with fauna composed mostly of crushed, partially pyritised, ammonites and bivalves; *Kosmoceras* spp. including *K.* (*Spinikosmokeras*) *acutistriatum* Buckman and *Binatisphinctes*		

comptoni (Pratt); *Meleagrinella* and nuculoids particularly common amongst bivalve fauna; locally common *Genicularia vertebralis*, rare belemnites and wood fragments also present — 7.64, 120.81

Coronatum Zone
Mudstone, silty, calcite-cemented, passing into cementstone; pale grey, extremely shelly with abundant partially pyritised nuculoid bivalves; shell plaster with belemnites, common wood fragments, crushed *Binatisphinctes comptoni* and *Kosmoceras* (the Comptoni Bed of the East Midlands) — 0.84, 121.65

Rhythmic, thinly bedded alternations of blocky, pale grey, moderately shelly, calcareous mudstone and fissile, brownish grey, very shelly, bituminous mudstone, mostly with pyritised fossils; traces of lamination preserved in the bituminous beds but mostly destroyed by bioturbation adjacent to contacts with the calcareous mudstones; non-ammonite fauna as in zone above with *Meleagrinella* especially common in the bituminous beds and nuculoids in plasters; *Kosmoceras* spp. common at many levels; passing rapidly down into shelly, very bioturbated, silty mudstone and then with bioturbation into — 5.39, 127.04

KELLAWAYS BEDS (see p.22 for details)

West Walton Beds

The West Walton Beds underlie the whole of the King's Lynn district at depths ranging from about 65 to 215 m below OD; they probably crop out in the floors of drift-filled valleys in the Lutton to Tydd St Giles area. The stratigraphy of the formation is poorly known at outcrop because it is largely covered by drift deposits. Throughout much of England, beds of equivalent age are present as limestones, sandstones and mudstones which represent marine shallow-water, reefal and lagoonal environments (Corallian Group).

At its maximum development under central Fenland, where its subcrop beneath drift deposits extends from Chatteris northwards to Wisbech and the south-western part of The Wash, the West Walton Beds consist of up to 15 m of mudstones, siltstones and silty mudstones with muddy limestone concretions at many levels. These overlie the Oxford Clay with lithological contrast and minor unconformity, the basal bed of the West Walton Beds being commonly very silty and burrowed down into the irregular surface of the Oxford Clay.

The full thickness of the formation was cored in the Gedney Drove End [4665 2939] and Tydd St Mary boreholes, and in the West Walton Highway Borehole [4913 1316] close to the boundary of the present district: the last-named serves as the type section in the absence of suitable exposures. The formation thins southwards and eastwards beneath the district, from 14.5m at Gedney Drove End, 14.1 m at West Walton Highway, 13.8m, at Tydd St Mary and 14.1 m at Hunstanton No.1, to 11.0 m at Wiggenhall and 10.6 m at Denver Sluice. This thinning is accompanied by a general increase in calcium carbon-

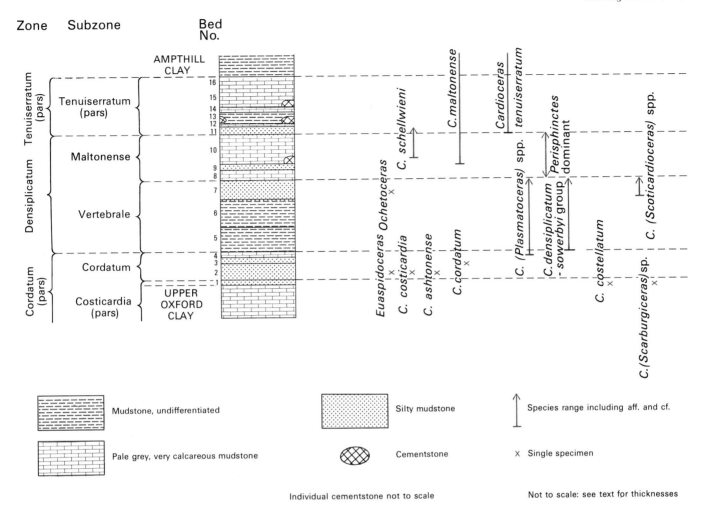

Figure 11 Generalised vertical section for the West Walton Beds of the district.

ate content, as demonstrated by the number of limestone horizons and an increase in silt content. It probably reflects the transition from a less to a more near-shore environment. North-westwards, the formation thickens to become 19 to 23 m thick beneath the Lincolnshire fens and wolds, before thinning again north of the River Humber as it approaches the positive area of the Market Weighton Axis.

The West Walton Beds contain a marine fauna similar to that of the Oxford Clay, with foraminifera, serpulids, bivalves and ammonites the most common elements, and also gastropods, belemnites, crustaceans, echinoids and a variety of trails and burrows. Plant debris and wood fragments are common at some levels. The formation falls within the Cordatum, Densiplicatum and Tenuiserratum zones (Table 4). It has been divided into 16 distinctive beds (WWB 1 to WWB 16) on the basis of a combination of lithological and faunal characters (Figure 11, after Gallois and Cox, 1977).

In the southern part of Fenland, the West Walton Beds are lithologically variable and were probably deposited in current-agitated shallow water close to a land area, the London Platform, fringed by patches of coral reef (Gallois, 1988, fig. 9). Coralline and oolitic limestones occur

at this stratigraphical level at Upware, Cambridgeshire, and mudstones with limestones rich in reefal debris occur at the same level at Elsworth, Cambridgeshire.

Details

West Walton Highway Borehole

The following sequence, which is the type section for the formation, was proved between 68.25 and 82.31 m in the West Walton Highway Borehole [4913 1316] . The main lithological and faunal features are summarised graphically in Figure 11. Comparison of its geophysical-log characters with those of the West Walton Beds sequences elsewhere in Norfolk and Lincolnshire, including those of the Hunstanton No.1 and Wiggenhall boreholes in the district, suggests that the predominant lithologies are laterally persistent. Minor variations in lithology probably occur over relatively short distances in shallow-water sediments such as these, particularly at the levels of the more silty horizons.

	Thickness	*Depth*
	m	m

Bed No. AMPTHILL CLAY (see p.33 for details)

WEST WALTON BEDS

WWB 16 Mudstone, medium to pale grey, silty, sparsely shelly with rare

		Thickness m	Depth m

myid and pectinid bivalves; *Pinna*, *Cardioceras tenuiserratum* and large *Perisphinctes*, all preserved as 'ghosts'; large serpulid-encrusted *Gryphaea* and small pyritised trails also present; locally containing small cementstone nodules; passing down with interburrowing into — 0.47 — 68.72

WWB 15 Mudstone, medium to pale grey, silty, with intense chondritic mottling; passing down into pale and very pale grey, very calcareous, slightly silty mudstone and cementstone; very sparsely shelly with myid bivalves, small oyster fragments encrusted by *Bullapora*, *Lopha*, *Pinna* and pyritised trails; interburrowed junction with — 1.52 — 70.24

WWB 14 Mudstone, pale and medium grey, intensely interburrowed with smooth-textured, pale grey mudstone and very gritty (shell debris), silty, medium grey mudstone; very irregular fracture; shelly with much broken bivalve debris, mostly oysters; *Lopha*, *Pholadomya* and pyritised ammonite 'ghosts', including *Cardioceras tenuiserratum* and *Perisphinctes*, also present; interburrowed junction with — 0.36 — 70.60

WWB 13 Mudstone, medium and pale grey, slightly silty to silty, passing locally into cementstone; very sparsely shelly with *Gryphaea* encrusted by *Bullapora*, rare pectinid bivalves, *Oxytoma* and *Cardioceras tenuiserratum*; passing down into — 0.85 — 71.45

WWB 12 Mudstone, medium and pale grey, very silty, shelly with large *Bullapora*-encrusted *Gryphaea*, *Chlamys* and partially pyritised *Cardioceras tenuiserratum* and *Perisphinctes*; interburrowed junction with — 0.12 — 71.57

WWB 11 Mudstone, dark grey, silty, foraminifera-spotted and with much silt grade bivalve debris; moderately shelly with bivalves including *Chlamys*, *Grammatodon* and *Plagiostoma* in partially leached calcite preservation; partially pyritised ammonites including *Cardioceras schellwieni* Boden, *C. tenuiserratum* (common in burrow clusters) and *Perisphinctes*; echinoid debris, *Dentalium*, wood fragments and pyritised trails also present; interburrowed junction with — 0.53 — 72.10

WWB 10 Mudstone, medium to pale grey, slightly silty and silty; burrow-mottled throughout with chondritic mottling in upper part; shelly in

part with calcitic and 'ghost'-preservation bivalves including small oysters, *Gryphaea* (encrusted by serpulids and *Bullapora*), arcid bivalves, *Astarte*, common *Chlamys*, *Gresslya*, *Myophorella*, *Pholadomya* and *Pleuromya*; partially pyritised ammonites including *Cardioceras maltonense*, *C.* cf. *schellwieni*, *C.* cf. *wrighti* Arkell and *Perisphinctes*; rare belemnites, *Dicroloma*, serpulids and pyritised trails also present; cementstone concretions formed around some *Gryphaea*; interburrowed junction with — 2.00 — 74.10

WWB 9 Mudstone, medium grey becoming darker with depth, silty with much comminuted shell debris; burrow-mottled, locally intensely foraminifera-spotted, plant-speckled in part; shelly and very shelly with arcid bivalves, *Astarte*, *Chlamys*, encrusted and bored *Gryphaea*, *Oxytoma*, poorly preserved *Cardioceras* including *C.* (*Subvertebriceras*) sp., *Perisphinctes*, small belemnites and pyritised trails; interburrowed junction with — 0.50 — 74.60

WWB 8 Mudstone, faintly greenish pale grey, slightly silty, passing locally into cementstone; intensely burrow-mottled with darker, more silty, burrowfills; sparsely shelly with poorly preserved bivalves including pyritised *Cardioceras* fragments and rare echinoid fragments; interburrowed junction with — 0.60 — 75.20

WWB 7 Mudstone, dark and medium grey, interburrowed, silty with much comminuted shell and plant debris, foraminifera and disseminated pyrite in the matrix; burrow and chondritic mottling; shelly with partially leached and/or partially pyritised calcitic preservation including *Chlamys*, encrusted *Gryphaea* fragments, *Lopha*, *Myophorella* and common *Pinna*, some in growth position; *Cardioceras* common including *C.* (*Plasmatoceras*) *popilianense* Boden, *C.* (*Scoticardioceras*) cf. *excavatum* (J Sowerby), *C.* (*S.*) cf. *expositum* (S S Buckman), *C.* (*S.*) cf. *serrigerum* (S S Buckman), *C.* (*Subvertebriceras*) cf. *densiplicatum* Boden; rare *Ochetoceras*, clusters of *Procerithium*, rare belemnites, common pyritised trails; oyster plaster at base — 1.40 — 76.60

WWB 6 Mudstone, pale grey with darker burrowfills, slightly silty with some silt and foraminifera-rich burrowfills; sparsely shelly with oyster fragments, *Pinna*, *Cardioceras*, crustacean fragments, pyrite knots

		Thickness m	Depth m

and trails; interburrowed junction with — 1.82 — 78.42

WWB 5 Mudstone, medium to pale grey, with rare cementstone concretions; slightly silty to silty, faintly foraminifera-spotted and plant-speckled in part; tiny pyritised trails common; burrow-mottled in part; sparsely shelly but with burrowfill concentrations of shell debris including·encrusted oysters, *Astarte*, *Chlamys*, *Myophorella*, *Pinna*, clusters of *Procerithium*, small belemnites (*Hibolites?*), serpulids and crustacean debris; *Cardioceras* including *C. (Subvertebriceras)* cf. *sowerbyi* Arkell, preserved as brown calcite films and pyritised nuclei; interburrowed junction with — 1.58 — 80.00

WWB 4 Mudstone, medium to dark grey, smooth-textured, very slightly silty, with rare cementstone concretions; very sparsely shelly with *Cardioceras* including *C. (Plasmatoceras)* sp; brown oxidised pyritised trails common; interbedded junction with — 0.40 — 80.40

WWB 3 Mudstone, medium to dark grey, silty, plant-speckled in part, especially towards base; burrow-mottled, faintly foraminifera-spotted; sparsely shelly with *Gryphaea*, *Procerithium* and *Cardioceras (C.)* cf. *cordatum* (J Sowerby); interbedded junction with — 0.50 — 80.90

WWB 2 Mudstone, medium grey becoming darker with depth, silty; chondritic and burrow mottling in part with some burrowfill concentrations of silt, plant debris and foraminifera; passing locally into cementstone; sparsely shelly with *Cardioceras* including *C.(C.) ashtonense* Arkell and *C. (C.)* aff. *costicardia* S S Buckman, in calcitic preservation and as pyritised nuclei; rare *Euaspidoceras*; pyritised trails common, together with irregular patches of pyrite film; interburrowed junction with — 1.05 — 81.95

WWB 1 Mudstone, medium to pale grey passing locally into cementstone; silty with much comminuted shell debris and foraminifera, faintly plant-speckled in part; sparsely shelly to shelly with *Cardioceras* including *C.(Subvertebriceras)* cf. *costellatum* S S Buckman and C. (*Scarburgiceras*) sp. in calcitic preservation and as pyritised nuclei; oysters and small belemnites common; strikingly interburrowed junction at base

with burrowfills of foraminifera-rich silt extending down into bed below — 0.36 — 82.31

UPPER OXFORD CLAY: pale grey calcareous mudstone

Gedney Drove End Borehole

A similar sequence, which can be summarised as follows, was proved between 85.30 and 99.80m in the Gedney Drove End Borehole [4665 2939]:

		Thickness m	Depth m

AMPTHILL CLAY (see p.33 for details)

WEST WALTON BEDS

Bed No.

WWB 16 Mudtone, pale and medium grey, with muddy limestone concretions; sparsely shelly with *Cardioceras tenuiserratum* (Oppel) and large *Perisphinctes*; bioturbated junction with — 0.30 — 85.60

WWB 11 to 15 Interbedded pale, medium and dark grey, slightly silty mudstones and silty mudstones; bioturbated with chondritic mottling and other burrows and pyritised trails; sparsely shelly with *Cardioceras schellwieni* Boden, common *C. tenuiserratum* and rare *Perisphinctes*; *Lopha*, *Pinna*, and *Gryphaea* encrusted by *Bullapora* — 9.80 — 95.40

WWB 1 to 10 (WWB 4 and 8 absent) Interbedded medium grey, silty mudstones and pale grey, slightly silty mudstones; bioturbated at many levels with several distinct burrowed surfaces; shelly at several levels with concentrations of ammonites, bivalves or gastropods, commonly as burrow linings; *Cardioceras* sp., *Cardioceras (Subvertebriceras)* sp., *C. (Cardioceras)* sp., *Perisphinctes*, *Chlamys*, *Gryphaea* encrusted by serpulids and *Bullapora*, *Lopha*, *Myophorella*, *Pholadomya*, *Pinna*, *Procerithium*, foraminifera and plant debris present; basal bed (WWB 1) occurs as burrow concentrations of foraminiferal silt penetrating the underlying Oxford Clay — 4.40 — 99.80

Tydd St Mary Borehole

The Tydd St Mary Borehole proved 15.45 m of West Walton Beds, between 55.20 and 70.65 m depth, consisting of a similar lithological and faunal sequence to that recorded at West Walton Highway and elsewhere in Fenland. The junction with the Oxford Clay is a burrowed surface on which the very silty mudstones of the West Walton Beds rest with marked lithological contrast on the smooth textured mudstones of the Oxford Clay.

Ampthill Clay

The Ampthill Clay crops out beneath the whole of the land area lying to the west of the Walpoles, in the floor of a broad drift-filled valley beneath Walker's Marsh and be-

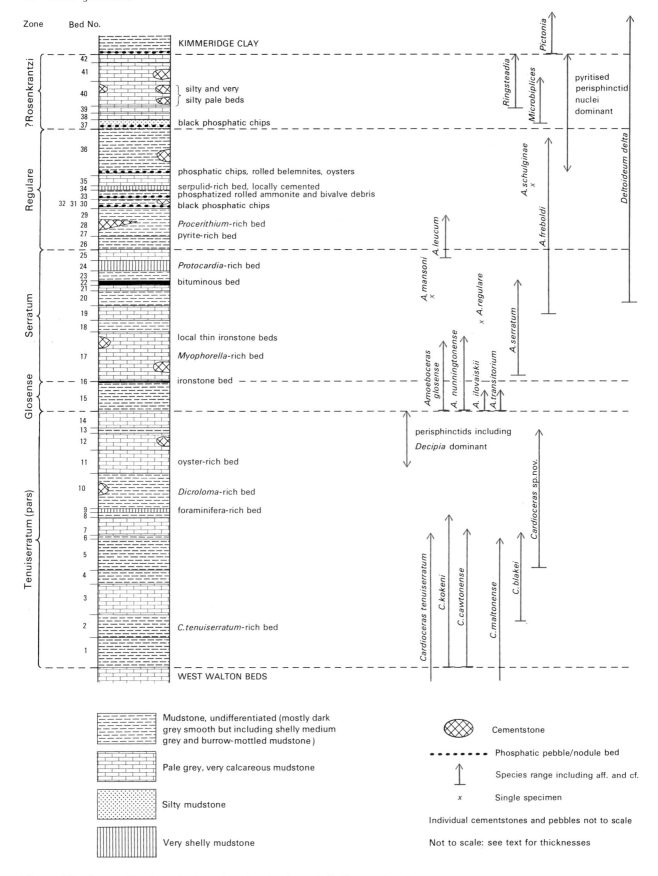

Figure 12   Generalised vertical section for the Ampthill Clay of the district.

neath the south-western part of The Wash. Its subcrop underlies the remainder of the district. Except for a small exposure [441 425] on the floor of The Wash near Roger Sand, the formation is everywhere covered by up to 30 m of Quaternary deposits. Its full thickness was cored in the Wash Feasibility Study boreholes at West Walton Highway and Gedney Drove End, and large parts of it were cored in other boreholes drilled for the study in the present district and in the Wisbech district. These cores have enabled the stratigraphy of the formation to be examined in detail and a standard sequence of 42 beds (AmC 1 to AmC 42) has been defined on the basis of a combination of lithological and faunal features (Gallois and Cox, 1977).

The generalised vertical section for the Ampthill Clay of the district, showing the lithologies, predominant faunal features, zonation, and bed numbers is shown in Figure 12. Descriptions of the individual beds, based largely on the sequences proved in the boreholes at Gedney Drove End, North Wootton and West Walton Highway, are given below.

The Ampthill Clay consists of shelly, soft, dark grey mudstones, slightly silty mudstones and pale grey, calcareous mudstones. Doggers and thin beds of muddy limestone occur at several levels, usually in association with the more calcareous mudstones. Rare thin beds of clay ironstone and kerogen-rich mudstone are also present. Pyrite occurs throughout the formation and bands of phosphatic nodules are common at several levels. The base of the formation has been taken at an upward lithological change from the medium and pale grey, silty calcareous mudstones of the West Walton Beds into medium and dark grey, slightly silty mudstones; this change is accompanied by a change in the fauna from rare small ammonites preserved as clay casts to abundant, more varied ammonites preserved in white calcite/aragonite.

The formation can be divided on the basis of gross lithology into a lower part (Beds AmC 1 to 15) which is slightly silty and forms a passage down into the West Walton Beds, a middle part (Beds AmC 16 to 29) in which smooth-textured mudstones are dominant and an upper part (Beds AmC 30 to 42) characterised by calcareous mudstones and silty mudstones (similar to those of the West Walton Beds) which contain several erosion surfaces overlain by phosphatic pebbles and encrusted by oysters.

The Ampthill Clay is composed almost entirely of clay- and silt-grade materials, mostly clay minerals, quartz and calcium carbonate, with small amounts of pyrite, phosphate, siderite and organic matter. The clay-mineral content varies in roughly inverse proportion to that of the quartz and calcium carbonate and is mostly in the range 25 per cent to 60 per cent; it consists of clay mica, probably largely illite, kaolinite, mixtures of expansible minerals and chlorite (Merriman and Strong *in* Gallois, 1979a). Quartz, where present, occurs almost entirely as crystalline silt-grade material and makes up mostly between 7 per cent and 22 per cent of the rock. Calcium carbonate contents vary from less than 15 per cent in the dark grey mudstones to between 30 per cent and 50 per cent in the pale grey mudstones. Concretions of muddy

limestone, with calcium carbonate contents of up to 90 per cent, commonly occur in the pale grey mudstones and the silty mudstones. These appear to have formed as early diagenetic concretions by the dissolution and migration of calcium carbonate from the macroscopic shelly fauna and, in the pale grey mudstones, from calcite mud which is probably largely derived from the disintegration of coccoliths, calcispheres, foraminifera, ostracods and macrofossils.

The Ampthill Clay thins southwards and eastwards in Fenland from more than 55 m (estimated) at Gedney Drove End to about 52 m at North Wootton and 36 m at Denver Sluice. This thinning is largely due to non-sequences in the upper part of the formation, but is in part due to overstep by the Kimmeridge Clay (Figure 13). Consequently, the lower and middle parts of the formation thin regularly south-eastwards, the upper part thins in a more rapid and irregular manner (Figure 14). This difference probably reflects the onset of minor earth-movements and folding close to the London Platform, at the margin of the Ampthill Clay basin of deposition. The thinnest sequences of the Ampthill Clay are likely to be in the south-eastern corner of the district closest to the London Platform, and the thickest in the north-west. The Ampthill Clay thickens rapidly north westwards from Gedney Drove End to 83.6 m in the Sibsey Borehole [361 502] and 93.7 m in the Nettleton Borehole [1185 9642] (Penn et al., 1986) in Lincolnshire. The formation again becomes attenuated northwards from there as it approaches another positive area, the Market Weighton Axis.

The Ampthill Clay contains a wholly marine fauna dominated by ammonites, bivalves and foraminifera, that suggests deposition in a shallow shelf sea. At many horizons the fauna is abundant but of low diversity. This feature, combined with the presence of common plant fragments, gastropods, oysters, crustaceans and echinoderms at certain levels suggests that, at times, the sea was sufficiently shallow for its circulation to be restricted and that land was probably not far distant. The phosphatic pebble beds that occur in the upper part of the formation indicate current-agitated conditions that were probably caused by a shallowing of the sea due to either eustatic changes or to reactivation of the nearby London Platform. In contrast, the smooth-textured clays with crushed, thin-shelled ammonites and bivalves that occur at several levels in the Ampthill Clay probably indicate more placid, deeper water and/or more offshore conditions.

The rich ammonite fauna is dominated by perisphinctids and the cardioceratid genera *Amoeboceras* and *Cardioceras*. Species of these latter have been used as the basis of the zonal scheme (Sykes and Callomon, 1979).

Details

A summary of the beds and thicknesses of those parts of the Ampthill Clay cored in boreholes drilled in the district for the Wash Feasibility Study are given in Table 5. The full thickness of the formation was also penetrated in the deep hydrocarbon-exploration boreholes at Hunstanton and Wiggenhall, but was not cored there. However, the geophysical log signatures are sufficiently distinctive to allow individual bed thicknesses to be identified with confidence (Penn et al., 1986, figs 12 to 14).

The generalised Ampthill Clay sequence for the district is given below. The lithological descriptions for each bed are common to all the' boreholes listed in Table 5. The faunas given are the combined records of the more common fossils from all the boreholes, with some additional observations on the ranges of the larger ammonites and oysters based on outcrop data from southern Fenland. The thicknesses are those of the North Wootton Borehole [6439 2457].

Thickness
m

KIMMERIDGE CLAY (see p.44 for details) (see p.44 for details)

AMPTHILL CLAY

Bed No

AmC 42 Mudstone, very pale grey, barren, with burrowfill concentrations of pyrite knots and pins; passing down into pale grey, slightly silty mudstone with medium grey burrowfills; sparsely shelly with *Oxytoma* and small oysters; minor erosion surface at base marked by striking colour change with interburrowing and scattered cream-coloured phosphatic nodules   0.20

AmC 41 Mudstone, medium grey, smooth-textured; moderately shelly with bivalves common, including *Anisocardia, Camptonectes, Oxytoma, Placunopsis, Protocardia, Thracia,* arcids and oysters; *Dicroloma,* rare serpulids, echinoid spines and brachiopods also present; *Ringsteadia* and partially pyritised perisphinctid nuclei common; burrowfills of shell debris common in upper part of bed; becoming slightly paler in lower part with rare 'cream-coloured' phosphatic nodules and prominent pale bed including cementstone dogger with *Deltoideum delta* (Wm Smith) and common *Thracia*; lowest few centimetres darker than above with comminuted plant debris and many pyrite knots; striking lithological change at base with interburrowed junction   1.90

AmC 40 Mudstone, pale and medium grey, silty with irregular fracture; well-developed chondritic mottling and other burrows occur throughout, a few partially phosphatised; cementstone bands and doggers at several levels; fauna dominated by oysters, many occurring as angular fragments in burrowfills and encrusted with *Bullapora; Chlamys, Ringsteadia,* perisphinctid nuclei, wood fragments and pyritised trails also present; weakly interburrowed junction with   2.00

AmC 39 Mudstone, pale and medium grey, smooth-textured with rare 'cream-coloured' phosphatic patches; shelly with *Camptonectes, Chlamys, Oxytoma,* oysters including *Nanogyra, Dicroloma,* serpulids and rare echinoid spines; partially pyritised perisphinctid nuclei and crushed *Ringsteadia?*; passing down into   0.90

AmC 38 Mudstone, pale grey, sparsely shelly with *D. delta, Oxytoma* and pectinids; partially pyritised perisphinctid nuclei and *Microbiplices*; pyritised trails; interburrowed junction with   0.75

AmC 37 Mudstone, medium grey, silty; intensely burrowed with small, angular, black, phosphatic chips and oyster fragments, some with encrusting *Bullapora* and serpulids, scattered throughout; shelly in part with *D. delta, Oxytoma, Trigonia, Microbiplices* and belemnites; widespread erosion surface at base marked by phosphatic nodule bed and interburrowing   0.33

AmC 36 Mudstone, medium slightly greenish grey, smooth-textured; alternating sparsely shelly and shelly with shell debris plasters; belemnites and serpulids common with *Isognomon, Modiolus, Oxytoma, Pinna, Protocardia, Thracia,* thin-shelled oysters, *Procerithium* and echinoids; partially pyritised perisphinctid nuclei and iridescent pyritised *Amoeboceras* including *A. freboldi* Spath; pyrite pins, knots and trails; widespread erosion surface at base marked by large oysters (*D. delta*), bored and rolled belemnites, black angular phosphatic pebbles and interburrowing   4.20

AmC 35 Mudstone, slightly greenish medium and pale grey; sparsely shelly with *D. delta, Pinna,* common *Thracia,* rare belemnites and *Amoeboceras* including *A. freboldi;* passing down into   1.12

AmC 34 Mudstone, medium grey, very shelly, packed with partially pyritised small gastropods and serpulids; rarer *Protocardia, Thracia* and crustacean fragments; rare *Amoeboceras* including *A. schulginae* Mesezhinov; passing down into   0.10

AmC 33 Mudstone, slightly greenish medium to dark grey with several large cream-coloured, partially phosphatised patches and local cementstones; sparsely shelly with *Astarte, Protocardia, Thracia depressa* J. Sowerby, *Amoeboceras freboldi* and *Perisphinctes;* widespread erosion surface at base marked by phosphatised oyster debris (including *D. delta*) and small angular phosphatic pebbles   0.48

AmC 32 Mudstone, medium slightly greenish grey; burrow-mottled with numerous pyritised trails; moderately shelly with *Amoeboceras* including *A. freboldi, Astarte, Protocardia* and small gastropods; passing down into   0.28

AmC 31 Mudstone, medium grey, silty, plant-speckled and with cream-coloured, phosphatic patches; shelly with plasters of *Astarte, Protocardia, Camptonectes* and other bivalves; *Amoeboceras,* small gastropods and rare serpulids also present; passing down into   0.20

AmC 30 Mudstone, medium slightly greenish grey, smooth-textured; sparsely shelly with *Amoeboceras* including *A.* cf. *freboldi* and *A. pectinatum* Mesezhnikov, belemnites, *Protocardia,* oysters and small gastropods; widespread minor erosion surface at base marked by oyster chips, rolled belemnites, small angular black phosphatic pebbles and, locally, by numerous *Modiolus* preserved in solid cream-coloured phosphate   0.25

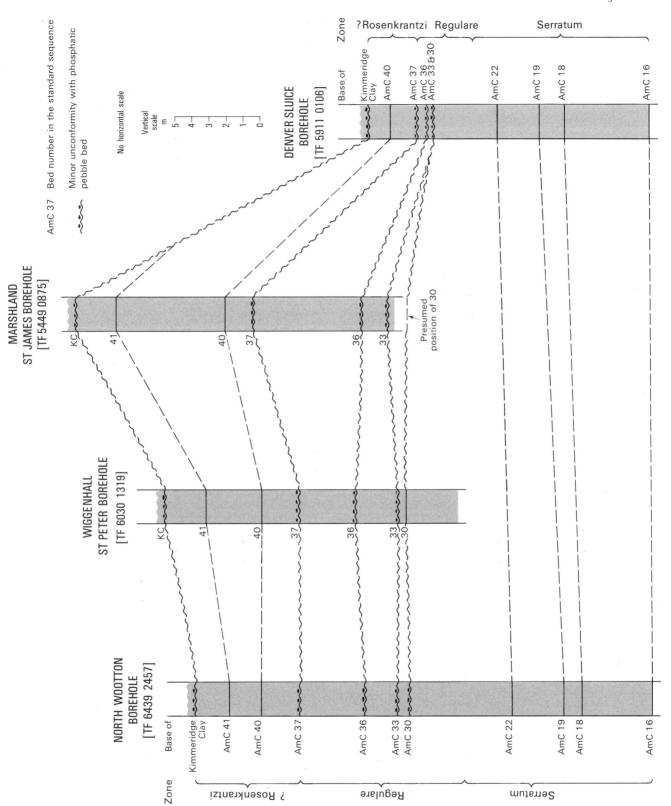

Figure 13   Attenuation and erosion surfaces in the Ampthill Clay.

Thickness
m

AmC 29 Mudstone, medium slightly greenish grey; sparsely shelly but with clusters of *Procerithium; Protocardia, Thracia*, small belemnites, serpulids, iridescent pyritised *Amoeboceras* including *A. freboldi* and phosphatised body-chamber fragments; pyritised trails also present; irregular junction with 0.25

AmC 28 Cementstone, densely cemented and locally composed largely of shells; rich in *Camptonectes, D. delta* and other oysters, *Protocardia, Thracia* and small *Amoeboceras;* lenticular bed of very variable thickness up to 0.70

AmC 27 Mudstone, medium and dark grey, slightly silty, plant-speckled and sooty-textured in part; oysters, including *D delta*, relatively common; *Thracia*, scattered serpulids, iridescent *Amoeboceras* including *A. freboldi* and *A.* cf. *leucum* Spath, and pyritised trails also present; passing down into 1.66

AmC 26 Mudstone, medium grey, slightly silty; very pyrite-rich with trails, knots and other burrowfills, all with prominent brown oxidation 'halos'; sparsely shelly with partially pyritised *Protocardia* common at some levels, *Thracia*, small belemnites, rare serpulids and *Amoeboceras* including A. cf. *leucum*; passing down into 1.40

AmC 25 Mudstone, medium and pale grey, slightly silty in part, mostly burrow-mottled; pyrite pins and trails abundant at some levels; sparsely shelly with nests of thick-shelled nuculoids, common *Thracia, Anisocardia*, rare belemnites, iridescent *Amoeboceras* including *A.* cf. *freboldi* and *A. leucum*, and *Perisphinctes;* base taken at highest *Protocardia* plaster of bed below 1.20

AmC 24 Mudstone, faintly greenish and brownish grey; slightly silty, sooty-textured and plant-rich in part; foraminifera-spotted in part; plasters and partial plasters of *Protocardia* at many levels; *Camptonectes* and belemnites; shell plaster at base 0.63

AmC 23 Mudstone, medium, grey, silty in part and with thin, brownish grey, weakly bituminous beds; shelly at some levels with *Protocardia* common, *Thracia*, belemnites, small gastropods, serpulids, rare fish fragments and faecal pellets; thin silty, foraminifera-spotted bed at base with bivalves in 'ghost' preservation and much fish debris; passing down into 0.82

AmC 22 Mudstone, medium brownish grey, fissile, bituminous; shelly with *Protocardia* and other bivalves in plasters; *Amoeboceras serratum* (J Sowerby), belemnites, small gastropods and abundant foraminifera; interburrowed junction with bed below 0.10

AmC 21 Mudstone, pale slightly greenish grey with darker burrowfills and rare cream-coloured phosphatic patches; shelly in part with *Pinna, Protocardia, Thracia*, oysters, serpulids, rare *Pentacrinus*, common *Perisphinctes* and some *Amoeboceras*, all

preserved as debris; pyrite pins and trails common; marked lithological change at base 0.64

AmC 20 Mudstone, medium grey, silty with much plant debris and rarer wood fragments; intensely burrow-mottled with many burrows pyrite-stained; thin beds of smoother, paler grey mudstone at several levels; sparsely shelly with *Camptonectes, Entolium, Pinna, Protocardia, Thracia*; huge oysters (*D. delta*) locally present; iridescent *Amoeboceras* including *A.* cf. *mansoni* Pringle and *A. serratum*, with rarer perisphinctids; passing down into 1.56

AmC 19 Mudstone, medium to pale grey, slightly silty with rare cream-coloured phosphatic patches; shelly in part with burrowfill concentrations of partially pyritised *Procerithium*, serpulids, *Pinna* and *Amoeboceras* fragments, scattered small oysters including *Nanogyra, Camptonectes, Thracia, Trigonia, Lingula*, echinoid spines, *Amoeboceras* including *A.* cf. *freboldi* and *A. regulare* Spath, with rarer *Perisphinctes;* shell plaster with ammonites and bivalves at base 0.81

AmC 18 Mudstone, medium to pale grey, smooth-textured, with cream-coloured phosphatic patches and some thin beds of foraminifera-spotted and sooty-textured mudstone containing plant debris; rare plasters of *Procerithium, Pinna* and *Camptonectes*, together with burrowfill concentrations of these as shell debris; sparsely shelly with *Camptonectes, Oxytoma, Protocardia*, serpulids, rare echinoid spines, *Amoeboceras* including *A.* cf. *serratum*, and pyritised trails; shell plaster at base 0.99

AmC 17 Mudstone, medium grey becoming pale grey with depth; smooth-textured, with silty burrowfills and wisps of shell debris; pyritised trails and burrowfills, some with oxidation 'halos', common at several levels; rare coffee-coloured phosphatic nodules; bed includes thin tabular siderite mudstone and thin lenticular shelly cementstone band; sparsely shelly with *Camptonectes, Grammatodon*, paired valves of *Myophorella, Oxytoma, Pinna, Thracia, Dicroloma*, echinoid spines and rare serpulids; pyritised ammonites including large *Perisphinctes* and *Amoeboceras* including *A.* cf. *glosense* (Bigot & Brasil), *A. nunningtonense* Wright and *A.* cf. *serratum*; probable minor but widespread erosion surface at base 4.10

AmC 16 Siderite mudstone, pale brown, dense, barren; irregular upper surface with shrinkage cracks, encrusting oysters and extensive boring and burrowing; lenticular bed apparently occurring as tabular masses up to 2 m across at outcrop; sharp lithological change at base up to 0.10

AmC 15 Mudstone, medium grey with pyrite pins and trails, some with oxidation 'halos'; cream-coloured phosphatic patches, some enclosing fauna, common in lower

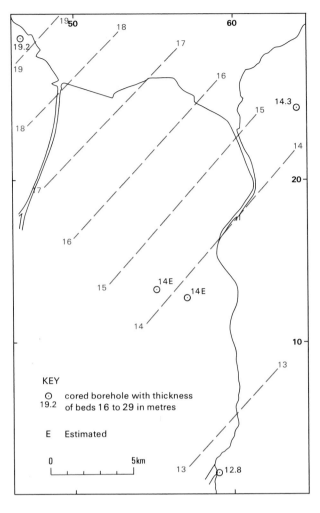

A   Beds AmC 16 to 29

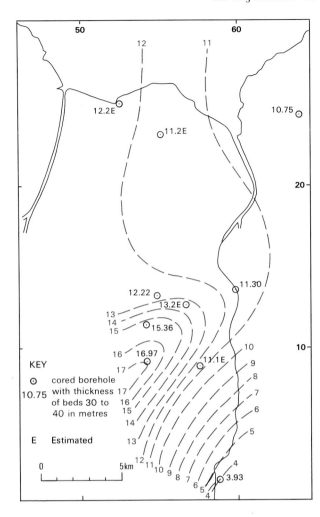

B   Beds AmC 30 to 40

Figure 14   Thickness variations in the upper part of the Ampthill Clay.

Table 5   Summary of the Ampthill Clay sequences proved in continuously cored boreholes in the district.

Borehole	Grid reference	AmC Beds cored, wholly or in part	Thickness m
Daseley's Sand (offshore)	5591 3189	40 to 42	3.52
Gedney Drove End	4665 2939	1 to 29	41.80
Walker's Marsh	5268 2530	36 to 42	12.28
Symington Farm	5499 2507	40 and 41	c. 6.0
Balaclava Farm	5549 2339	33 to 42	10.41
Ongar Hill	5767 2444	?36 to 40	c. 6.7
North Wootton	6439 2457	1 to 42	50.94
Terrington St John	5401 1435	41 and 42	0.90
Tydd St Mary	4307 1737	1 to 4	c. 6.3
West Walton Highway	4913 1316	1 to 16	24.04
North Runcton	6404 1264	41 and 42	4.27

Thickness
m

part; shelly with fauna dominated by small partially pyritised, iridescent perisphinctid nuclei (*Decipia?*) and *Amoeboceras* including *A. ilovaiskii* (Sokolov) and *A.* cf. *transitorium* Spath, some concentrated in burrowfills; larger *Amoeboceras* including *A. glosense, A. nunningtonense* and *A.* cf. *transitorium; Entolium, Pleuromya, Thracia, Pentacrinus* columnals and rare serpulids also present; equivalent of the younger Long Stanton fauna of Torrens and Callomon (1968); passing down into   2.72

AmC 14 Mudstone, medium and pale grey, slightly silty, almost barren, with pyrite knots and trails; rare cream-coloured phosphatic patches; becoming paler and more calcareous with depth and with incipient cementstone bands; *Camptonectes*, small oyster fragments, *Dicroloma*, and perisphinctid fragments and inner whorls including *Decipia*; at outcrop fauna dominated by *Decipia* spp., older Long Stanton fauna of Torrens and Callomon (1968); passing down into   1.22

AmC 13 Mudstone, pale grey becoming medium grey with depth; gritty and with irregular fracture due to included shell debris; pyrite pins and trails; burrowfill concentrations and partial plasters of comminuted shells and small fossils including *Camptonectes*, oysters, *Dicroloma, Procerithium* and perisphinctid nuclei; large oysters including, at outcrop, serpulid-encrusted *Gryphaea dilatata* J Sowerby common in lower part of bed; interburrowed junction at base   0.75

AmC 12 Mudstone, pale and very pale grey, calcareous, passing locally into cementstone; smooth fracture tending to subconchoidal; very sparsely shelly with *Camptonectes, Pinna, Thracia*, small oysters, perisphinctid fragments, *Cardioceras* sp. nov. and pyritised trails; including, at outcrop, fossiliferous cementstone doggers rich in *Decipia* and allied forms; becoming medium grey with paler burrowfills in lower part; passing down into   1.25

AmC 11 Mudstone, medium and pale grey, slightly silty in part; sparsely shelly but with many large *Gryphaea*, some smaller oysters, *Entolium, Grammatodon, Thracia, Dicroloma, Cardioceras* sp. nov., perisphinctid fragments and pyrite pins and trails; base taken at lowest *Gryphaea* plaster   1.80

AmC 10 Mudstone, medium grey, gritty textured due to shell debris; passing down into pale grey mudstone with impersistent weak cementstone; shelly in upper part, becoming sparsely shelly with depth and including *Chlamys, Entolium, Myophorella, Oxytoma*, fragmentary and small oysters, some with encrusting *Bullapora*, rare *Lingula, Cardioceras* sp. nov., perisphinctid fragments and pyrite pins and trails;

Dicroloma, many with associated guilielmites structures, common in lower part; passing down into   2.90

AmC 9 Mudstone, medium grey, slightly silty with irregular fracture; foraminifera-spotted throughout; moderately shelly with *Thracia*, other bivalve debris, *Dicroloma, Cardioceras* sp. nov. and perisphinctid fragments; passing down into   0.15

AmC 8 Mudstone, medium to dark grey, slightly silty; shelly in part with burrowfill concentrations of serpulids, *Dicroloma* with guilielmites structures and bivalve and ammonite debris; scattered perisphinctid fragments common; *Cardioceras kokeni* Boden and *C.* sp. nov. in plasters; passing down into   0.45

AmC 7 Mudstone, medium to pale grey, slightly silty, with many small pyritised trails; irregular fracture due to scattered shell debris; generally sparsely shelly but with many serpulids; small oysters, *Entolium, Oxytoma, Dicroloma* with associated guilielmites structures, *Cardioceras kokeni* and *C.* sp. nov. also present; passing down into   1.42

AmC 6 Mudstone, medium grey, slightly silty but with smoother texture than bed above; shelly with crushed, small thin-shelled *Cardioceras* preserved in white calcite and often partially pyritised, including *Cardioceras blakei* Spath, *C. cawtonense* (Blake & Hudleston) and *C.* sp. nov.; scattered serpulids, bivalve debris and perisphinctid fragments; passing down into   0.28

AmC 5 Mudstone, medium grey with some slightly darker and paler bands; in part silty and with sooty texture, particularly in upper half of bed; rare plant-speckled, brownish grey, faintly bituminous layers; rare cream-coloured phosphatic nodules; sparsely shelly in upper part but with small belemnites common; moderately shelly in lower part with *Grammatodon, Thracia, Dicroloma*, rare serpulids and crustacean fragments, and crushed small white *Cardioceras blakei, C. cawtonense, C. kokeni, C maltonense* (Young and Bird), *C. tenuiserratum* (Oppel), *C. (Subvertebriceras)* sp. and *C.* sp. nov., some with guilielmites structures; pyrite knots with oxidation 'halos' common in lower part of bed; passing down into   3.10

AmC 4 Mudstone, medium grey, silty, plant-speckled; sparsely shelly with bivalve debris, *C. kokeni, C. maltonense, C. tenuiserratum*, rare small belemnites and wood fragments; passing down into   0.80

AmC 3 Mudstone, medium to pale grey, slightly silty to silty, burrow-mottled, finely plant-speckled in part; sparsely to very sparsely shelly with *Cardioceras tenuiserratum* very common at some levels in grey 'ghost' preservation; scattered bivalve debris including *Thracia; C. cawtonense, C. kokeni* and *C. maltonense* preserved partly as 'ghosts' and partly in white calcite; passing down into   2.70

Thickness
m

AmC 2   Mudstone, medium and dark grey,
slightly silty to silty, burrow-mottled and
plant-speckled; foraminifera-spotted in
part; moderately shelly with shells mostly
in white or weakly iridescent calcite
preservation; some shell debris
concentrated in burrowfills; *Thracia,
Cardioceras* aff. *blakei, C. cawtonense?, C.
kokeni, C.maltonense, C. tenuiserratum* and
rare belemnites; passing down into   2.40

AmC 1   Mudstone, medium grey, slightly silty
becoming paler and more silty in lower
part; silty burrowfills with plant specks
and burrowfill shell concentrations
common at some levels; moderately
shelly with bivalves and ammonites
preserved in white calcite, including
*Grammatodon, Thracia, Cardioceras
cawtonense, C. kokeni, C. maltonense, C.
tenuiserratum, C. (Subvertebriceras)* sp. and
rare belemnites; junction taken at the
lowest white or iridescent calcitic fossils;
these are coincident with a downward
change to paler, more calcareous and
more silty mudstone   3.15

WEST WALTON BEDS (see p.25 for details).

Kimmeridge Clay

The Kimmeridge Clay crops out beneath about one third
of the King's Lynn district, but is almost everywhere overlain by Quaternary deposits. The highest part of the formation is exposed along the eastern edge of Fenland between South Wootton and West Winch, and small exposures occur in the floor of some of the deeper channels
in The Wash in the Roger Sand and Ferrier Sand areas.
The formation underlies the Clenchwarton, King's Lynn,
Terrington and Tilney areas where it is overlain by up to
70 m of Quaternary deposits.

Because of the paucity of exposure, little work had
been done on the stratigraphy of the Kimmeridge Clay
of the district until the present survey. Although temporary sections have become available from time to time
in the exposed area, they have usually been too deeply
weathered to be stratigraphically useful. However, the
full thickness of the formation has now been cored in the
North Runcton [6404 1624] and North Wootton
[6439 2457] boreholes, and large parts of it were cored
in boreholes in the Fenland part of the district and in
The Wash (Figure 15) for the Wash Feasibility Study.
All these boreholes were drilled on or close to the outcrop. In the subcrop area, parts of the formation were
cored in the Geological Survey boreholes at Hunstanton [6857 4078], Gayton [7280 1974] and Marham
[7051 0803], and in oil-shale exploration boreholes
drilled between 1916 and 1923. The full thickness of
the formation has been penetrated in deep boreholes
drilled in or close to the district for hydrocarbon exploration, notably at North Creake (Kent, 1947) and Hunstanton.

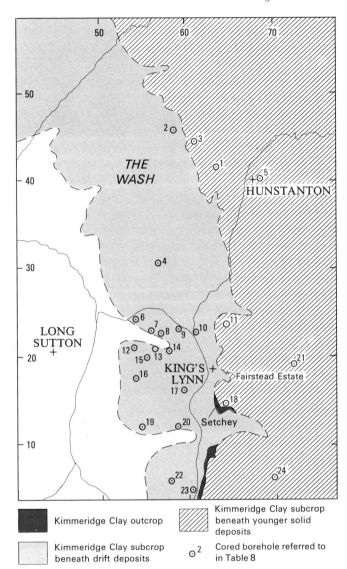

Figure 15   Distribution of the Kimmeridge Clay in the
district showing key localities.

The limits of the Kimmeridge Clay subcrop in the region can be approximately determined from deep boreholes at Trunch [TG 2933 3455] (BGS, 1975), East Ruston [TG 353 268] (Hamilton Bros., 1971), Somerton
[TG 4607 2120] (Conoco, 1969), Saxthorpe [TG 1226
3013] (Duntex Petroleum, 1970), Culford [TL 831 710]
(for water, 1890), Great Ellingham [TM 0262 9847] (Superior Oil, 1965), Rocklands [TL 9952 9670] (Norris Oil,
1969), Lowestoft [TM 538 926] (for water, 1902), Four
Ashes [TM 0223 7186] (Superior Oil, 1965), Lakenheath
[TL 748 830] (Superior Oil, 1965) and Breckles [TL 286
390] (Norris Oil, 1969), which proved Cretaceous rocks
resting on pre-Kimmeridgian rocks (Figure 16). The absence of the formation over much of Norfolk is probably
largely due to erosion close to the edge of the London
Platform during the late Jurassic and early Cretaceous.

The Kimmeridge Clay is made up of soft shelly mudstones, calcareous or kerogen-rich (oil shales) in part,

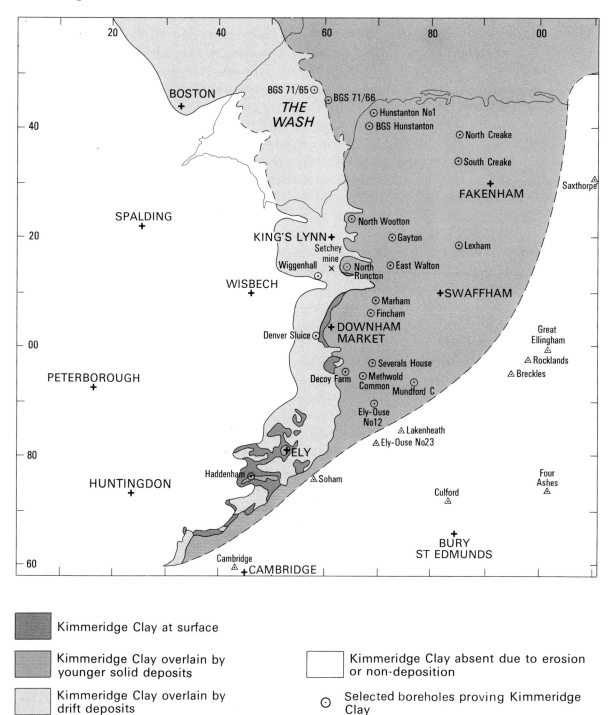

Figure 16 Outcrop and subcrop of the Kimmeridge Clay in Norfolk.

with small amounts of silty mudstone, siltstone and muddy limestone. Pyrite is present throughout and phosphatic pebble beds occur at a few horizons. The mudstones are generally shelly and contain a rich marine fauna dominated by bivalves and ammonites, but with gastropods, serpulids, brachiopods, echinoderms and vertebrates common at some levels. Much of the faunal and lithological detail of the formation is destroyed at

outcrop by weathering and the mudstones give rise to a heavy grey to yellowish brown clayey subsoil rich in selenite. The junction with the Ampthill Clay is taken at a lithological and faunal change that occurs at a minor, but widespread unconformity. This break in sedimentation is marked in the present district by a sharp lithological change from the very pale grey, highly calcareous mudstones of the upper Ampthill Clay to a medium grey silty mudstone with phosphatic pebbles. This change is readily recognisable in borehole cores and is accompanied by an upward change in the ammonite assemblages, from one dominated by *Amoeboceras* and *Ringsteadia* to one in which *Pictonia* and *Rasenia* predominate, and by distinct signatures in geophysical logs.

The large amount of continuous core available from boreholes has enabled the stratigraphy of the Kimmeridge Clay of the district to be studied in detail. As a result, a standard sequence of 49 distinctive beds (KC 1 to KC 49) has been defined (Gallois and Cox, 1976; Cox and Gallois in Gallois, 1979a) on the basis of a combination of lithological and faunal characters (Figures 17 A and B).

The Kimmeridge Clay of the district is thin in comparison with that of large parts of southern England. At its most complete in the district, beneath the northern part of The Wash, the formation is about 130 m thick; it thins southwards to 93 m at the southern limit of the district at North Runcton and to about 60 m at Denver in the adjacent Wisbech district. Elsewhere in England it is mostly 200 to 500 m thick. These variations in thickness are due mainly to differences in rates of sedimentation, the lowest rates being close to the stable massif of the London Platform. The thickness of the Kimmeridge Clay sequence was further reduced in Norfolk by late Jurassic erosion, but the effects of this were small in the present district, for although the Sandringham Sands overstep the formation in a southerly direction, only the highest three beds (Beds KC 47 to 49) have been removed between the northern Wash and North Runcton.

Despite these large differences in thickness the sequence in the King's Lynn district can be matched in both lithological and faunal detail with those proved in the Warlingham Borehole, Surrey (Worssam and Ivimey-Cook, 1971) and in the type section at Kimmeridge, Dorset (Cox and Gallois, 1981). The more prominent lithological markers such as the silty Beds KC 5, 8, 15 and 24, the thick calcareous Beds KC 18, 30 and 44 and the oil-shale-rich bands in Beds KC 29, 32, 36, 42 and 46 can be readily recognised in all three sections. Faunal ranges and many faunal marker bands are similar in each section.

Comparison of the thicknesses and lithologies of groups of beds in the formation within the present district shows that the attenuation occurs without intraformational erosion, without change in the relative proportions of the various lithologies, and with no obvious evidence of an approach to a shoreline (Figure 18).

The Kimmeridge Clay is made up of a complex sequence of small-scale rhythms. In the lower part of the formation these rhythms consist of thin silts or silty mudstones overlain by dark grey mudstones which in turn are overlain by pale grey calcareous mudstones (Type A of Gallois, 1979a). In their middle and upper parts, the rhythms consist of brownish grey, kerogen-rich mudstones overlain by dark grey mudstones and then by pale grey, calcareous mudstones (Type B), the last-named sometimes including thin beds or doggers of muddy limestone (cementstone). Many of the individual rhythms and limestones can be correlated throughout the region. Superimposed on this rhythmic sequence are broader lithological changes, from more to less calcare-

ous and from more to less kerogen-rich, which can be regarded as larger-scale rhythms and which can be correlated throughout the English Kimmeridge Clay.

The various lithologies are made up of a relatively small number of siliciclastic, bioclastic, biogenic and chemogenic components. The clastic materials are clay minerals (mostly illite and kaolinite), quartz, calcium carbonate, in the form of microscopic and macroscopic shell debris, and phosphatised fossil debris. The biogenic component consists mostly of calcareous fossils, and of kerogen composed largely of diagenetically modified lipids probably derived in part from marine algae such as dinoflagellates. Diagenetically formed pyrite, phosphate and cementstone concretions make up the relatively small chemogenic component. The typical composition of some of the commoner Kimmeridge Clay lithologies is summarised below and a selection of analyses are given in Table 6.

(i) dark grey mudstone — clay minerals 45 to 65 per cent, quartz 10 to 30 per cent, calcium carbonate 5 to 20 per cent depending upon shell content, kerogen <2 per cent.

(ii) medium grey mudstone — clay minerals 35 to 55 per cent, quartz 10 to 15 per cent, calcium carbonate 20 to 35 per cent, kerogen <1 per cent.

(iii) pale grey mudstone — clay minerals 25 to 45 per cent, quartz 8 to 15 per cent, calcium carbonate 25 to 55 per cent, kerogen <1 per cent.

(iv) cementstone — clay minerals 10 to 20 per cent, quartz 2 to 6 per cent, calcium carbonate 60 to 90 per cent, kerogen <1 per cent.

(v) oil shale — clay minerals 20 to 40 per cent, quartz 10 to 15 per cent, calcium carbonate 10 to 25 per cent, kerogen 10 to 45 per cent.

More than 98 per cent by volume of the Kimmeridge Clay of the district is composed of soft mudstones. The most obvious lithological difference within them is colour, and a number of shades can be recognised ranging from very dark grey to very pale grey. These colours are largely related to calcium carbonate contents and although it has commonly been suggested that the colours of many grey Mesozoic mudstones are largely controlled by disseminated carbonised plant debris, this is not the case with the Kimmeridge Clay. Lithological variations within the mudstones can also be detected by the two related features of texture and the nature of their fracture.

The two most distinctive Kimmeridge Clay lithologies that can be recognised in weathered sections at outcrop in the district are muddy limestones and oil shales.

Six types of limestone are present in the Kimmeridge Clay in the district; in the past, most of these have been described as 'cementstone' (muddy limestone suitable for cement manufacture).

The six types are:
1. Muddy limestone — the most common type of cementstone
2. Silty limestone — the second most common type of cementstone; well-developed bands occur in the Lower Kimmeridge Clay

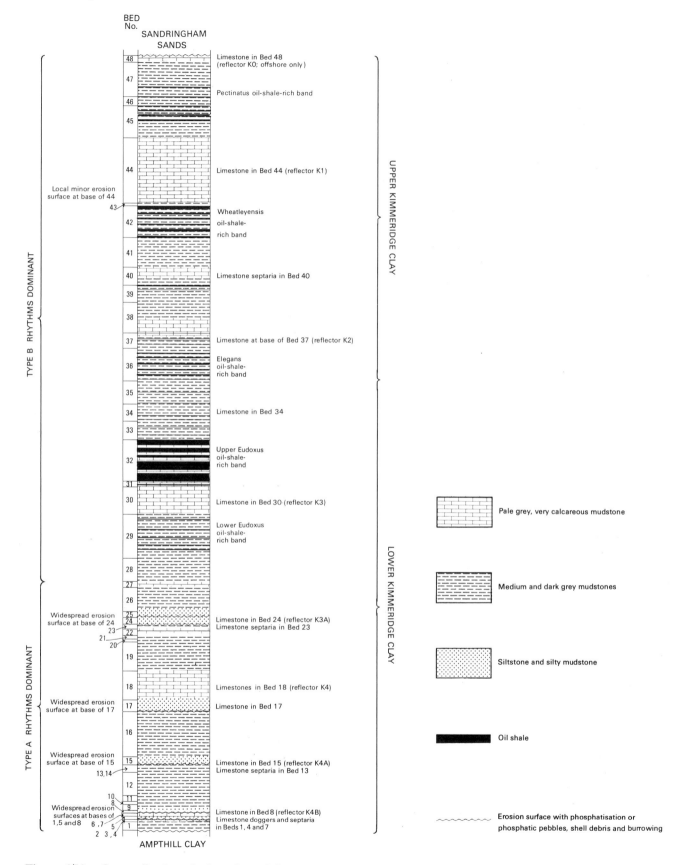

Figure 17A Generalised vertical section of the Kimmeridge Clay of the district showing lithologies and marker bands.

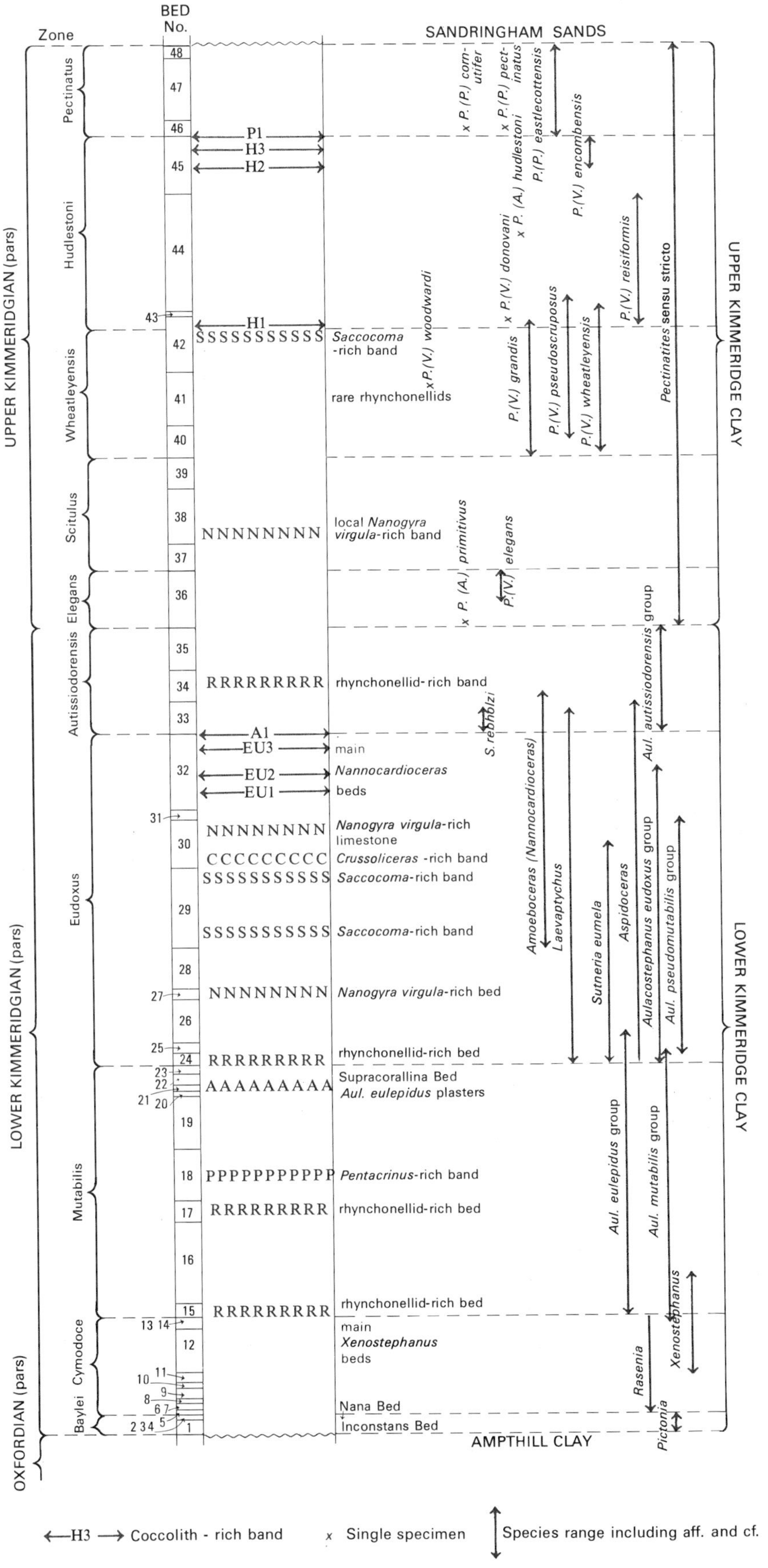

Figure 17B Generalised vertical section of the Kimmeridge Clay of the district showing faunal ranges and marker bands.

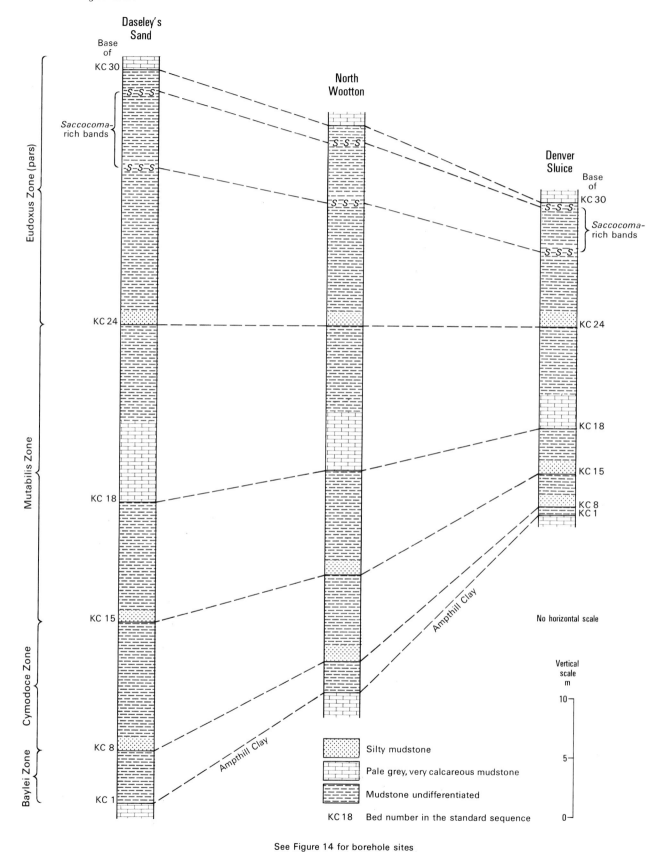

Figure 18   Attenuation in the Kimmeridge Clay between Daseley's Sand and Denver Sluice.

Table 6 Selected analyses of typical Kimmeridge Clay lithologies from the district.

Borehole	Depth m	Clay minerals present (<2μm fraction)	Whole rock			Other minerals detected (j.d.) just detectable ? denotes uncertainty	Lithology
			Total clay %	Quartz %	Calcite %		
North Wootton	31.85	Illite>kaolinite≥chlorite	23	7	54	Aragonite, pyrite	Mudstone, pale, grey, barren, slightly silty
North Wootton	36.10	Illite>kaolinite	20	9	18	Aragonite (15%), gypsum	Oil shale, shelly
North Wootton	45.70	Illite≥kaolinite≥chlorite	46	20	—	Gypsum, pyrite, feldspar	Oil shale, barren
North Wootton	47.85	Illite≥kaolinite>chlorite	18	9	66	Aragonite, pyrite	Cementstone, pyritic with oil shale
North Wootton	64.00	Illite≥kaolinite	27	12	20	Pyrite, gypsum, aragonite	Oil shale, shelly
North Wootton	68.20	Illite>kaolinite≥chlorite	35	13	26	Aragonite, pyrite, (?) dolomite	Oil shale/mudstone, interburrowed, shelly
North Wootton	69.15	Illite>kaolinite>chlorite	17	4	61	Aragonite, pyrite	Cementstone, muddy, shelly in part
North Wootton	75.80	Illite>kaolinite≥chlorite	35	12	32	Gypsum (j.d.) pyrite, aragonite, dolomite	Mudstone, medium grey, silty
North Wootton	76.10	Illite≥kaolinite	33	12	13	Pyrite, gypsum, aragonite, (?) dolomite	Oil shale, barren, fissile
North Wootton	98.30	Illite>kaolinite, chlorite, smectite	13	5	65	Aragonite	Cementstone, weakly cemented, slightly silty
North Wootton	100.12	Illite, smectite, chlorite	7	2	90	—	Cementstone, densely cemented with calcite veining
Emorsgate	37.20	Illite>kaolinite>chlorite	47	14	—	Gypsum, pyrite	Mudtone, dark grey, fissile
Emorsgate	47.14	Illite>kaolinite≥chlorite	50	16	6	Gypsum, aragonite, pyrite, (?) apatite, (?) dolomite	Mudstone, dark grey, sparsely shelly
Terrington St John	21.13	Illite>kaolinite≥chlorite, smectite	25	8	47	Aragonite, pyrite, dolomite	Mudstone, pale grey, silty, almost barren
Terrington St John	35.14	Illite>kaolinite≥chlorite	45	12	18	Gypsum, aragonite, pyrite, (?) dolomite	Mudstone, medium grey, sparsely shelly

3. Bituminous limestone — restricted to two bands
4. Septarian nodules — at many levels and generally widely spaced
5. Shelly limestone (shell coquina)
6. Coccolith limestone — occurs as very thin (1 to 50 mm thick) bands at 11 levels.

The stratigraphical distribution of the 16 limestone horizons recorded in cored boreholes in the Kimmeridge Clay in the region is summarised in Table 7. However, it should be noted that the reliability of the information is uneven because some stratigraphical horizons have been cored as many as twelve times, whereas others have been cored only twice. Widely spaced doggers are likely to have been missed where they occur at stratigraphical horizons that have been sampled only a few times. The continuous tabular cemented bed at the base of the Sandringham Sands is included in Table 8 for completeness because it forms an important seismic reflector which marks the top of the Kimmeridge Clay. The coccolith limestones have been omitted from Table 7 because of their thinness. Nine of the 16 horizons were recorded as single occurrences (mostly widely spaced doggers or septarian nodules): three of these (in Beds KC 8, 17 and 24) can be correlated with doggers at outcrop elsewhere in England and are clearly persistent horizons. The remaining 7 horizons are probably persistent tabular beds or closely spaced doggers.

The limestones which form persistent reflectors in the seismic (sparker) traverses carried out in The Wash are thought to be continuous tabular beds or lines of closely spaced doggers. These are numbered K0 to K4 in Table

Table 7 Distribution of limestones in the Kimmeridge Clay in Norfolk.

Stratigraphical horizon KC Bed No.	No. of times stratigraphical horizon penetrated in boreholes	Total no. of limestones recorded	Maximum no. of limestone bands in any one borehole	Seismic reflector	Probable lithology
Base Sandringham Sands	6	6	1	TK	calcareous sandstone
48	2	2	1	KO	muddy
44	6	6	1	K1	muddy
40	5	1	1		septarian
37	2	2	1	K2	bituminous
34	4	1	1		bituminous
30	3	3	2	K3	muddy
24	9	1	1	K3A	silty
23	9	1	1		septarian
18	10	8	2	K4	muddy
17	10	1	1		silty
15	10	3	1	K4A	silty
13	9	1	1		septarian
8	12	1	1	K4B	silty
7	12	1	1		septarian
4	11	3	1		muddy
1	12	1	1		muddy

Source of data: Wash Feasibility Study boreholes and IGS boreholes 71/65; 71/66; Skegness, Hunstanton, Gayton and Marham. Coccolith-rich bands omitted.

7 and Figure 17A (K0 = first seismic-reflection signal from the Kimmeridge Clay) in accordance with the geological interpretations of the sparker traverses (Wingfield et al., 1979). Their outcrops beneath Quaternary deposits are shown on the 1 to 50 000 geological map; these boundaries are based on seismic reflection data in the offshore area and on boreholes in the land area. Limestones K0, K3 and K4 are known only from boreholes; K1 and K2 have been recorded in excavations in the King's Lynn and West Winch areas.

Persistent tabular beds or closely spaced doggers of limestone occur in the Mutabilis (Bed KC 18), Eudoxus (Bed KC 30), Scitulus (Bed KC 37), Hudlestoni (Bed KC 44) and Pectinatus (Bed KC 48) zones. Each of these can be recognised on the basis of either fauna or lithology. The Bed KC 18 limestones commonly enclose large *Aulacostephanus* with smooth body chambers (*A. mutabilis* (J. Sowerby)) and small, finely ribbed forms (*A. eulepidus* (Schneid)); the Bed KC 30 limestones and adjacent mudstones commonly contain coarsely ribbed *Aulacostephanus* of the *eudoxus* group, the perisphinctid *Crussoliceras*, *Aspidoceras*, large forms of the oyster *Nanogyra virgula* (Defrance) and common serpulids. The limestones in Beds KC 37, 44 and 48 all contain similar pectinatitid and bivalve faunas, but the first is a tabular, calcite-cemented oil shale, the second consists of large doggers and lies only 1 to 2m above thick oil shales with *Saccocoma* (Bed KC 42), and the third only occurs beneath The Wash.

Thin coccolith-rich limestones have been recorded at 11 horizons in the Kimmeridge Clay, 8 of which have been proved in boreholes in the district (Gallois and Medd, 1979). They are recognisable in borehole cores by their pale colour and low density, and at outcrop by their almost white weathering patina. Each of these limestones is laterally persistent throughout much of the English Kimmeridge Clay. The thickest bed recorded to date (the White Stone Band of Dorset) is about 0.9 m thick and composed of over 100 coccolith-rich laminae; in Norfolk the limestones are mostly less than 0.1 m thick and some occur as a few laminae less than 10 mm thick in total. The coccolith component of the limestone varies from 30 to 90 per cent of the calcium carbonate contents; this is a minimum value because, in addition to the recognisable coccolith debris present, most bands contain minute calcite fragments and recrystallised calcite which are probably of coccolith origin. The coccolith contents of the limestones vary between 2 and 27 species, but most are dominated by a single species. In the Upper Kimmeridge Clay this is usually *Ellipsagelosphaera britannica* (Stradner). Most of the coccoliths are well preserved and with the scanning electron microscope many of the coccolith-rich bands appear to be composed exclusively of coccoliths. In every occurrence except one, the coccolith-rich limestones are interlaminated with oil shales. This has led to the suggestion that these lithologies are derived largely from algal blooms composed of coccoliths (the limestones) and dinoflagellates (the oil shales) (Gallois, 1976).

Thin seams of oil shale occur throughout the Eudoxus to Pectinatus zones inclusive, but are concentrated in five main oil-shale-rich bands in the Eudoxus (2), Elegans, Hudlestoni and Pectinatus zones. The North Runcton Borehole [6404 1624], drilled to examine the oil shales, proved 80 seams ranging from 1 to 47 cm in thickness (see p.181). The oil shales are noticeably low in density, have a faint bituminous smell when freshly broken and are the only Kimmeridge Clay lithology to have a brown streak.

Assays have shown that mudstones containing only a few per cent of kerogen and yielding as little as 5 gal oil/ton shale can be recognised as oil shale in hand specimen.

The oil shales are usually much altered at outcrop due to the interaction of the weathering products of their high pyrite content with their shelly fauna. When weathered, they are usually fissile, parting along bedding planes encrusted with secondary gypsum [$CaSO_42H_2O$] crystals and powdery natrojarosite [$NaFe_3(SO_4)_2(OH)_6$]. At some levels the oil shales are rich in phosphatised fish debris and faecal pellets, and this may be the source of the phosphorous that enables the brilliant blue mineral vivianite [$Fe_3(PO_4)_28H_2O$] to form in some of the weathered oil shales. The organic (kerogen) part of the oil shales is surprisingly resistant to weathering and where the oil shale is only sparsely shelly it can be split to reveal well-preserved calcareous and pyritic fossils.

The formation contains a rich marine fauna dominated by bivalves and ammonites (Plate 2). Crushed, but otherwise well-preserved ammonites are common throughout the sequence and, because they occur in assemblages of rapidly evolving forms, they provide an excellent basis for a zonal scheme (Ziegler, 1962; Cope, 1967; 1978). The ammonite faunas are dominated by perisphinctaceans, the zonal scheme being based on species of *Pictonia, Rasenia, Aulacostephanus* and *Pectinatites* (Figure 17B). Other ammonites are common in the *Rasenia* and *Aulacostephanus* zones, notably *Amoeboceras, Aspidoceras* (with *Laevaptychus*) and *Sutneria*, and these assist with zonal recognition and subdivision. In the overlying *Pectinatites* zones, no ammonite other than *Pectinatites* has been recorded from the district. *Gravesia*, which has been collected at this stratigraphical level elsewhere in England and which forms an important link with Kimmeridgian sequences throughout Europe, has not been found in the district. The highest part of the Kimmeridge Clay, characterised in southern England by species of *Pavlovia* and *Virgatopavlovia*, is absent due to uplift and erosion in the late Jurassic, but phosphatised, water-worn fragments of these ammonites occur in the basal pebble bed of the overlying Sandringham Sands.

Thin marker bands containing flood occurrences of coccoliths, serpulids, brachiopods, certain species of bivalve, and crinoids occur at several levels and when combined with the rhythmic variation in lithology that occurs throughout the formation, they provide an additional means of subdivision. Many of the faunal changes coincide with lithological changes that can be recognised throughout southern England; this suggests that they reflect widespread events that affected the whole of the basin of deposition. In boreholes, the zonal boundaries and the positions of selected marker bands can be rapidly identified using a combination of a small number of core specimens and geophysical logs.

A number of faunal marker bands have been identified in boreholes in the district (Figure 17B); a few have been recorded at outcrop in the spoil from the deeper temporary excavations and drains. Among the bivalves the small oyster *Nanogyra virgula* (Defrance) ranges from the Mutabilis to the Wheatleyensis zones but is especially abundant in three thin beds, two in the Eudoxus Zone

and one in the basal Wheatleyensis Zone. The highest band is overlain by only thin Recent deposits in the area between King's Lynn and the River Nar and might be expected to occur in excavations there. In the Upper Kimmeridge Clay of the same area, the bivalves *Isocyprina minuscula* (Blake), *Protocardia morinica* (de Loriol), '*Astarte*' and small oysters, the inarticulate brachiopod *Lingula* and the limpet-type gastropod *Pseudorhytidopilus latissima* (J Sowerby), are not diagnostic of any particular stratigraphical level.

Crinoids are rare in the Kimmeridge Clay except at one level in the Mutabilis Zone where pentacrinoid columnals are abundant, and at four levels (in the Eudoxus (2), Autissiodorensis and Wheatleyensis zones) where pyritised plates of the free-swimming *Saccocoma* occur in great abundance. Each *Saccocoma*-rich band occurs in oil shale and, when fresh, the tiny brass-coloured plates stand out in marked contrast to the dull brown of the matrix. The highest of the four bands has been recorded at outcrop at Setchey [6268 1452]. There, the pyrite is largely destroyed by weathering, but the shape of the *Saccocoma* plates has been preserved in secondary gypsum overgrowths.

Rhynchonellid brachiopods are common in the muddy siltstones (Beds KC 15, 17 and 24) in the Mutabilis and Eudoxus zones. These beds are covered by thick drift deposits everywhere in the district, but loose blocks of the siltstones have been recorded from the local boulder clay. Rhynchonellids have also been recorded from the Wheatleyensis Zone (Bed KC 41) in boreholes in the district, but not at outcrop. Too few specimens have been recorded for their stratigraphical distribution to be accurately known.

Details

There is no natural exposure of Kimmeridge Clay in the district at the present time nor is there any working clay pit. However, small parts of the sequence have been exposed from time to time in the deeper drains and other excavations and fragments of the more resistant Kimmeridge Clay lithologies, notably oil shales, cementstones and some of the harder mudstones persist in the spoil from these for many years. When compared with the standard sequence described below and summarised in Figure 17A, these fragments can provide useful stratigraphical information. The positions of the sections described below, and their relationship to the Kimmeridge Clay outcrop and its subcrop beneath drift deposits are shown in Figure 15. Beds KC 1 to 35 are overlain by 6 to 70 m of Quaternary deposits and are known only from boreholes; Beds KC 36 to 41 are present at shallow depths beneath Recent deposits between North Lynn and Saddle Bow, and should occur from time to time in drain spoil; Beds KC 42 to 47 crop out between South Wootton and West Winch and are exposed in temporary excavations from time to time.

South Wootton to King's Lynn

A small excavation dug for an electricity substation in 1972 at Chapel Lane, South Wootton [6405 2318], close to the junction with the Sandringham Sands, proved about 1 m of weathered mudstone and oil shale with fragments of finely ribbed pectinatitid ammonites (probably Bed KC 46 or 47).

Spoil from a 2 m-diameter sewerage tunnel which passed beneath the River Great Ouse at North Lynn [604 227] in 1972 included large doggers of densely calcite-cemented oil shale, together with mudstones and thin oil shales. The cementstone was a single impersistent bed that followed the axis of the tunnel for most of its length. The mudstones and oil shales contained *Isocyprina*, *Protocardia*, *Lingula* and pectinatitid ammonites which suggest that the stone band is that which marks the base of the Scitulus Zone (Bed KC 37).

Spoil from sewerage trenches [642 194] dug for Fairstead Estate, King's Lynn in 1966 showed the basal bed of the Sandringham Sands underlain by up to 1 m of softened dark grey clay with thin oil shales. These shales include fragments of a soft, very pale brownish grey, coccolith-rich limestone with *Pectinatites* (*P.*) *eastlecottensis* (Salfeld) (base of Bed KC 46) which can be correlated with the White Stone Band of Dorset.

King's Lynn to West Winch

Between Lynn sugar-beet factory [6050 1730] and St Germans Bridge [6084 1571], the spoil from the Fenland Flood Relief Channel shows its floor to have been cut in Upper Kimmeridge Clay. Loose doggers of cemented oil shale (Bed KC 37) occur near the sugar-beet factory. Between St Germans Bridge and Wiggenhall St Mary Magdalen railway bridge [6000 1025], the channel is cut in glacial deposits.

Contemporary photographs of the small opencast pit [6268 1452] dug for oil shales in about 1920 at Setchey suggest that it was in an oil-shale-rich band (Bed KC 42) in the Wheatleyensis Zone and that the overburden included the pale calcareous mudstones and cementstones of Bed KC 44. This suggestion was supported by the observations of Pringle (unpublished MSS, 1919) who obtained samples from the pit in 1917 via a testing works at Chiswick. The fauna of the oil shales included *Saccocoma* and pectinatitid ammonites, and enabled Pringle to correlate the Setchey seams with the Blackstone of Dorset. In 1975, the Geological Survey made a temporary excavation in the northern side of the pit to examine the weathering profile of the shales and to obtain samples for chemical analysis. The following section was measured:

KIMMERIDGE CLAY
Bed No.

		Thickness m
	Pleistocene deposits: cryoturbated flint gravel, ferruginous sand and sandy clay (derived largely from the Kimmeridge Clay)	1.9
KC 44	Pale and medium grey, deeply weathered calcareous clays with a line of cementstone doggers in lower part of bed; shelly in part with rotted bivalves and ammonites; sharp planar contact with bed below disturbed by burrows	0.16
KC 43	Very dark grey, apparently barren clay softened by weathering 1.6	
KC 42	Oil shale, dark brownish grey with brown streak; fissile, shelly with abundant *Isocyprina minuscula* and rarer *Protocardia*, small oysters, *Pseudorhytidopilus latissima* and fragments of *Pectinatites*; fish debris and faecal pellets locally common; *Saccocoma* abundant at one level; interbedded in 5 to 30 cm-thick seams with sparsely shelly dark grey clays seen	1.7

Plate 2 Fossils from the Kimmeridge Clay (all × 1, except Figs 5b and 7b × 2).

1 *Pectinatites* (*Pectinatites*) *eastlecottensis* (Salfeld); KC 46; Pectinatus Zone; BGS Skegness Borehole, 120.78 m (BGS BDF5533)

2 *Pectinatites* (*Virgatosphinctoides*) *pseudoscruposus* (Spath); KC 41; Wheatleyensis Zone; BGS Gayton Borehole, 92.15 m (BGS BDB9151)

3 *Pseudorhytidopilus* ['*Discinisca*'] *latissima* (J Sowerby); KC 42; Wheatleyensis Zone; BGS Hunstanton Borehole, 120.37 m (BGS BDD3739)

4 *Aulacostephanus* juveniles; KC 35; Autissiodorensis Zone; BGS North Wootton Borehole, 53.00 m (BGS BDK7418)

5a–b plaster of *Amoeboceras* (*Nannocardioceras*) *krausei* (Salfeld); KC 32; Eudoxus Zone; BGS North Wootton Borehole, 60.68 m (BGS BDK7539)

6 plaster of *Isocyprina minuscula* (Blake); KC 39; Scitulus Zone; BGS Marham Borehole, 89.94 m (BGS BDC1321)

7a–b pentacrinoid columnals; KC 18; Mutabilis Zone; BGS North Wootton Borehole, 99.06 m (BGS BDK8164)

8 *Aulacostephanus eudoxus* (d'Orbigny); KC 32; Eudoxus Zone; BGS North Wootton Borehole, 61.62 m (BGS BDK7555)

9 ammonite aptychal plate (*Laevaptychus*); KC 26; Eudoxus Zone; BGS North Wootton Borehole, 86.65 m (BGS BDK7977)

10 *Nicaniella extensa* (Phillips) ['*Astarte supracorallina* d'Orbigny']; KC 22 (Supracorallina Bed); Mutabilis Zone; BGS North Wootton Borehole, 89.06 m (BGS BDK8030)

11 *Sutneria eumela* (d'Orbigny); KC 29; Eudoxus Zone; BGS North Wootton Borehole, 74.70 m (BGS BDK7768)

12 plaster of *Rasenia* spp.; KC 12; Cymodoce Zone; BGS North Wootton Borehole, 112.98 m (BGS BDK8407)

13 *Nanogyra virgula* (Defrance); KC 29; Eudoxus Zone; BGS North Wootton Borehole, 73.63 m (BGS BDK7754)

The oil shales were well jointed and yielded such large quantities of water that it became impossible to deepen the excavation.

The oil-shale seams that were worked in the Setchey opencast pit crop out in the banks of the Puny Drain [626 145], about 100 m west of the pit. There is currently no exposure, but large amounts of deeply weathered oil shale are present. Most of the calcareous fossils and all the pyritised fossils in the shale are completely rotted.

In his account of the oil-shale workings, Forbes-Leslie (1917a, p.18) referred to a mine [625 145] that was overcome by an inrush of oil-covered water. It is unlikely that this mine ever worked oil shale because it seems to have been abandoned for financial reasons during the construction stage. The layout of the mine, as shown on the Ordnance Survey 25-inch scale map of 1925, consisted of two shafts and an engine house. One of these shafts was open until about 1970 when it was backfilled by the local landowner, who measured the depth at that time as approximately 15 m. The small amount of spoil still present consists largely of dark grey shelly mudstone with thin oil-shale seams. The faunal assemblage, mostly *Isocyprina*, *Protocardia* and *Pectinatites* fragments, together with the depth of the shaft, suggests that the seams encountered were those of the Elegans Zone (Bed KC 36). Forbes-Leslie (1917a, pp.16–17) recorded at least three 6 ft [1.83 m]-thick oil-shale seams in the Kimmeridge Clay of the Setchey area, but no comparable seam was recorded in the present survey (see p.181). Details of the stratigraphical distribution of oil shales in the area, together with details of their potential oil yields, are given in Chapter Seven.

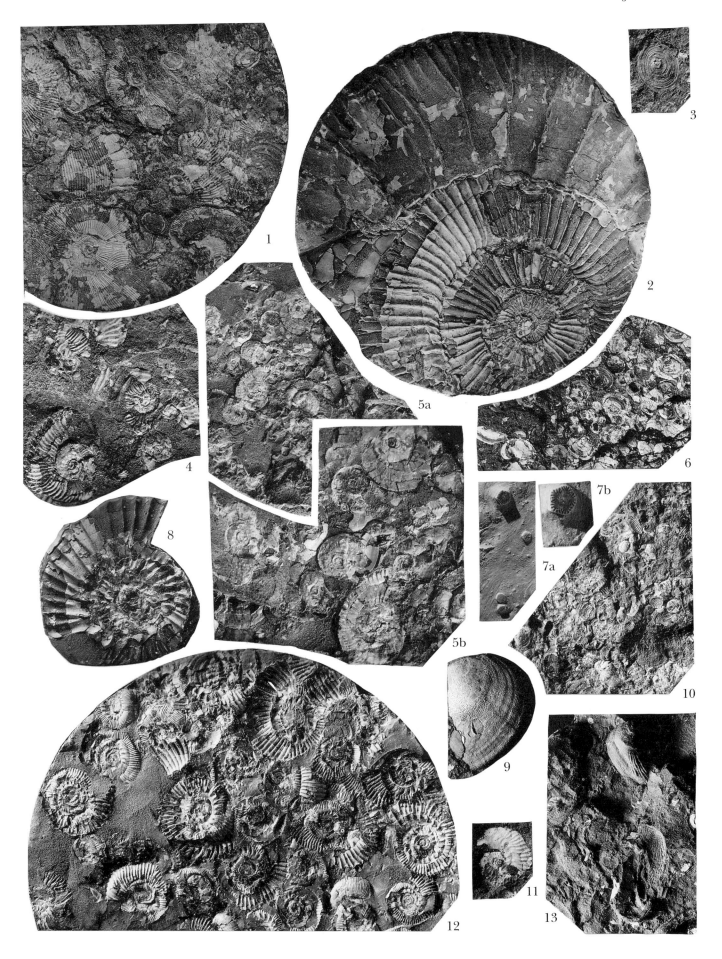

Boreholes

The full thickness of the Kimmeridge Clay was continuously cored in the Wash Feasibility Study borehole at North Wootton [6439 2457] and the BGS oil-shale exploration borehole at North Runcton [6404 1624]. Parts of the sequence were cored in other Wash Feasibility Study boreholes and in stratigraphical boreholes drilled by the Geological Survey in 1970 to 1972. Graphic sections and gamma-ray logs for the Feasibility Study boreholes have been published elsewhere (Gallois, 1979a). The ranges of strata penetrated in the cored boreholes in the district, and their thicknesses, are summarised in Table 8.

The following bed-by-bed description of the standard stratigraphical sequence for the district is based on the combined evidence from all these boreholes. The thicknesses for Beds KC 1 to 46 are those of the North Wootton Borehole [6439 2457] and for Beds KC 47 to 49 those of Geological Survey Borehole 71/65 [5850 4676] in The Wash.

		Thickness m
Bed No.		
KC 49	Oil shale, sparsely shelly with some *Isocyprina* plasters and three thin coccolith-rich bands; interbedded with dark grey, sparsely shelly mudstone	4.49
KC 48	Cementstone, medium grey, muddy limestone with secondary calcite veins; barren; lenticular bed	0.11 to 0.22
KC 47	Mudstone, predominantly dark and medium grey with several thin interbeds of pale grey mudstone and, in upper part, oil shale; generally sparsely shelly with *Isocyprina minuscula* (Blake), *Protocardia morinica* (de Loriol), *Pseudorhytidopilus latissima* (J Sowerby) and *Lingula ovalis* J	

Sowerby common at some levels; *Camptonectes* cf. *morini* (de Loriol), *Grammatodon, Modiolus autissiodorensis* (Cotteau), *Pleuromya, Oxytoma* and small oysters also present; fragments of finely ribbed perisphinctid ammonites including *Pectinatites* (*P.*) *eastlecottensis* (Salfeld); base taken at top of coccolith-rich band — 4.33

KC 46 Mudstone, dark and medium grey, thinly interbedded with fissile, shelly oil shales which include several thin bands of pale brownish grey, coccolith-rich limestone; equivalent of White Stone Band of Dorset; fauna as Bed KC 47 but with fish debris and faecal pellets common in the oil shales; *Pectinatites* (*P.*) *eastlecottensis* common throughout; rarer *P.*(*P.*) *cornutifer* (Buckman) and *P.*(*P.*) *pectinatus* (Phillips) base taken at base of shelly oil shale which marks the lower limit of *P.*(*P.*) *eastlecottensis* — 0.35

KC 45 Mudstone, dark and medium grey with thin oil-shale interbeds common in upper part; prominent pale grey band in middle part; sparsely shelly except in lower part; scattered *Dicroloma, Lingula, Isocyprina, Protocardia, Thracia* and small oysters; *Pectinatites*, including *P.* (*Virgatosphinctoides*) *encombensis* Cope, scattered throughout; colour change at base — 4.70

KC 44 Mudstone, pale and medium grey, highly calcareous; sparsely shelly with well-preserved ammonites including *Pectinatites* (*Virgatosphinctoides*) *reisiformis* Cope and rarer *P.* (*Arkellites*) *hudlestoni* Cope, *P.*(*V.*) *donovani* Cope and, in the lower part, *P.*(*V.*) *pseudoscruposus* (Spath); epizoic oysters common and other bivalves

Table 8 Summary of Kimmeridge Clay sequences proved in continuously cored boreholes in the district.

Borehole (* indicates offshore)	Grid reference	KC beds cored wholly or in part	Thickness m	No. in Figure 15
IGS 72/77*	6313 4835	45 to 49	3.18	1
IGS 71/65*	5850 4676	34 to 44	31.50	2
IGS 71/66*	6117 4475	42 to 47	22.38	3
Daseley's Sand*	5591 3189	1 to 30	64.50	4
Hunstanton	6857 4078	37 to 47	26.39	5
Walker's Marsh	5268 2530	1 to 10	10.22	6
Symington Farm	5499 2507	1 to 12	14.1	7
Admiral's Farm	5651 2486	2 to 16	17.70	8
Ongar Hill	5767 2444	1 to 13	13.4	9
Vinegar Middle	6092 2505	12 to 19	19.30	10
North Wootton	6439 2457	1 to 47	99.65	11
New Common Marsh Farm	5299 2308	14 to 19	20.4	12
Balaclava Farm	5549 2339	1 to 12	8.5	13
Pierrepont Farm	5748 2322	1 to 16	18.99	14
Racecourse Road	5510 2194	19 to 28	18.8	15
Emorsgate	5321 2006	6 to 18	25.9	16
Tilney All Saints	5791 1779	18 to 30	35.15	17
North Runcton	6404 1624	1 to 46	92.79	18
Terrington St John	5401 1435	1 to 18	31.6	19
The Grange	5807 1329	18 to 25	20.5	20
Gayton	7280 1974	37 to 45	14.86	21
Hook Drain	5721 0827	1 to 24	25.67	22
Stowbridge	6022 0640	16 to 36	44 81	23
Marham	7051 0803	37 to 45	17.25	24

Thickness
m

including *Pleuromya*; *Dentalium* and fish fragments; persistent tabular cementstone band near base of bed; pyritised pins; base marked by sharp colour change with evidence of minor erosion in some sections — 6.35

KC 43 Mudstone, medium and dark grey; sparsely shelly with fragments of bivalves including oysters, fish debris, *Dicroloma* and ammonites including *Pectinatites* (*V.*) *pseudoscruposus*, *P.*(*V.*) *reisiformis* and *P.*(*V.*) *wheatleyensis* (Neaverson); base taken at top of oil-shale seam — 0.35

KC 42 Oil shale, fissile, shelly, foraminifera-spotted, with plasters of *Isocyprina* and fragmentary ammonites including *Pectinatites* (*V.*) *grandis* (Neaverson), *P.*(*V.*) *pseudoscruposus*, *P.*(*V.*) *reisiformis* (highest part only) and *P.*(*V.*) *wheatleyensis*; pyritised radial plates of *Saccocoma* common in one band in upper part of bed; interbeds of dark and medium grey, sparsely fossiliferous mudstone occur throughout; *Isocyprina*, small oysters and *Dentalium* locally common; *Protocardia*, *Opis*, 'Chemnitzia', *Dicroloma*; *Pseudorhytidopilus*, *Lingula* and fish fragments also present; the junction of the Hudlestoni and Wheatleyensis zones falls within this bed; base of bed taken at base of oil-shale seam — 3.63

KC 41 Mudstone, dark grey, smooth-textured with pyrite 'halos'; sparsely shelly with bivalves including '*Astarte*' and well-preserved ammonites including *Pectinatites* (*V.*) *grandis*, *P.*(*V.*) *pseudoscruposus*, *P.*(*V.*) *wheatleyensis* and *P.*(*V.*) *woodwardi* (Neaverson), some with epizoic oysters and some with partial infilling of cream-coloured phosphate; rare wood fragments and rhynchonellids; colour change at base — 2.97

KC 40 Mudstone, pale grey in upper part becoming medium and dark grey, brownish grey and silty textured with depth, with some burrowfills of oil shale; sparsely shelly in part with *Lingula*, *Isocyprina minuscula*, *Modiolus autissiodorensis*, *Nanogyra virgula* (Defrance), *Protocardia* and ammonites including *Pectinatites* (*V.*) *grandis*, *P.*(*V.*) *pseudoscruposus* and *P.*(*V.*) *wheatleyensis*; base taken at top of thin oil-shale seam — 1.58

KC 39 Oil shale, brownish grey, shelly with *Isocyprina* plasters, *Modiolus*, *Protocardia*, *Thracia* and oysters; passing down into pale grey and brownish grey, smooth-textured mudstone; colour change at base — 0.92

KC 38 Mudstone, medium and dark grey, sooty-textured in part with much comminuted plant debris; foraminifera-spotted in part; sparsely shelly with *Grammatodon*, *Protocardia* oysters, *Pseudorhytidopilus*, *Lingula* and fragments of *Pectinatites* including *P.*(*Virgatosphinctoides*) sp.; rare thin interbeds of oil shale; bed crowded with *Nanogyra virgula* locally present near base — 2.35

KC 37 Oil shale with interbeds of dark and medium grey mudstone; shelly in part with *Isocyprina* plasters and burrowfill concentrations of other bivalves including '*Astarte*' and *Protocardia*; rarer *Dentalium*, *Dicroloma*, *Pseudorhytidopilus* and *Pectinatites*; base taken at base of densely calcite-cemented oil shale — 2.75

KC 36 Mudstone, dark, medium and pale grey interbedded and with thin interbeds of oil shale; shelly in part with '*Astarte*', *Camptonectes*, *Inoceramus*, *Isocyprina*, *Nanogyra virgula*, *Protocardia*, *Dicroloma* and *Pectinatites* including *P.*(*V.*) *elegans* Cope and *P.* (*Arkellites*) *primitivus* Cope; base taken at base of oil shale immediately above highest *Aulacostephanus* — 3.35

KC 35 Mudstone, dark and medium grey, sparsely shelly, with *Aulacostephanus* (*Aulacostephanoceras*) *autissiodorensis* (Cotteau), *A.* (*Aulacostephanoceras*) cf. *volgensis* (Vischniakoff) and *A.* (*Aulacostephanus*) cf. *fallax* Ziegler; several thin interbeds of brownish grey, fissile, shelly mudstone; iridescent *Aulacostephanus* spat debris common in upper part of bed; '*Astarte*' and oysters common throughout; this bed includes highest *Aulacostephanus* and lowest *Pectinatites?* fragments; base is taken at change from sparsely to very shelly mudstones — 3.75

KC 34 Mudstone, medium and dark grey, blocky, with a few thin beds of oil shale and weakly bituminous shale (quasi oil shale); some thin beds of beef and calcite-coated surfaces; shelly and very shelly with large *Aulacostephanus* (*Aulacostephanoceras*) *autissiodorensis* common in pyrite preservation with calcite and gypsum overgrowths; *A.* (*Aulacostephanoceras*) aff. *rigidus* Ziegler and *A.*(*Aulacostephanoceras*) *volgensis* also present; plasters of *Amoeboceras* (*Nannocardioceras*) spp., including *A.*(*N.*) *krausei* (Salfeld) and a distinctive form with very fine rectiradiate ribbing identified as *A.*(*N.*) aff. *anglicum* (Salfeld), in lowest part of bed; *Nanogyra virgula* (Defrance), *Lopha* and rhynchonellid brachiopod including *Rhynchonella* cf. *subvariabilis* (Davidson) locally common; base taken at change to dark grey mudstones coinciding with upper limit of range of *Aspidoceras* — 2.70

KC 33 Mudstone, dark grey, moderately shelly, with some thin beds of oil shale crowded with iridescent *Amoeboceras* (*Nannocardioceras*) *krausei* and rarer *A.*(*N.*) cf. *anglicum*; *Aspidoceras* (*Aspidoceras*) cf. *longispinum* (J de C Sowerby), *A.*(*A.*) cf. *sesquinodosum* (Fontannes); *Aulacostephanus* (*Aulacostephanoceras*) cf. *autissiodorensis*, *A.* (*Aulacostephanoceras*) cf. *jasonoides* (Pavlow), *A.* (*Aulacostephanoceras*) aff. *kirghisensis* (d'Orbigny), *A.* (*Aulacostephanoceras*) cf. *volgensis*,

Thickness
m

Laevaptchus, Sutneria cf. *rebholzi*
(Berckhemer), oysters and other small
bivalves also present; base taken at top of
prominent oil-shale seam   2.55

KC 32 Oil shale, fissile, shelly, with plasters of
Amoeboceras (*Nannocardioceras*) cf. *anglicum*
and *A.*(*N.*) cf. *krausei*, the latter dominant
in the upper part of the bed, and larger
Amoeboceras (*Amoebites*) including aff.
quadratolineatum (Salfeld), commonly with
foraminifera and ammonite dust debris;
including a few thin beds of pale and
medium grey, sparsely shelly mudstone
with *Aulacostephanus* (*Aulacostephanoceras*)
cf. *eudoxus* (d'Orbigny), *A.*
(*Aulacostephanus*) cf. *rigidus, Laevaptychus*,
small bivalves including oysters, and
Dicroloma; base taken at base of prominent
oil-shale seam   6.25

KC 31 Mudstone, pale grey, burrow-mottled,
sparsely and moderately shelly with large
Nanogyra virgula; interbedded with oil
shale, brownish grey, fissile, shelly,
including *Aspidoceras, Sutneria, Protocardia*
and lowest *Nannocardioceras* plaster;
serpulids locally common in lower part of
bed; base taken at base of prominent oil-
shale seam   0.80

KC 30 Mudstone, pale and medium grey, blocky,
shelly, rubbly, slightly silty-textured;
Nanogyra virgula common and very
common, often large in size, in places
forming a *N. virgula*-rich soft 'limestone'
in upper part of bed; persistent band of
cementstone doggers in middle part of
bed; fauna includes *Amoeboceras*
(*Nannocardioceras*) cf. *anglicum, Aspidoceras*
(*Aspidoceras*) cf. *iphericum* (Oppel), *A.*
(*Aspidoceras*) *sesquinodosum*,
Aulacostephanus (*Aulacostephanoceras*) cf.
eudoxus, A. (*Aulacostephanus*) cf.
pseudomutabilis (de Loriol) *A.*
(*Aulacostephanoceras*) cf. *undorae* (Pavlow),
Laevaptychus, Sutneria sp., small bivalves
including '*Astarte*', *Entolium, Grammatodon,
Isocyprina, Protocardia, Thracia*, with fish
fragments and *Dicroloma; Crussoliceras*
plasters occur in middle part of bed
forming good marker band; base taken at
top of prominent oil-shale seam   4.45

KC 29 Mudstone, pale and medium grey,
moderately shelly, hackly fracture, burrow-
mottled, interbedded with oil shale,
brownish grey, fissile, shelly, intensely
foraminifera-spotted; *Amoeboceras*
(*Amoebites*) spp. including aff. *elegans* Spath,
A. (*Nannocardioceras*) cf. *anglicum*,
Aspidoceras spp. including *A.* (*Aspidoceras*)
longispinum, Aulacostephanus
(*Aulacostephanoceras*) cf. *eudoxus, A.*
(*Aulacostephanus*) cf. *pseudomutabilis, A.*
(*Aulacostephanoceras*) *mammatus* Ziegler, *A.*
(*Aulacostephanoceras*) cf. *volgensis*,
Laevaptychus, Sutneria cf. *cyclodorsata*
(Moesch); *Sutneria eumela* (d'Orbigny)

common; small bivalves including
Grammatodon Nanogyra virgula and other
oysters, *Oxytoma, Palaeoneilo?, Plicatula,
Posidonia, Protocardia* and *Thracia*; fish
fragments, *Dicroloma* and *Lingula*; serpulids
common particularly in upper part of bed;
two bands with *Saccocoma*, one near top of
bed and one at base, form widespread
marker bands; base taken at base of oil
shale   7.35

KC 28 Mudstone, medium to dark grey, burrow-
mottled, silty, in part cemented, shelly with
rubbly and hackly fracture; *Amoeboceras*
(*Amoebites*) spp., *Aspidoceras* sp.,
Aulacostephanus (*Aulacostephanoceras*) cf.
eudoxus, A. (*Aulacostephanoceras*) aff.
mammatus, A. (*Aulacostephanus*)
pseudomutabilis, Laevaptychus, Sutneria spp.;
small bivalves, often fragmentary,
including tiny '*Astarte*', *Grammatodon,
Protocardia* and small oysters; *Lingula,
Dicroloma*, rare small belemnites; passing
down into   3.30

KC 27 Mudstone, medium and dark grey, shelly,
burrow-mottled, rubbly and hackly fracture;
extremely rich in *Nanogyra virgula*, with
Aspidoceras sp., *Aulacostephanus*
(*Aulacostephanus*) *pseudomutabilis* and
Laevaptychus; passing down into   0.80

KC 26 Mudstone, dark and medium grey, slightly
silty; moderately and sparsely shelly with
thin, more shelly beds often rich in
'*Astarte*'; oysters common; brownish grey
quasi oil-shale bed in lower part;
foraminifera-spotted in burrow
concentrations; *Aspidoceras* spp. including
cf. *iphericum* and cf. *sesquinodosum*,
Aulacostephanus (*Aulacostephanoceras*) cf.
eudoxus, A. (*Aulacostephanoides*) cf. *mutabilis*
(J de C Sowerby), *A.* (*Aulacostephanus*)
pseudomutabilis, A. (*Aulacostephanoceras*)
volgensis, Laevaptychus, Dicroloma; highest *A.*
(*Aulacostephanites*) *eulepidus* (Schneid) in
lowest part of bed; passing down into   3.70

KC 25 Mudstone, dark grey, fissile, very shelly with
plasters of *Aulacostephanus*
(*Aulacostephanites*) *eulepidus* and *A.*
(*Aulacostephanoides*) *linealis* (Quenstedt);
also *A.* (*Aulacostephanoceras*) cf. *eudoxus, A.*
(*Aulacostephanus*) cf. *pseudomutabilis*,
Aspidoceras sp., *Laevaptychus, Nanogyra
virgula*, rhynchonellid brachiopods; in
places silty and indistinguishable from
KC 24 below   0.50

KC 24 Mudstone, dark grey, shelly, locally very
shelly at base; slightly silty throughout
becoming very silty at base, partially calcite-
cemented with rare cementstone doggers;
Aspidoceras sp. locally very common,
Aulacostephanus (*Aulacostephanites*) *eulepidus,
A.* (*Aulacostephanoides*) *linealis, A.*
(*Aulacostephanoides*) *mutabilis, Laevaptychus*,
rare *Aulacostephanus* of the *eudoxus* group
and *Sutneria* sp.; rhynchonellid
brachiopods (*Rhynchonella?*) common;
bivalves including *Nanogyra* and other
oysters, *Entolium, Grammatodon*; very rare

		Thickness m

Saccocoma; interburrowed junction and widespread erosion surface at base marking incoming of *Aspidoceras*, *Aulacostephanus* of the *eudoxus* group and *Sutneria* — 1.25

KC 23 Mudstone, medium grey, moderately shelly with *Aulacostephanus* (*Aulacostephanites*) *eulepidus*; as KC 22 but less shelly; passing down into — 0.45

KC 22 Supracorallina Bed: mudstone, pale to medium grey, tough, shelly and intensely shelly in part, with myriads of tiny crushed '*Astarte supracorallina* d'Orbigny' [now *Nicaniella extensa* (Phillips)]; partially calcite-cemented; including also some almost barren pale beds; *Aulacostephanus* (*Aulacostephanites*) *eulepidus*, A (*Aulacostephanites*) cf. *peregrinus* Ziegler; oyster fragments including *Nanogyra* and *Grammatodon*; passing down into — 1.00

KC 21 Mudstone, medium to dark grey, with some paler bands; mostly sparsely shelly with *Aulacostephanus* (*Aulacostephanites*) cf. *peregrinus*, common *Nicaniella* and *Entolium*; passing down into — 0.65

KC 20 Mudstone, medium slightly brownish grey, faintly bituminous, fissile, shelly, with plasters and debris of *Aulacostephanus* (*Aulacostephanites*) *eulepidus*; A. (*Aulacostephanoides*) cf. *linealis*, small oysters, *Protocardia*, fish fragments; locally with thin oil-shale beds; passing down into — 0.15

KC 19 Mudstone, medium grey, sparsely to moderately shelly with a few slightly more shelly bands with *Aulacostephanus* (*Aulacostephanites*) *eulepidus*, A. (*Aulacostephanoides*) cf. *linealis*, A. (*Aulacostephanoides*) cf. *mutabilis*, bivalve fragments including *Entolium* and *Thracia*; silty in lower part with plant debris; shell chips with *Bullapora*, echinoid spines, oysters, a *Nicaniella* plaster and *Pleuromya*; passing down into — 4.90

KC 18 Mudstone, pale grey, blocky, mostly sparsely shelly, but locally shelly in lower part; some very pale bands and persistent cementstone at one level, locally at two; locally silty or with silty burrowfills, gritty in part with broken shell debris, pyrite pins and trails; *Aulacostephanus* (*Aulacostephanites*) *eulepidus*, A. (*Aulacostephanoides*) cf. *mutabilis*, very rare small aptychi cf. *Laevaptychus*, indeterminate perisphinctid; *Entolium*, *Isocyprina* (*Venericyprina*) cf. *compressa* Cox common, *Lopha*, *Nanogyra* (typically of small size with large attachment area), *Parallelodon*, *Pholadomya* cf. *acuticosta* J. de C. Sowerby, *Pinna*, *Pleuromya*, *Dicroloma*; pentacrinoid columnals form widespread marker band; echinoid spines and other fragments common, rare rhynchonellid brachiopods; crustacean claws and pyritised wood; passing down into — 4.60

KC 17 Mudstone, medium and dark grey, interburrowed, moderately shelly; becoming very shelly, silty and intensely burrow-mottled in lower part; *Aulacostephanus* (*Aulacostephanites*) cf. *eulepidus*, A. (*Aulacostephanoides*) cf. *linealis*, A. (*Aulacostephanoides*) *mutabilis*; bivalves, mostly broken, including '*Astarte*', *Lopha*, *Nanogyra*, *Thracia*; rhynchonellid brachiopods common; interburrowed base with minor erosion surface — 1.10

KC 16 Alterations of mudstone, dark grey, very smooth, almost barren; quasi oil shale, dark brownish grey, fissile, shelly; and mudstone, dark grey fissile, moderately shelly, foraminifera-spotted; in part sooty-textured due to comminuted plant debris; *Amoeboceras* (*Amoebites*) sp., *Aulacostephanus* (*Aulacostephanoides*) aff. *desmonotus* (Oppel), A. (*Aulacostephanites*) *eulepidus* (well-preserved, iridescent specimens of a more coarsely ribbed variety particularly abundant), A. (*Aulacostephanites*) cf. *peregrinus*; bivalves including *Nanogyra* and other oysters *Entolium*; in places containing a shelly oyster bed with *Bullapora* and belemnites; rare *Xenostephanus* sp. in lower part of bed; passing down into — 6.80

KC 15 Mudstone, medium to pale grey, becoming silty to very silty towards base; sparsely to moderately shelly with *Aulacostephanus* (*Aulacostephanoides*) cf. *mutabilis*, *Rasenia* (*Semirasenia*) cf. *moeschi* (Oppel), *Anisocardia*, *Pholadomya acuticosta*, *Pleuromya* and oysters; very shelly oyster-rich bed with rhynchonellid brachiopods at base, locally cemented into doggers; interburrowing and chondritic mottling marking minor, but widespread, erosion surface at base — 1.15

KC 14 Mudstone, medium grey, slightly silty, shelly, with body-chamber fragments of large *Rasenia*, *Rasenia* (*Semirasenia*) cf. *moeschi*, *Xenostephanus*?, '*Astarte*', oysters, *Isognomon*?, *Thracia* and rare rhynchonellid brachiopods; passing down into — 0.75

KC 13 Mudstone, medium and pale grey, slightly silty, sparsely shelly, with rare cementstones *Rasenia* (*Rasenoides*) cf. *lepidula* (Oppel), *Aulacostephanus* (*Aulacostephanoides*) aff. *mutabilis*, *Xenostephanus* sp. and *Thracia*; passing down into — 0.90

KC 12 Mudstone, medium and dark grey, shelly; pale silty, shelly bed in middle part; well-preserved *Xenostephanus* common in upper part; *Rasenia* aff. *cymodoce* (d'Orbigny), R. aff. *erinus* (d'Orbigny), R. (*Rasenioides*) cf. *lepidula*, R. (*Eurasenia*) aff. *trifurcata* (Reinecke), R. (*Involuticeras*) sp. and *Lingula*; interburrowed junction — 3.70

KC 11 Mudstone, medium and dark grey, paler in upper part, interburrowed; sparsely to moderately shelly with oysters and *Rasenia* spp. including R. cf. *anglica* Geyer and R. aff. *lepidula*; very foraminifera-spotted in part; interburrowed base — 0.55

		Thickness m

KC 10 Mudstone, pale grey, sparsely shelly, locally foraminifera-spotted; *Rasenia* spp. including *R.* aff. *anglica* and *R.* aff. *erinus*; in places a basal shelly oyster-ammonite bed with serpulids; striking chondritic mottling and interburrowing at base 0.30

KC 9 Mudstone, medium to dark grey, smooth, sparsely to moderately shelly; well-preserved, fine-ribbed *Rasenia*, including *R* (*Semirasenia*) aff. *askepta* Ziegler, with *R.* aff. *anglica* and *R.* (*Rasenioides*) cf. *paralepida* Schneid; passing down into 0.95

KC 8 Mudstone, medium grey, silty and very silty; partially calcite-cemented and locally well-cemented to form doggers; shelly with leached calcite shells and pyrite ghosts; *Amoeboceras* (*Amoebites*) spp. including *A.* aff. *cricki* (Salfeld) locally common; large encrusted *Rasenia* and *Pictonia?* with numerous small *Rasenia*; strikingly interburrowed base marking widespread minor erosion surface with local phosphatisation, rare soft, pale brown phosphatic nodules and hard black phosphatic chips 0.47

KC 7 Mudstone, very pale grey, almost barren, locally intensely interburrowed with silt from above; rare cementstones; passing down into 0.29

KC 6 Mudstone, dark grey, almost barren, finely laminated, sooty-textured, silty; rare *Amoeboceras*; passing down into 0.29

KC 5 Mudstone, medium and dark grey, partially calcite-cemented; rare large *Pachypictonia?* and bivalves infilled with soft, pale brown phosphate; strikingly interburrowed and silty at base with hard, black phosphatic angular chips and similarly preserved ammonite fragments marking widespread erosion surface 0.13

KC 4 Mudstone, very pale grey, almost barren, with cementstone doggers; passing down into 0.07

KC 3 Mudstone, medium grey, shelly to moderately shelly with *Dicroloma*, small *Gryphaea* and other oysters, *Oxytoma* and *Thracia*; passing down into 0.85

KC 2 Mudstone, medium and pale grey with rare cementstones; very sparsely shelly with pyrite trails and pins and rare pyritised perisphinctid nuclei; *Dicroloma*, *Deltoideum delta* (Wm Smith), *Pinna*, *Placunopsis* and *Thracia*; passing down into 0.50

KC 1 Mudstone, strikingly interburrowed, pale and dark grey; sparsely shelly with *Entolium* and *Modiolus*; minor but widespread erosion surface at base marked by burrowing and rare phosphatic nodules, with some bivalves and ammonites (including *Pictonia*) preserved in soft, pale brown phosphate 0.30

AMPTHILL CLAY (see p.30 for details)

FIVE
Cretaceous

The late Jurassic–early Cretaceous rocks of Norfolk crop out in a series of low escarpments that run south from Hunstanton via the eastern part of King's Lynn to Denver, a distance of about 40 km. At its maximum, between King's Lynn and Gayton, the outcrop has a width of about 10 km (Figure 19). Much of this part of the sequence is composed of alternations of loose sand and sandy clay that give rise to low, gently rounded features with few natural exposures. Between Hunstanton and the Babingley River, these features are well displayed and provide considerable aid in geological surveying. South of the Babingley River, the outcrop is overlain by extensive tracts of Quaternary deposits and features are less well developed.

Between Hunstanton and Gayton, the Lower Chalk, capped by the more resistant Melbourn Rock, forms a prominent feature that separates the open, gently undulating topography of the Middle and Upper Chalk from the more dissected and wooded Lower Cretaceous. This feature is largely free from drift deposits and was used in the area south and east of Dersingham to define the eastern limit of the present district.

As with many other parts of England, William Smith, in his pioneer geological maps of the counties, was the first to recognise the main divisions of the Jurassic and Cretaceous rocks of the district. On his geological map of Norfolk (1819) he delineated the 'Oaktree Clay' [Kimmeridge Clay] and the 'Golt Brick Earth' [Gault] separated by 'Sand' (Table 9). Rose (1862, pp.234–236) noted that the beds between the Kimmeridge Clay and the Gault (in south-west Norfolk) and between the Kimmeridge Clay and the Red Chalk (in north-west Norfolk) could be divided into three parts, namely 'loose white sand', a ferruginous sandstone known locally as 'Carstone' (or 'Carr Stone' = Fen Stone) and a 'breccia' (the Carstone of Hunstanton Cliffs). Teall (1875, pp.16–19) recognised three slightly different divisions, comprising the white sands and the Carstone, but with an intervening clay, and these were confirmed by the Geological Survey for those parts of Old Series Sheets 65 and 69 (published 1883 and 1885 respectively) between Hunstanton and King's Lynn. The names Sandringham Sands and Snettisham Beds [Clay] were therefore added to that of the Carstone (Whitaker and Jukes-Browne, 1899, pp.6, 16). In proposing the name Sandringham Sands for the lowest of their three divisions, Whitaker and Jukes-Browne adopted with slight modification a term that had been suggested by Harmer (1877, pl.1).

Whitaker and Jukes-Browne (1899, pp.5, 8) described the Sandringham Sands as being probably over 100 ft (greater than 30 m) thick and consisting mostly of light-coloured, sharp and silvery sands which, in their upper part, are iron-stained and in places cemented into a flag-gy brown stone. The overlying Snettisham Clay, although poorly exposed, was recognised as a continuous bed from Hunstanton (where it had been mistakenly identified by Rose and other early workers as Kimmeridge Clay) as far south as Ashwicken (Figure 19).

The Sandringham Sands invited natural lithological comparison with parts of the Lower Greensand of southern England and all the early workers except William Smith appear to have assumed the two formations to be of similar age. Lamplugh (in Whitaker and Jukes-Browne, 1899, p.22) was clearly aware that the Sandringham Sands were older than the Lower Greensand, but it was not until Spath (1924, p.79) assigned the overlying Snettisham Clay to the Lower Barremian on the basis of its ammonite fauna, that the Sandringham Sands became universally recognised as early Cretaceous in age. They were tentatively assigned to the Berriasian (≈ Ryazanian) to Hauterivian stages. At that time few fossils, other than the rare bivalves and bits of wood recorded by Rose (1862, p.230), had been found in the Sandringham Sands and it was generally assumed that the basal beds of the formation were the earliest Cretaceous rocks in eastern England.

Because of the paucity of exposure and the demise of the few brickpits that had yielded fossils, little new information was gained between the 1880s, the time of the Whitaker and Jukes-Browne survey, and 1961 when the first good sections became available through the lower part of the Sandringham Sands (at West Dereham in south-west Norfolk). Ammonites indicative of a late Jurassic to early Cretaceous age were obtained from the basal beds (Casey, 1961a; 1963). Subsequent exposures in the cuttings for the King's Lynn Bypass (1964–65) and in gas-pipeline trenches (1967–68) in the West Winch area provided additional faunal and lithological details of the lower part of the Sandringham Sands and showed that this part of the formation underwent rapid lateral variation at outcrop. These various sections, together with the present survey and continuously-cored boreholes drilled through the complete late Jurassic–early Cretaceous sequences at Skegness, in The Wash, at Hunstanton, Gayton and Marham to supplement the survey work, enabled new lithological (Casey and Gallois, 1973) and palaeontological (Casey, 1973) classifications to be devised.

The recognition by Casey (1961a) that the Jurassic–Cretaceous boundary, assumed by all previous workers except William Smith to be represented by the unconformity at the junction of the Kimmeridge Clay and Sandringham Sands, lay within a relatively complete, ammonite-bearing sequence stimulated much discussion concerning the position of this boundary in other sequences in the Boreal province. William Smith had noted in the marginal comments on his 1819 map of Norfolk that the 'Portland Rock' was occasionally found in the 'Sand be-

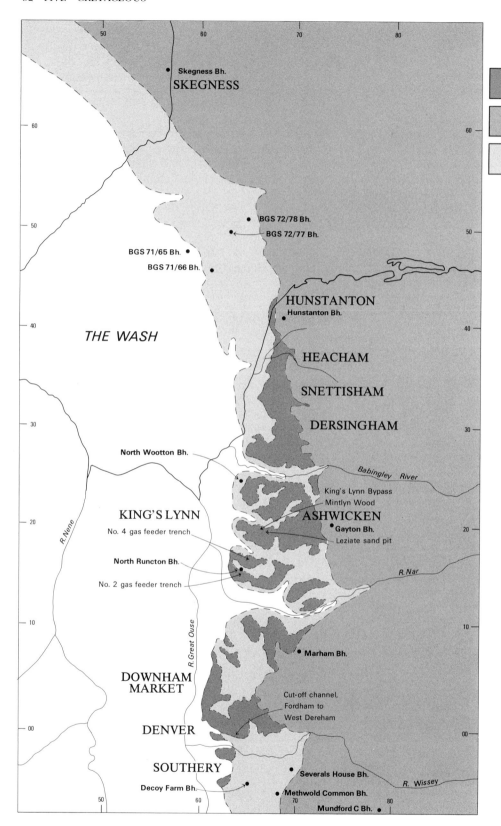

Figure 19 Distribution of the top Jurassic–Lower Cretaceous rocks showing boreholes and key localities.

Table 9 Evolution of the nomenclature of the top Jurassic–Lower Cretaceous rocks of the district. Discussion of the ages of the various formations is given in the text.

Wm Smith, 1819	Rose, 1835; 1862	Teall, 1875	Whitaker and Jukes-Browne, 1899	Present work	Modern stages	
Chalk	Chalk	Chalk	Chalk	Chalk	Maastrichtian to Cenomanian	Upper Cretaceous
Golt Brick Earth	Red Chalk / Gault	Red Chalk / Gault	Red Chalk / Gault	Red Chalk / Gault	Albian	Lower Cretaceous
Sand beneath the Golt	'breccia'	Carstone	Carstone	Carstone		
			erosional gap	Sutterby Marl[1]	Aptian	
				Skegness Clay[1]		
	Carstone	'clay'		Roach	Barremian	
			Snettisham Clay	Snettisham Clay		
				Dersingham Beds	Hauterivian	
	'loose white sand'	'white sands'	Sandringham Sands	Sandringham Sands[2]	Valanginian to Volgian (pars)	
Oaktree Clay	Kimmeridge Clay	Kimmeridge Clay	Kimmeridge Clay	Kimmeridge Clay	Kimmeridgian	Upper Jurassic (pars)

1 Submarine outcrops beneath The Wash only.
2 See Table 10 for subdivisions of Sandringham Sands.

neath the Golt', but the significance of this perceptive observation was overlooked for almost 150 years.

The position of the Jurassic–Cretaceous boundary has provoked considerable discussion in recent years. The nomenclature for the youngest Jurassic and the oldest Cretaceous stages followed here (Table 10) is that given for the Boreal faunal realm by Casey et al. (1977), in which the system boundary is placed at the junction of the Volgian and Ryazanian stages.

Whitaker and Jukes-Browne's (1899) definition of the Sandringham Sands included all the beds between the Kimmeridge Clay and the Snettisham Clay. Lamplugh (in Whitaker and Jukes- Browne, 1899, pp.16–17) drew attention to a bed of ferruginous sandstone in the upper part of the Sandringham Sands which he thought appeared to pass northwards into the lower part of the Snettisham Clay. Kent (1947, p.17) suggested that the disappearance of the Snettisham Clay when traced southwards from its type area was due to truncation by younger Cretaceous strata. The present work suggests that the distribution of the Snettisham Clay results from a complex situation involving both overstep and facies change along the outcrop. In the Hunstanton Borehole, clays that include Snettisham Clay lithologies rest on clean white sands of the Sandringham Sands. When

traced southwards via Heacham, Snettisham and Dersingham, the lower part of these clays appears to pass laterally into a complex rhythmic sequence of sand, silt and clay with a persistent bed of ferruginous sandstone (that noted by Lamplugh) at its base. This last bed rests on the same clean white sand throughout its outcrop. The rhythmic complex and the Snettisham Clay have therefore been grouped under the name Dersingham Beds (Casey and Gallois, 1973).

The stratigraphical results of the Hunstanton Borehole and the field survey suggested that the Snettisham Clay was not unconformably overlain by the Carstone throughout its outcrop (as has previously been supposed by Whitaker and Jukes-Browne, 1899) but that it was comformably overlain by an equivalent of the Lincolnshire Roach (Swinnerton, 1935). The presence of the Roach (Barremian in age) in Norfolk was subsequently confirmed by trenches dug on Hunstanton beach.

The Lower Cretaceous sequence in south-east Lincolnshire has long been known, from the work of Swinnerton (1935), to be more complete than that in Norfolk due to the easterly overstep of the Carstone. In the Skegness Borehole the Roach is separated from the Carstone by two Aptian formations, the Skegness Clay and the Sutterby Marl. Thin representatives of both formations are

Table 10 The subdivisions and zones of the Sandringham Sands.

Formation	Member		Zone (Casey, 1973; Casey et al., 1977)	Stage and Substage	System
Dersingham Beds			not zoned	Hauterivian to Lower Barremian	CRETACEOUS (pars)
Sandringham Sands	Leziate Beds		not zoned	Valanginian	
			Paratollia spp.	Valanginian	
			Peregrinoceras albidum	Upper Ryazanian–Berriasian	
	Mintlyn Beds		*Bojarkia stenomphala*	Upper Ryazanian–Berriasian	
			Lynnia icenii	Upper Ryazanian–Berriasian	
			Hectoroceras kochi	Lower Ryazanian–Berriasian	
	remanié fauna in overlying nodule bed		*Praetollia (Runctonia) runctoni*	Lower Ryazanian–Berriasian	
	Runcton Beds		*Subcraspedites (Volgidiscus) lamplughi*	Upper Volgian	JURASSIC (pars)
	remanié fauna in overlying nodule bed		*Subcraspedites (Subcraspedites) preplicomphalus*	Upper Volgian	
	not recorded in Norfolk		*Subcraspedites (Swinnertonia) primitivus*	Upper Volgian	
	Roxham Beds		*Paracraspedites oppressus*	Middle Volgian	
Kimmeridge Clay			*Pectinatites pectinatus*	Kimmeridgian	

Key to predominant lithology

⌇⌇⌇⌇ erosion surface and phosphatic nodule/pebble bed

∴∴∴∴ Sandstone/sand

— — — Mudstone

also present in boreholes in the central part of The Wash, but both have been removed by the unconformity at the base of the Carstone before the Norfolk coast is reached. At Hunstanton beach the Carstone rests on the Roach. The stratigraphical relationships of the late Jurassic–early Cretaceous formations between Skegness and the River Little Ouse are shown diagrammatically in Figure 20.

For many years the relationship of the Carstone to the Lower Greensand (Woburn Sands) of the Buckinghamshire to Cambridgeshire area was poorly understood. The l9th century geological maps of Norfolk classified both deposits as 'Lower Greensand' and it was generally assumed that they were contemporaneous. A number of authors noted that the two distinctive lithologies appeared to occupy mutually exclusive areas, the Carstone being confined to Norfolk and the Woburn sands to more southerly counties. Recent work in the Ely district (Gallois, 1988) has shown that there is a considerable area of overlap between the outcrops of the Woburn Sands and the Carstone, and that the Carstone (Albian) progressively cuts down into the Woburn Sands (Aptian) in a northerly direction.

Because of its striking lithology and colour and its excellent exposure in the cliffs at Hunstanton, the youngest Lower Cretaceous formation in the district, the Red Chalk, has attracted much attention in the past. It was first recorded by William Smith (1815, p.10) and its stratigraphical relationship as the lateral equivalent of the Cambridgeshire Gault was appreciated as early as l846 by Sedgwick.

Southwards from Hunstanton, the Red Chalk becomes progressively more argillaceous until, in the area to the south of the Babingley River, it is replaced by the Gault. Little has been written about this last-named formation

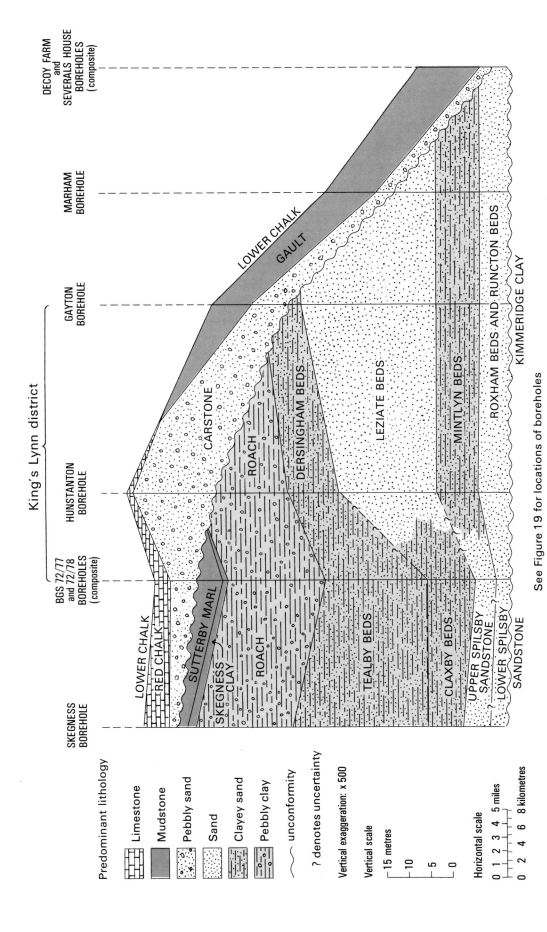

Figure 20 Stratigraphical relationships of the late Jurassic–Lower Cretaceous sequences of Norfolk.

in the district and the Geological Survey memoir (Whitaker et al., 1893), describing the Norfolk part of Old Series Sheet 65 and Jukes-Browne's review in the Cretaceous rocks of Britain (Vol. 1, 1900), remained the most comprehensive accounts until the description of the stratigraphy of the Gault in cored boreholes in the region was published (Gallois and Morter, 1982).

The position of the outcrop of that part of the Chalk which is included in the present survey differs little from that shown on William Smith's map of Norfolk (1819) and on Old Series Sheets 65 and 69. Some of the more important exposures were described by Jukes-Browne (1903) and the stratigraphical sequence has been reviewed and updated by Peake and Hancock (1961).

SANDRINGHAM SANDS

The Sandringham Sands comprise up to about 50 m of predominantly arenaceous sediments which rise from the fringing marshes of The Wash near Heacham and form an almost continuous narrow outcrop from there to Denver (Figure 19). The upper part of the formation consists of clean white and grey sands that were formerly extensively worked for glass sand in the Dersingham to King's Lynn area. These sands give rise to the heathland scenery of the Sandringham area, from which the formation takes its name. Much of the outcrop is covered by Pleistocene deposits and, as the sediments are poorly consolidated, natural exposures are rare, small and generally deeply weathered.

At outcrop between Heacham and Ashwicken, the Sandringham Sands are conformably overlain by the Dersingham Beds (Figure 20). To the south of Ashwicken, the Carstone rests directly on the Sandringham Sands and the unconformity at its base progressively cuts out the latter in a southerly direction.

The Sandringham Sands have a relatively small subcrop that does not extend far beyond their outcrop. They were proved in the deep hydrocarbon-exploration boreholes at Hunstanton, North Creake and South Creake and are probably present at Lexham. Sandringham Sands were absent from the Breckles, Great Ellingham, Lakenheath, Rocklands, and BGS Soham boreholes, and from the more southerly boreholes (Nos. 1 to 20) drilled for the site investigation of the Ely-Ouse Water Transfer Scheme tunnels. When traced north-westwards from their outcrop into The Wash, the Sandringham Sands become rapidly more clayey and pebbly. At most stratigraphical levels the lithologies that characterise the Norfolk outcrop can no longer be recognised in BGS Boreholes 72/77 and 72/78 (8.5 and 9.0 km. from Hunstanton respectively) are reached.

The Sandringham Sands can therefore be regarded as a mass of predominantly sandy sediment with a maximum proven area of distribution of about 12 km (The Wash to the South Creake Borehole) by 55 km (Hunstanton to the River Little Ouse). The extent of the subcrop north of Hunstanton beneath the North Sea is not known. Thick sandy Lower Cretaceous sequences have been proved in many of the hydrocarbon-exploration

boreholes in the southern North Sea (for example, in the Wash No. 1 Borehole), but no core is available at this stratigraphical level and the palaeontological data obtained from the rock cuttings are inadequate to enable the sequences to be classified in detail.

The original depositional extent of the Sandringham Sands, even on the land area, can no longer be determined because of their later erosion by younger formations. Patches of sand, which may in part be correlatives of the Sandringham Sands, occur along the western margin of the London Platform as far south as Buckinghamshire. Only in The Wash, where the formation passes laterally into the Lincolnshire sequence, can the original depositional limit be determined with confidence.

The Sandringham Sands have been divided in the present survey into four members (Casey and Gallois, 1973) on the basis of lithological characters that were recognised in the field by means of exposures (mostly temporary excavations, Figure 21), soil and subsoil brash and topographical features. The outcrop of the formation is mostly drift-free in the area between Heacham

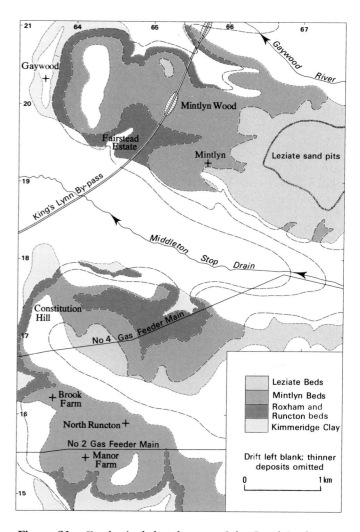

Figure 21 Geological sketch map of the Sandringham Sands outcrop in the King's Lynn area showing the positions of key sections.

and the River Nar and, although the relief is low (5 to 30 m above OD), features are well developed. Within that area, fragments of the more durable lithologies, notably phosphatic pebbles, ironstones and thin clay seams, are commonly exposed in drainage ditches.

The interpretation of such data needs, however, to be treated with caution because the effects of late Pleistocene cryoturbation and solifluction are widespread in the district. Fortunately, the geological structure is extremely simple; dips are mostly less than 1/2° eastwards to north-eastwards and only two small faults (at Mintlyn and Roydon Common) have been recorded, with the result that the features can usually be interpreted with confidence. A typical section through part of the Sandringham Sands outcrop between West Winch [631 167] and Middleton [658 175], based on the field survey and a gas pipeline trench, is shown in Figure 22.

The first complete sections through the lower part of the Sandringham Sands became available in 1961 in the Cut-Off Channel of the Fenland Flood Relief Scheme between Fordham [TL 639 995] and West Derham [TL 662 996] in the Wisbech (Sheet 159) district. The stratigraphy revealed was more complex than had hitherto been described. The basal beds yielded a Middle Volgian (late Jurassic) fauna and the Jurassic–Cretaceous boundary was shown to lie within the lower part of the formation in a condensed sequence of glauconitic sands with phosphatic pebble beds (Casey, 1961a).

The rich ammonite faunas obtained from the Sandringham Sands in Norfolk and from the Spilsby Sandstone of Lincolnshire, enabled Casey (1973) to establish the first British zonal scheme for the beds at this (Volgian and Ryazanian) stratigraphical level. The zonation and nomenclature of the Sandringham Sands is summarised in Table 10. The recognition of the stratigraphical importance of the Sandringham Sands, and the availability of unweathered, fossiliferous marine material prompted additional palaeontological studies. Ager (1971) has described brachiopods and Creber (1972) gymnospermous wood from the basal beds of the Sandringham Sands, and Kelly (1977, 1984–92) and Pinckney (1978) have described the bivalves and belemnites respectively of the lower part of the Sandringham Sands and the Spilsby Sandstone.

Roxham Beds

The basal member of the Sandringham Sands consists of 3 to 6 m of poorly consolidated grey and yellowish green, locally glauconitic sands with disseminated pyrite and, at some levels, pyritic nodules. The formation takes its name from the section in the Cut-Off Channel at Roxham Farm [TL 638 995], West Dereham, where Casey (in Casey and Gallois, 1973) recorded 3.5 m of glauconitic slightly clayey sand with a basal pebbly sandstone.

The Roxham Beds have been mapped from their most northerly occurrence, on the banks of the Babingley River [652 255], to their disappearance beneath Methwold Fens in the Ely (Sheet 173) district. Temporary sections at King's Lynn, North Runcton, Middleton and Wormegay [655 119] have exposed parts of the formation and have suggested that it remains lithologically uniform at outcrop throughout the present district. This uniformity has been confirmed in the subcrop by boreholes at Hunstanton, North Wootton, Gayton and Marham (Figure 23). When traced north-westwards from the Babingley River, beneath The Wash, the Roxham Beds are partly cut out by erosion at the base of the overlying Mintlyn Beds (e.g. Borehole 72/77), but continue with unchanged lithology into Lincolnshire as the lower part of the Spilsby Sandstone.

At the base of the Roxham Beds, the junction with the Kimmeridge Clay is marked by an erosion surface overlain by up to 1.5 m of densely calcite-cemented, fine-grained, grey sandstone crowded in its lower part with pebbles of black chert (lydite) and black phosphate. Many of the latter are water-worn moulds of pavloviid ammonites and myid bivalves derived from the Kimmeridge Clay (Plate 3). Vein quartz and other pebbles are less common. The basal sandstone is intensely bioturbated throughout with Ophiomorpha and other burrows common, and with Ophiomorpha and Skolithos burrows filled with sand penetrating several tens of centimetres into the underlying Kimmeridge Clay (Plate 3). The sandstone forms a fining-upward unit in which very pebbly sand passes up into shelly sand and then into sparsely shelly, very fine-grained sand with abundant Ophiomorpha (Plate 3).

This sandstone persists throughout the Sandringham Sands outcrop and subcrop in the district and forms a marked feature and spring-line at outcrop. It is present in every borehole in the district that has penetrated this stratigraphical level. Its thickness is variable, and the cement may locally be patchy. The common occurence of large rounded boulders (up to 2 m across) of this sandstone in the local Chalky-Jurassic till suggests that the sandstone may locally be present as concretionary masses rather than a tabular bed. It has been traced as a prominent seismic reflector beneath The Wash (Wingfield et al., 1979) and has been proved at the base of the Spilsby Sandstone on the Lincolnshire side of The Wash in boreholes (e.g. Skegness Borehole) and at outcrop (Beds A and B of Swinnerton, 1935). The geographical source of the sandstone erratics in the till is not entirely clear. Some have probably been derived from the west Norfolk outcrop, and many more from former outcrops on the floor and eastern side of the Wash, from areas now overlain by Recent deposits. Erratics of almost identical lithology, but of probable Lincolnshire origin, have been recorded in the drift at Leziate (Casey, 1973, pl. 3, fig. 4). They contain preplicomphalus Zone ammonites, a younger fauna than that recorded from the sandstone at outcrop in Norfolk (see below).

The basal phosphatic pebble bed of the Roxham Beds was correlated by Casey (1973), on the basis of its distinctive lydite cherts and derived phosphatised pavloviid ammonites, with the Upper Lydite Bed, a phosphatic pebble bed that marks the base of the Portland Beds in Buckinghamshire and Wiltshire. This correlation is, however, possibly misleading because the Upper Lydite Bed is not a marker bed in the chronostratigraphical sense. Its age varies from kerberus to oppressus zone (early to late Middle Volgian) between Wiltshire and Norfolk, and from op-

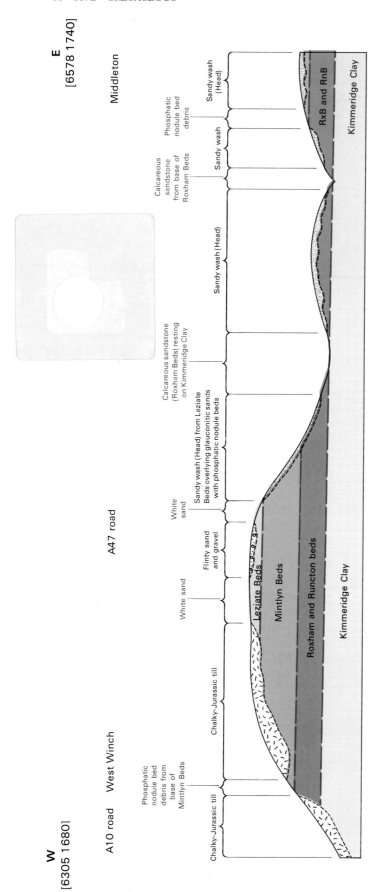

W
[6305 1680]

A10 road   West Winch

Phosphatic
nodule bed
debris from
base of
Mintlyn Beds

Chalky-Jurassic till

A47 road

White sand

Chalky-Jurassic till

Flinty sand
and gravel

White
sand

Sandy wash (Head) from Leziate
Beds overlying glauconitic sands
with phosphatic nodule beds

Calcareous sandstone
(Roxham Beds) resting
on Kimmeridge Clay

Sandy wash (Head)

Calcareous
sandstone
from base of
Roxham Beds

Sandy wash

Phosphatic
nodule bed
debris

Sandy wash
(Head)

Middleton

E
[6578 1740]

Leziate Beds

Mintlyn Beds

Roxham and Runcton beds

Kimmeridge Clay

RxB and RnB

Kimmeridge Clay

Vertical exaggeration x 18 approx

Notes refer to strata exposed in trench

See Figure 21 for line of trench

Figure 22   Geological sketch section along the No. 4 Gas Feeder Main trench between West
Winch and Middleton.

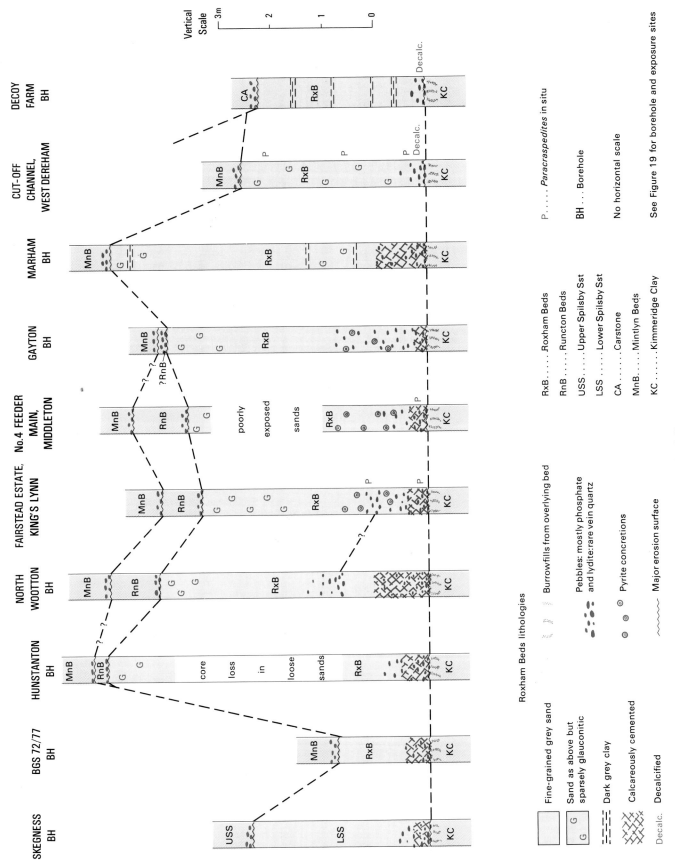

Roxham Beds lithologies

Fine-grained grey sand	Burrowfills from overlying bed
G Sand as above but sparsely glauconitic	Pebbles: mostly phosphate and lydite:rare vein quartz
Dark grey clay	Pyrite concretions
Calcareously cemented	Major erosion surface
Decalc. Decalcified	

P..... *Paracraspedites* in situ

BH ... Borehole

No horizontal scale

See Figure 19 for borehole and exposure sites

RxB..... Roxham Beds
RnB..... Runcton Beds
USS..... Upper Spilsby Sst
LSS..... Lower Spilsby Sst
CA...... Carstone
MnB..... Mintlyn Beds
KC...... Kimmeridge Clay

Figure 23   Correlations between the Roxham Beds sequences in Norfolk.

pressus to *primitivus* zone between Norfolk and north Lincolnshire.

The most prolific source of indigenous fossils from the Roxham Beds has been the basal calcareous sandstone. Pyritised fossils occur sparingly in the overlying loose sands. At Roxham Farm, West Dereham, the sandstone was decalcified by weathering to a loose pebbly sand with pyritic nodules containing moulds of bivalves and ammonites. These were identified by Casey (1973, p.199) as including the Middle Volgian zonal index ammonite *Paracraspedites oppressus* Casey, together with *P.* cf. *bifurcatus* Swinnerton, *P. stenomphaloides* Swinnerton and *Glottoptychinites? trifurcatus* Swinnerton. The Roxham Beds sands in trenches at Fairstead Estate [6430 1945], King's Lynn, also yielded pyritised, but poorly preserved, *Paracraspedites*. Elsewhere at outcrop in the district, the pyritic fossils above the basal bed crop out above the water table and have been destroyed by oxidation. As a result, the basal sandstone is the only consistently fossiliferous horizon. An unweathered exposure [6507 1709] of this bed near Fair Green, Middleton, yielded abundant *Paracraspedites* spp., the belemnite *Acroteuthis*, the large terebratulid brachiopod *Rouillieria ovoides* (J Sowerby) (*Terebratula rex* auctt.) and numerous bivalves. The same bed was also encountered in drainage trenches at Wormegay [655 119], but was decalcified and much of its fauna destroyed. Unweathered basal Roxham Beds sandstone was also present in the Hunstanton, Gayton and Marham boreholes, but the core diameters were too small for much of the fauna to have been sampled. Many of the erratic boulders of basal sandstone in the till in the cuttings of the King's Lynn Bypass and in the overburden at Leziate Sand Pits yielded a prolific fauna of ammonites (Casey, 1973), bivalves (Kelly, 1977; 1984–92) belemnites (Pinckney and Rawson, 1974; Pinckney, 1978) and brachiopods (Ager, 1971).

Details

Babingley River to River Nar

The Roxham Beds are deeply weathered everywhere at outcrop and commonly occur only as loose, brown, grey or green sands in ditches, or where deeply ploughed. Even the basal calcareous sandstone breaks down to a loose pebbly sand. The junction with the Kimmeridge Clay can be traced by means of a poorly developed feature and an impersistent spring line.

Spoil from foundation trenches [6430 1935] at Fairstead Estate, King's Lynn, included the basal sandstone, loose glauconitic sands and a few pyritised sandstone concretions. These last-named enclosed poorly preserved *Paracraspedites*. The basal sandstone was exposed, beneath the water table and unweathered, in the No. 4 Gas Feeder Main trench [6504 1709] near Fair Green, Middleton. It consisted of 0.3 m of very pebbly, shelly sandstone that passed up into sparsely shelly, intensely bioturbated sandstone with common *Ophiomorpha*. The junction with the Kimmeridge Clay was disturbed by cryoturbation and the excavators, but appeared to be slightly irregular. Pyritised burrow infillings of sand and pebbly sand were common in the highest 0.3 m of the clay. The diverse fauna of the sandstone includes common *Paracraspedites* spp., *Acroteuthis*, *Rouillieria ovoides*, and common *Camptonectes*, *Entolium*, *Modiolus*, *Moyophorella*, *Oxytoma* and *Pleuromya*.

Large, richly fossiliferous erratic blocks of the basal sandstone have been recorded from a number of excavations in the district,

Plate 3 Details of the basal cemented bed of the Roxham Beds (Sandringham Sands).

A. Partially rounded to well-rounded erratic boulders of the basal beds of the Roxham Beds in Chalky-Jurassic till close to the Roxham Beds outcrop. Snowre Hall, Fordham [TL 621 995].

B. Almost complete section through the cemented basal bed showing concentration of phosphatic and other pebbles at base, and gradation upward into well-sorted bioturbated sand. Snowre Hall; ×0.35.

C. Detail of *Ophiomorpha* burrows in upper part of bed shown in B. Snowre Hall; ×1.

D. Phosphatised pavlovid ammonites in the basal bed of the Roxham Beds. Snowre Hall; ×1.

E. Section through the basal Roxham Beds pebble bed showing the rich calcareous fauna in which belemnites (be) and bivalves are dominant. The phosphatic pebbles contain derived bivalves (bi), foraminifera and other small fossils. North Runcton Borehole 8.90 m; ×1.

F. Detail of basal pebble bed showing graded bedding, shelly fossils, and (upper right) a large bivalve shell that has acted as a trap for small phosphatic pebbles. Snowre Hall; ×0.6.

notably in the cuttings [660 216 to 666 228] of the King's Lynn Bypass between Reffley Wood and Rising Lodge, in the overburden at Leziate Sand Pits [672 194] and in the No. 4 Gas Feeder Main trench [631 167] at West Winch. The ammonites from some of these have been described by Casey (1973), the belemnites by Pinckney (1978), the bivalves by Kelly (1977) and the brachiopods by Ager (1971). All but one of these erratics contain *Paracraspedites* spp. indicative of the *oppressus* Zone, and they were probably derived from local outcrops in Norfolk or beneath The Wash. A single block from Leziate contains *Subcraspedites* indicative of the *preplicomphalus* Zone, and was derived from the more northerly part of the Lincolnshire outcrop or from an as yet undescribed horizon within the upper part of the Roxham Beds (see below). The *oppressus* Zone blocks have yielded the brachiopods *Rhynchonella* aff. *subvariabilis* (Davidson) and *Rouillieria ovoides*; the bivalves *Anisocardia* sp. nov., *Arctotis* cf. *intermedia* Bodylevsky, *Camptonectes morini* (de Loriol), *Codakia? crassa* (J Sowerby), *Corbicellopsis claxbiensis* (Woods), *Cucullaea* (*Dicranodonta*) *vagans* Keeping, *Entolium orbiculare* (J Sowerby), *Falcimytilus suprajurensis* Cox, *Grammatodon* sp., *Iotrigonia* sp. nov., *Isognomon* cf. *cuneatum* Zakharov, *Liostrea plastica* (Trautshold), a lucinid, *Mesomiltha kostromensis* (Gerasimov), *Modiolus* sp., *Musculus* (*M.*) *fischerianus* (d'Orbigny), *Myophorella* (*M.*) *intermedia* (Fahrenkohl), *M.* sp., *Nanogyra thurmanni* (Etallon), *Neocrassina* (*Lyapinella*) *asiatica* Zakharov, *N.* (*Pressastarte*) sp. nov., *Nicaniella claxbiensis* (Woods), *N. mnewnikensis* (Milaschwitch), *N.* sp., an ostreid, *Oxytoma octavia* (d'Orbigny), *Pinna cf. constantini* de Loriol, *Pinna* sp., *Plagiostoma* sp., *Pleuromya* cf. *uniformis* (J Sowerby), *Pleuromya* sp., *Protocardia* (*P.*) *concinna* (von Buch), *Pseudolimea arctica* Zakharov and *Senis* aff. *petschorae* (Keyserling); indeterminate gastropods, and the cephalopods *Paracraspedites* spp. and *Acroteuthis* sp..

Boreholes

The full thickness of the Roxham Beds was cored in IGS Borehole 72/77 in the mouth of The Wash, and in the BGS Hunstanton, North Wootton, North Runcton, Gayton and Marham boreholes, but because of the poorly cemented nature of the sands large parts of the middle and upper parts of the sequence were lost at most of these localities. Borehole 72/77

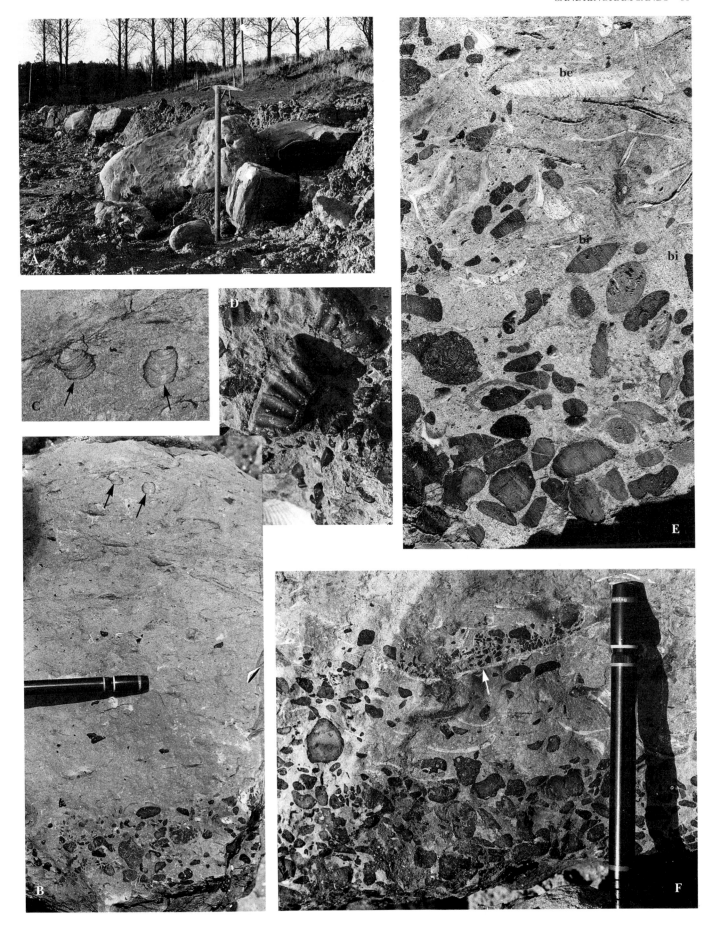

proved 0.52 m of densely calcite- and pyrite-cemented, fine-grained sandstone with abundant black phosphatic pebbles in its lower part, some enclosing casts of ammonites or bivalves derived from the Kimmeridge Clay, resting on an irregular, burrowed surface of Kimmeridge Clay. The only identifiable fossils preserved were *Entolium* and *Myophorella*, but the lithology is sufficiently distinctive for it to be clear that this is the basal bed of the Roxham Beds of the Norfolk outcrop. Seismic profiling between the borehole and the Lincolnshire and Norfolk coasts strongly suggests that this bed is continuous beneath The Wash. The remaining part of the Roxham Beds, 1.2 m thick and presumed to be loose sands, was lost during coring.

The Hunstanton Borehole proved a total of 6.1 m of Roxham Beds. This was made up of 0.3 m of densely calcite- and in part pyrite-cemented, very shelly, pebbly sandstone overlain by 0.2 m of sparsely shelly, bioturbated sandstone with common *Ophiomorpha*. These basal beds were overlain by 5.0 m of soft, fine-grained, pale grey and greenish grey glauconitic sand with traces of bioturbation including *Ophiomorpha*, overlain by 0.6 m of calcite- and pyrite-cemented sandstone with pectinid bivalves, rare belemnites, *Ophiomorpha* and scattered small (2 to 4 mm diameter), well-rounded, black phosphatic pebbles. The lithology and fauna of the basal bed is the same as that of the basal bed at outcrop and in erratics in Norfolk. The higher cemented bed is similar in lithology and fauna to the basal bed and is of particular interest in that it has not been recorded at any other locality. It is likely that cemented patches occur at several stratigraphical levels within the Roxham Beds, but that these become decalcified at outcrop. Cemented horizons such as this in the upper part of the member could be the source of the *preplicomphalus* Zone faunas that have been recorded in erratic blocks at Leziate.

The basal sandstone was 0.97 m thick in the North Wootton Borehole and was overlain by up to 5 m of soft, pale and dark grey and greenish grey, glauconitic sand and clayey sand. The basal, cemented bed contained the usual pebbles and rich bivalve fauna; the higher beds yielded only a few poorly preserved bivalve fragments and the ubiquitous *Ophiomorpha* burrows. The middle part of the sequence included a phosphatic nodule bed (0.15 m thick) overlain by 0.55 m of dark green highly glauconitic sands; both beds appear on lithology to have been displaced, by labelling error, from the overlying Runcton Beds. With these beds removed, the Roxham Beds may, therefore, be only about 5.3 m thick at North Wootton.

The North Runcton Borehole was sited on glacial deposits (8.87 m thick) which rested on the basal sandstone of the Roxham Beds (0.38 m thick). As elsewhere, this bed was densely cemented, shelly and pebbly in its lower part.

The full thickness of the Roxham Beds was cored in the Gayton and Marham boreholes, where 4.9 m and 5.9 m respectively of the member were proved. The sequences are graphically summarised in Figure 22. The basal sandstone was well developed in both sections and is overlain by loose grey and greenish grey, sparsely fossiliferous, glauconitic, clayey, fine-grained sands with a few pyritic concretions.

Runcton Beds

The Runcton Beds consist of up to 1.5 m of dark green, clayey, highly glauconitic sands with phosphatic nodule beds and phosphatic pebble beds[1] at several levels. They contain bivalves and ammonites in clay-cast and phosphatic preservation, but few of these survive weathering at outcrop. The sands are intensely bioturbated throughout and no other sedimentary feature has been recorded in them either at outcrop or in boreholes. Where observed in temporary sections, the contact with the underlying Roxham Beds is erosional and marked by a sharp colour contrast and either a phosphatic nodule-bed that rests on an irregular burrowed surface, for example at North Runcton, or by channelling, for example at West Dereham.

The Runcton Beds, which are themselves overlain unconformably by the Mintlyn Beds, are a highly condensed sequence and the sedimentary breaks bounding them and within them occur in such a complex manner that the sequence is laterally very variable in detail. They have been mapped almost continuously between the Babingley River and the River Nar, but because of the narrowness of their outcrop they have been combined for convenience with the Roxham Beds (with which they have Jurassic faunal affinities) on the geological map. Phosphatic debris from the pebble beds that bound the Runcton Beds occurs as persistent soil fragments and drain spoil.

The present-day distribution of the Runcton Beds appears to be that of a narrow subcrop elongated parallel to the outcrop. In the northern part of the district the outcrop is overlain by the Recent deposits of Wolferton Marshes and The Wash. It presumably continues as far north as the latitude of Hunstanton (proved in the Hunstanton Borehole), but does not extend much farther northwards (absent in Borehole 72/77). To the south (West Dereham) and east (Gayton and Marham boreholes) the formation is absent due to overstep by the Mintlyn Beds, and the subcrop is narrow.

The Runcton Beds attain their maximum thickness of about 1.5 m, estimated from the geological survey, at outcrop in the North Runcton area. Complete sections through them were exposed in temporary excavations at Brook Farm [6370 1630] and Manor Farm [6515 1555], North Runcton, and at Hardwick [635 174]. At all three localities, the base of the Runcton Beds is marked by a phosphatic pebble bed that rests on a slightly irregular, burrowed surface of Roxham Beds. The green, glauconitic sands of the Runcton Beds form a marked colour contrast with the uniformly grey Roxham Beds, and a change of permeability between the slightly clayey Runcton Beds and

---

1 The distinction between phosphatic nodules and phosphatic pebbles is commonly overlooked in the description of Mesozoic strata. In argillaceous formations such as the Ampthill Clay, Kimmeridge Clay and Gault, phosphatic nodules occur in situ as accretions around fossils and trace fossils. When intraformational erosion occurs such nodules commonly become incorporated in younger strata as waterworn pebbles or simply as exhumed masses from which the adjacent soft sediment has been removed by winnowing. Additional phosphatisation can occur during this process, but its effects are generally slight and the difference between in-situ phosphatic *nodules* and derived phosphatic *pebbles* is usually obvious. In the Sandringham Sands the situation is more complex. Phosphatisation and winnowing by currents appear to have occurred almost continuously so that, whilst individual phosphatic nodules and pebbles can be recognised, much of the phosphatic material in the sands has a complex origin. Irregular phosphatic masses up to l0 cm across are common at some levels and these may contain several generations of pebbles and more than one phase of phosphatisation of the matrix. Materials of composite origin such as this are described as 'pebble/nodule beds' in the present account. Where recognisable, 'pebbles' and 'nodules' are distinguished.

the relatively clean Roxham Beds gives rise to a line of seepage at their junction. At Hardwick, the overlying basal Mintlyn Beds pebble bed has removed the top part of the Runcton Beds and the formation there is less than 1 m thick. Northwards from Hardwick, the formation continues as about 1 m of glauconitic sand with phosphatic pebble/nodule beds at several levels. It was proved as loose debris in excavations at King's Lynn and North Wootton Common, and in the North Wootton Borehole.

The Runcton Beds are absent in the central part of The Wash, where the basal Mintlyn Beds pebble bed rests on Roxham Beds (Figure 24). On the Lincolnshire side of the Wash, Casey (1973, fig. 2) has correlated the Runcton Beds with the bulk of the Lower Spilsby Sandstone of Lincolnshire on faunal evidence. To the south of North Runcton, the Runcton Beds have been recorded as loose debris in temporary sections at Wormegay [655 119], Wimbotsham [618 052], Downham Market [612 041], Denver [615 005] and West Dereham [TL 645 997]. At this last locality they occur as *remanié* phosphatic pebbles within a thick pebble/nodule bed complex at the base of Mintlyn Beds.

The fauna of the Runcton Beds is sparse. Derived *Subcraspedites* (*S.*) cf. *sowerbyi* Swinnerton, indicative of the former presence of the Upper Volgian *Subcraspedites* (*S.*) *preplicomphalus* Zone, occur in the basal pebble bed of the formation at North Runcton (Casey, 1973, p.200). No evidence of the underlying *Subcraspedites* (*Swinnertonia*) *primitivus* Zone has yet been recorded in Norfolk. *Subcraspedites* (*Volgidiscus*) *lamplughi* Spath, the zonal index of the *lamplughi* Zone (Casey, 1973, p.200), the youngest Jurassic zone in Britain, occurred in soft brown phosphatic preservation (clearly indigenous) in the trenches at Manor Farm and Brook Farm (Casey *in* Casey and Gallois, 1973). The same ammonite is commonly a derived component of the basal Mintlyn Beds pebble bed in Norfolk and Lincolnshire, indicating that beds of *lamplughi* Zone age were formerly widespread.

Details

North Runcton Area

The thickest observed sequence recorded to date is that in the No. 2 Gas Feeder Main trench at Manor Farm, North Runcton [6515 1555], where the formation consisted of 1.15 m of dark green, glauconitic clayey sand with bivalves and ammonites in clay-cast and phosphatic preservation. The following section (after Casey and Gallois, 1973, p.6) is a composite based on sections recorded over about 100 m of trench: the lowest nodule bed was measured and collected in situ below the water table, but was seen as spoil:

	Thickness m
MINTLYN BEDS	
Basal phosphatic nodule/pebble bed	0.10 to 0.15
RUNCTON BEDS	
Sand, bright green, glauconitic, clayey with brown phosphatised clay wisps; clay casts and 'ghosts' of fossils	0.15
Phosphatic nodule bed, brown, fragile; mostly moulds of ammonites and bivalves in matrix as	
above; the ammonites, including *Subcraspedites lamplughi* Spath and *S.* spp. nov., tend to fall apart at sutures	0.03 to 0.10
Sand, dark green, clayey, glauconitic	0.75
Phosphatic nodules/pebbles, black, rough, some in clotted masses; in situ but below water level in trench; *Subcraspedites* cf. *sowerbyi* Spath loose in this mode of preservation	0.15
ROXHAM BEDS (part only)	
Sand, grey-green, faintly glauconitic, slightly clayey, loose; collected from below water level in trench.	

A similar, but laterally more variable sequence, was recorded in three trenches about 5 m apart at nearby Brook Farm [6370 1630], North Runcton. These showed the non-sequences bounding and within the Runcton Beds to be complex and laterally variable over short distances (Figure 25). The following composite section was measured:

	Thickness m
MINTLYN BEDS (see p.72 for details)	
Phosphatic nodule/pebble bed at base resting on irregular surface of	
RUNCTON BEDS	
Sand, fine-grained, greenish grey, weathering to brownish grey, glauconitic, slightly clayey	0.33
Nodule bed: impersistent band of widely spaced, soft, medium brown phosphatic casts of small bivalves and a *Subcraspedites*	up to 0.02
Sand, fine-grained, dark green, glauconitic; clayey throughout, with wisps of glauconitic clay; small patches of secondary ferruginous cement	0.33
Nodule/Pebble bed: composite bed, ferruginously cemented in part, consisting of an upper (0.08 m-thick) and a lower (0.10 m-thick) bed separated by a secondary ferruginous pan; upper bed composed of large, irregular, dark bluish grey phosphatic nodules mixed with aragonitic shell debris and soft, crumbly, medium brown phosphatic nodules; irregular base	0.12 to 0.18
ROXHAM BEDS	
Sand, dark grey (pale above water table), fine-grained, faintly glauconite-speckled, with prominent burrowfills of dark green clayey sand extending down from overlying bed; well-sorted, clean; faintly purplish black in part due to comminuted plant debris	seen 1.16

These same beds were probably briefly exposed in the trench for the No. 4 Gas Feeder Main between the King's Lynn to Norwich road [643 169] and Middleton [650 171] (Figure 22). Because of the loose, waterlogged nature of the sands, the pipeline was emplaced shortly after excavation and no section was recorded. The spoil included glauconitic, greenish grey, fine-grained, clayey sands with phosphatic nodule beds within and at their base, including, in the middle part of the bed, soft, medium brown, phosphatised bivalves (as seen at Brook Farm and Manor Farm).

Boreholes

The Runcton Beds appear to be present in the Hunstanton Borehole only as a *remanié* deposit that consisted of 0.27 m of ferruginously cemented fine-grained sandstone crowded with

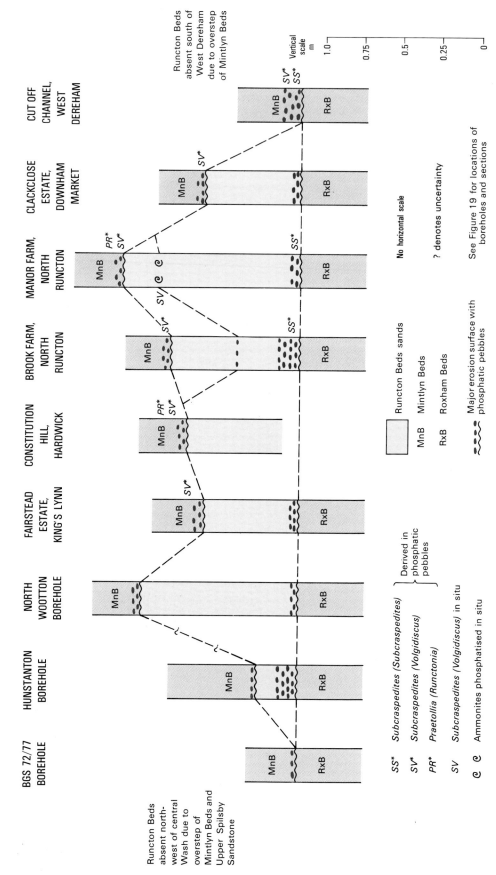

Figure 24   Correlations between the Runcton Beds sequences in Norfolk.

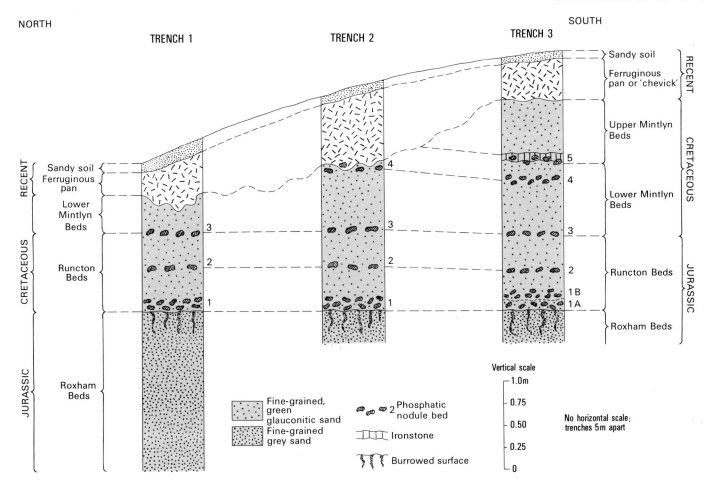

Figure 25 Correlation of trench sections in Jurassic–Cretaceous boundary beds at Brook Farm, West Winch.

phosphatic pebbles. There is no faunal evidence to differentiate this bed from the basal Mintlyn Beds phosphatic pebble bed that overlies it, but many of the pebbles in the lower bed are lithologically similar to those of the Runcton Beds and their size and concentration suggests in-situ winnowing rather than transportation.

The North Wootton Borehole proved between 1.0 and 1.7 m of Runcton Beds (uncertainties in the core labelling make a precise determination impossible) that consisted of soft, bright green and dark green, fine- and medium-grained, intensely bioturbated, clayey, highly glauconitic sands with ferruginously cemented phosphatic pebble beds at the base and near the top, and with uncemented phosphatic pebble beds at two other levels.

Mintlyn Beds

The Mintlyn Beds consist of grey, greyish green and green glauconitic sands and clayey sands (Plate 4) with thin beds and lines of doggers of clay ironstone (siderite mudstone) and thin (mostly less than 15 cm thick) beds of green glauconitic clay. Erosion surfaces marked by phosphatic pebble beds occur at two or more levels in the lower part of the formation. The high iron content and low permeability of the Mintlyn Beds cause them to weather to poorly drained, dirty orange-brown sand with much secondary limonite, commonly in the form of iron-pan.

In the few large exposures recorded in the Mintlyn Beds, the ironstone bands and clay seams appear to be even- and parallel-bedded over long distances. Small-scale sedimentary structures are rare at outcrop because of the deep weathering, but in borehole cores bioturbation, planar lamination and rarer cross-lamination are picked out by clay wisps and concentrations of glauconite, shell debris, carbonaceous sand and clean white sand. Indeterminate burrow mottling occurs throughout the member, with *Chondrites* and *Rhizocorallium* common at several levels. Minor erosion surfaces with cf. *Skolithos* and other burrows commonly mark the junctions of two lithologies, and several of the ironstones are partially phosphatised to form hardgrounds, each with a burrowed and bored upper surface that carries a concentration of shells and, in a few instances, phosphatic pebbles. Body fossils are rare except in the ironstones and a few pyritic concretions, and as casts in some of the thicker clay seams. Plant scraps and bits of coalified wood are common throughout.

The formation crops out almost continuously between Wolferton and West Dereham (in the Wisbech district), mostly on low ground. The higher parts of the Mintlyn Beds are largely covered by loose sandy wash derived from the overlying Leziate Beds in the present district.

Only on the interfluves at the widest part of their outcrop, between the Babingley River and the River Nar, can the Mintlyn Beds be readily studied. Even there, they are deeply weathered and give rise to poorly drained soils rich in secondary iron. The member takes its name from that area, from the former parish of Mintlyn, near King's Lynn, where much of the sequence was well exposed in the Mintlyn Wood cutting [652 200] of the King's Lynn Bypass (Figure 21).

Eastwards from their outcrop, the Mintlyn Beds are present in the same lithologies in the Hunstanton, Gayton and Marham boreholes (Figure 26). Their original depositional thickness appears to have been somewhat variable in their known area of preservation. They thin northwards and eastwards from an estimated thickness of about 20 m at King's Lynn to 11.4 m in the Hunstanton Borehole, 12.9 m in the Gayton Borehole and 14.0 m in the Marham Borehole. At outcrop, the progressive southerly overstep of the Carstone across the Sandringham Sands brings the Carstone to rest on the Mintlyn Beds at Ryston Park, Denver [624 014], and then cuts out the Mintlyn Beds between there and Southery in southwest Norfolk.

The northerly and westerly limits of the Mintlyn Beds are poorly known. A thin representative of the formation, overlain by Claxby Beds, was recorded in the central part of The Wash in Borehole 72/77 (Figure 26).

Throughout their outcrop in the King's Lynn district the Mintlyn Beds rest unconformably on the Runcton Beds. In most sections the erosive nature of the basal pebble bed of the Mintlyn Beds is clearly demonstrated by its irregular, commonly channelled contact with the underlying beds. The phosphatic pebbles of both the basal Runcton Beds pebble bed and the basal Mintlyn Beds pebble bed occur as gritty, glauconitic, irregular masses cemented by secondary phosphate, and both contain derived *Subcraspedites*. However, the pebbles from the two horizons can be distinguished in soil fragments because they differ in colour and texture. The Runcton Beds pebbles/nodules are dense, uniformly black or bluish black (and remain so on weathering), with rare worn *Subcraspedites* fragments, whereas the basal Mintlyn Beds pebbles/nodules are commonly dark brown or black, but contain paler and darker patches (due to the incorporation of earlier pebbles), have a creamy white-weathering patina that accentuates the colour of the glauconite grains, yield relatively common, worn *Subcraspedites* and other ammonites, and commonly contain iridescent shell debris.

In the North Runcton and Hardwick areas the basal nodule bed of the Mintlyn Beds yielded fragments of phosphatised ammonites that Casey (1973, p.242) considered to have affinities with both the Jurassic (Upper Volgian) genus *Subcraspedites* (*Volgidiscus*) and the Cretaceous (Lower Ryazanian) genus *Hectoroceras*. He named these *Runctonia* gen. nov., but subsequently (*in* Casey et al., 1977, p.16) renamed them *Praetollia* (*Runctonia*) and suggested, on the basis of comparison with the late Jurassic to early Cretaceous faunas of north-west Europe, the Russian Platform, the Subarctic Urals, Northern Siberia and Greenland, that they are the oldest Cretaceous fauna yet recognized in Britain. The base of the Mintlyn Beds in

Norfolk (and its equivalent in Lincolnshire, the mid-Spilsby nodule bed) has therefore been taken to mark the base of the Cretaceous in eastern England (Casey, 1973).

The age of the preserved Mintlyn Beds varies throughout the length of their outcrop. At West Dereham, the formation consists of 6.5 m of glauconitic and pyritic sands with bands of fossiliferous clay ironstone, unconformably overlain by the Carstone. The ironstones have yielded a fauna rich in *Hectoroceras* and include over 25 species of bivalve, lignite logs up to 1 m long, and bones and teeth of large marine reptiles (Casey, 1971). These beds were assigned by Casey (1973) to the *Hectoroceras kochi* Zone (Table 10). When traced northwards, this lower part of the Mintlyn Beds (the Hectoroceras Beds) becomes strikingly attenuated due to erosion at the base of the overlying bed. At North Runcton (4 m), Hardwick (1 m) and Mintlyn Wood (1 m), the Hectoroceras Beds are overlain with minor unconformity by sands with *Lynnia* (= *Surites* (*Lynnia*) of Casey, 1973). There is no exposure of these beds between Mintlyn Wood and North Wootton. Drainage ditches on the south side of the Babingley Valley at North Wootton [6524 2536], which yielded large, off-white phosphatic nodules containing worn *Subcraspedites* (basal Mintlyn Beds nodule bed) in the present survey, are probably close to the locality which yielded phosphatised *Hectoroceras* cf. *kochi* Spath to Whitaker during the original (1883) geological survey. This suggests (Casey, 1973, p.201) that at Whitaker's locality the Hectoroceras Beds are represented only by a *remanié* fauna at the base of the Mintlyn Beds.

Plate 4   Selected Lower Cretaceous lithologies, all of which weather at outcrop to ferruginous sandy clay or clayey sand. All photographs at natural size.

A. Bioturbated clayey, glauconitic fine-grained sand with phosphatic pebbles. View normal to bedding, Specimen no. BDD 3426, Hunstanton Borehole, 96.37 m (316 ft 2 in), low Mintlyn Beds.

B. Bioturbated clayey, glauconitic fine-grained sand densely cemented in part by phosphatic ironstone; prominent burrow at left is lined with glauconitic clayey sand and filled with clean brown sand. View parallel to bedding, Specimen no. BDC 1049, Marham Borehole, 66.27 m (217 ft 5 in), low Mintlyn Beds.

C. Bioturbated sparsely pebbly, faintly sandy, oolitic (limonite) silty clay. View normal to bedding, Specimen no. BDD 2853, Hunstanton Borehole, 53.70 m (176 ft 2 in), high Dersingham Beds.

D. Faintly sandy, silty clay with burrowfill concentrations of limonite ooids and densely cemented by phosphatic ironstone. View normal to bedding, Specimen no. BDD 2758, Hunstanton Borehole, 49.05 m (160 ft 11 in), Roach.

E. Bioturbated, pebbly, intensely oolitic sandstone with limonite/'chamosite' ooids and ferruginous cement; pebbles dominantly ironstone. View normal to bedding, Specimen no. BDD 2361, Hunstanton Borehole, 21.03 m (69 ft 0 in), Carstone.

F. Bioturbated intensely pebble, oolitic sandstone with ferruginous cement as E; pebbles mostly ironstone and vein quartz. View normal to bedding, Specimen no. BDD 2357, Hunstanton Borehole, 20.57 m (67 ft 6 in), Carstone.

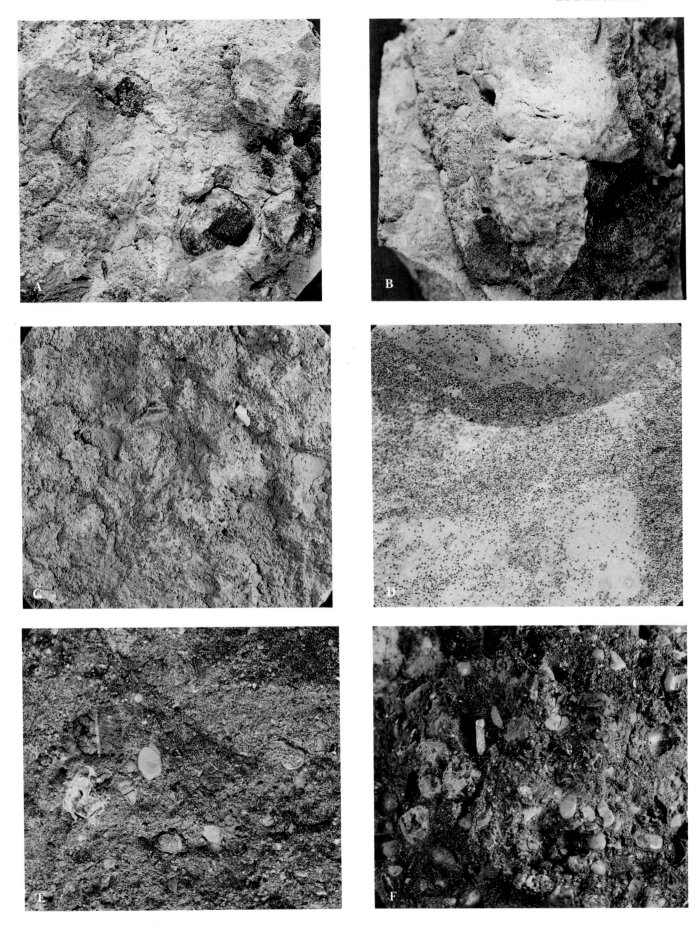

As yet the Hectoroceras Beds are unknown in England outside Norfolk. They are absent beneath the central part of The Wash (IGS Borehole 72/77) and in Lincolnshire (e.g. Skegness Borehole). However, *Hectoroceras* occurs widely in the Boreal province and has been recorded from East Greenland (Spath, 1947), Western Siberia (Klimova, 1969), the Subarctic Urals (Golbert et al., 1972), Canada (Wiedmann, 1968) and the Central Graben of the North Sea (Birkelund et al., 1983).

The beds above the Hectoroceras Beds form the bulk of the Mintlyn Beds at outcrop in the present district. Exposures are rare, but spoil from the ironstone bands and irregularly shaped limonitic concretions secondarily derived from them, and also glauconitic sand and clay, occur in many ditches dug across the poorly drained Mintlyn Beds outcrop.

The most complete section recorded to date in the lower and middle parts of the member is that exposed in 1964 in the King's Lynn Bypass cutting at Mintlyn Wood [6508 1987 to 6530 2010]. There, 8.75 m of glauconitic sands and clayey sands with four thin (0.10 to 0.20 m) bands of fossiliferous clay ironstone were recorded. When fresh, the ironstones are especially fossiliferous and yielded a rich ammonite (Plate 5) and bivalve fauna including, in ascending order, species of the ammonites *Praetollia* (*Runctonia*), *Hectoroceras, Borealites, Lynnia* and *Bojarkia* (= *Surites* (*Bojarkia*) of Casey, 1973) indicative of the *kochi, icenii* and *stenomphalus* zones of the Ryazanian (Casey, 1973; Casey et al., 1977). The bivalves have been identified by Kelly (1984); they include species of *Anisocardia, Camptonectes, Corbula, Cucullaea* (*Dicranodonta*), *Girardotia, Myophorella, Neocrassina* (*Lyapinella*), *Neocrassina* (*Pressastarte*), *Nicaniella, Pinna, Pleuromya, Protocardia, Thracia* and ostreids. None of this diverse bivalve fauna appears to be diagnostic of any particular bed within the Mintlyn Beds. Indeterminate gastropods and wood fragments were also present. Phosphatic nodules accompanied by evidence of erosion occur at two levels, in the *kochi* and *icenii* zones, and the sequence is clearly condensed.

The fossiliferous part of the Mintlyn Beds in the cutting is overlain by about 4 m of sparsely fossiliferous sand, clayey sand and ironstone that did not yield any ammonite. A second cutting [6563 2089], 800 m north-north-east of the Mintlyn Wood cutting and estimated from the field survey to be about 2 to 3 m stratigraphically above the highest beds exposed at Mintlyn Wood, proved a thin clay ironstone with *Peregrinoceras* spp., indicative of the *albidum* Zone (Casey, 1973, p.200).

The highest part of the Mintlyn Beds is poorly exposed and has not yielded any fauna. A specimen of *Paratollia*, indicative of a Valanginian age, was recorded in the upper part of the formation in Borehole 72/77 in the central part of The Wash, thereby suggesting that the Ryazanian–Valanginian boundary falls within the top part of the Mintlyn Beds. Ammonites are so rare in the Valanginian rocks in Norfolk and Lincolnshire that no attempt has yet been made to zone this stage. Elsewhere in the Boreal Province, zones have been established at this stratigraphical level largely on the basis of species of *Platylenticeras, Polyptychites* and *Dichotomites*.

Details

Wolferton to North Runcton

The Mintlyn Beds outcrop is covered by thick drift deposits between Wolferton and the Babingley River; their most northerly outcrop is on the south side of the river valley where brown and green, glauconitic, clayey sands with much secondary ironstone pan give rise to poorly drained land between Wootton Carr and Castle Rising. The spoil from a drainage ditch [6524 2536 to 6523 2510] running northwards from the Carr yielded large off-white, glauconite-speckled, phosphatic nodules with worn *Subcraspedites*, and small barren, bluish black nodules, both from the basal bed of the formation. Whitaker (*in* Whitaker and Jukes-Browne, 1899, p.7) collected a phosphatised *Hectoroceras kochi* from this or an adjacent locality. Stratigraphically higher sections in the ditch show thinly interbedded dark grey clays and green sands, that weather to pale grey and white respectively, overlain by ferruginous pan and sandy flinty wash. Phosphatic nodules from the basal bed of the Mintlyn Beds have also been thrown out from an excavation [6380 2429] near North Wootton station and a drain [6383 2498] at the Common, but no section was seen at either locality.

Green glauconitic sands and clays, and slabs of fossiliferous ironstone have been revealed from time to time in the deeper excavations in the built-up areas of North Wootton, South Wootton, Gaywood and Fairstead Estate, King's Lynn, but in most of these areas the Mintlyn Beds are deeply oxidised and commonly covered by weathered, brown, clayey sands and bleached white sand. Spoil from several trenches at Fairstead Estate [643 197] included phosphatic nodules from the basal bed and several rotted ironstone bands, one with phosphatic pebbles, within a sequence of grey and green, glauconitic sands and clays.

The largest and most stratigraphically revealing section yet recorded in the Mintlyn Beds is that in the cutting [6508 1987 to 6530 2010] dug for the King's Lynn Bypass at the western end of Mintlyn Wood. The beds dip north eastwards at less than one degree. The following composite section was recorded in 1965 (Casey and Gallois, 1973, pp.13–14).

	Thickness m
MINTLYN BEDS	
Clay-ironstone, sandy, slightly glauconitic, buff-coloured, with *Neocrassina* and other bivalve fragments; limonitic 'box-stone' weathering	seen 0.10
Sand, slightly clayey, buff-coloured	0.55
Clay-ironstone 'box-stones', as above	0.10
Sand, slightly clayey, buff and brown, passing down into	1.50
Sand, clayey, grey, passing down into	1.80
Sandy clay, glauconitic, dark grey, with vivid green sandy bands	1.60
Clay-ironstone, sandy, buff and putty-coloured, with grains of glauconite, shrinkage cracks and splintery fracture; limonitic crust where in weathering zone; a few bivalves and *Bojarkia* spp. including *B. stenomphala* (Pavlow)	0.15 to 0.20
Very clayey sand and sandy clay, glauconitic, blue-green, with bits of carbonised wood	0.45 to 0.60
Clay-ironstone, putty-coloured, in lenses up to 1 m long; some pyrite; shrinkage cracks; full of moulds of fossils including *Neocrassina, Myophorella, Bojarkia stenomphala* and allied species; bits of carbonised wood common	up to 0.20
Very clayey sand, glauconitic, grey-green, with bits of black wood and *Chondrites*-type burrows infilled	

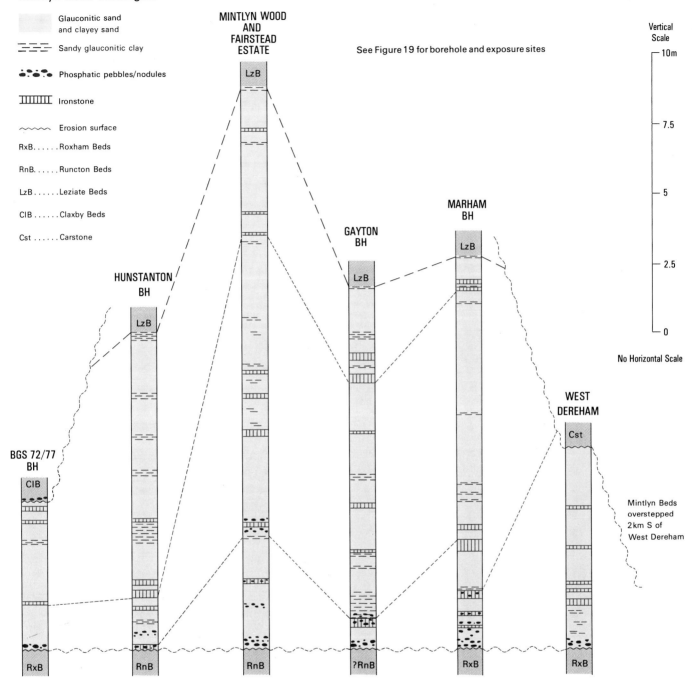

Figure 26 Correlations between the Mintlyn Beds sequences in Norfolk.

	Thickness m
with pale coloured sand; 200 mm-thick band of green laminated sand 300 mm above the base	1.00 to 1.15
Clay-ironstone; putty-coloured, tabular bed; fewer fossils than in ironstone above; shrinkage cracks	0.15 to 0.20
Sandy clay and clayey sand, blue-green and dark grey, glauconitic; beds become less sandy downwards and pass from sand with clay laminae to clay with sand laminae; lowest 150 mm a stiff clay with line of black, pyrite-coated wood fragments at base; clay-ironstone 'doggers' 0.6 m and 1.45 m above base; *Chondrites*; ammonite 'ghosts' in bottom 150 mm	3.50
Phosphatic nodules; black, water-worn, resting on irregular surface of bed below; *Surites* (sensu Casey, 1973) spp. common, rare *Bojarkia* sp.	0.05
Clay-ironstone, sandy, putty-coloured, patchily fossiliferous; *Ophiomorpha*-type burrows; knobs of semi-phosphatised, sandy and glauconitic clay-ironstone cemented to upper surface, in places enveloping nodules of the bed above; moulds of *Thracia* in same material lie on upper surface; *Surites* (sensu Casey, 1973) sp. nov.	0.10 to 0.15
Clayey sand, glauconitic, blue-green	0.40
Phosphatic nodules, small, black and brown, well-rolled, crowded in dark green, clayey sand, indurated in places; fragments of *Hectoroceras*	0.15 to 0.20
Clayey sand, hard, brown	0.15
Clayey sands, buff and green, becoming greener and more clayey with depth	0.50

Spoil from the trenches at nearby Fairstead Estate (see above) subsequently showed the basal nodule bed of the Mintlyn Beds to be less than 1 m below the lowest bed proved in the cutting. A second cutting [6563 2089] on the Bypass showed about 0.6 m of weathered, yellowish green, clayey sand overlain by 0.3 m of fossiliferous clay ironstone that yielded *Peregrinoceras* sp. nov., *P.* cf. *albidum* sp. nov. and *P.* cf. *pseudotolli* (Neale), indicative of the *albidum* Zone (Casey, 1973, p.200).

The ironstones from both cuttings contain a rich bivalve fauna mostly preserved as moulds. These have been identified by Kelly (1977) and include *Anisocardia* sp. nov., *Camptonectes* (*Boreionectes*) *cinctus* (J Sowerby), *Camptonectes morini* (de Loriol), *Corbicellopsis claxbiensis* (Woods), *Corbula* sp., *Cucullaea* (*Dicranodonta*) *vagans* Keeping, *Discoloripes* sp. nov., *Girardotia* sp. nov., *Myophorella keepingi* (Lycett), *Neocrassina* (*Lyapinella*) sp. nov. (*Astarte* cf. *saemanni* Spath *non* de Loriol), *N.* (*Pressastarte*) sp. nov., *Nicaniella claxbiensis* (Woods), *N.* sp., ostreids indet., *Pinna* sp., *Pleuromya* cf. *uniformis* (J Sowerby), *Protocardia* sp., and *Thracia phillipsii* (Roemer). None of these species is diagnostic of any particular bed within the Mintlyn Beds. Indeterminate gastropods and wood fragments are also present.

The junction of the Mintlyn Beds and the Leziate Beds has been exposed at times in the most westerly part [6663 1962] of Leziate sand-pits, adjacent to Mintlyn Wood. At the time of the survey, it was below the water table and the Mintlyn Beds were seen only as green glauconitic clay and sand spoil.

The basal beds of the Mintlyn Beds were exposed in the floor of a large borrow-pit [635 174] at Constitution Hill, Hardwick, where the following section was measured:

1a–c *Bojarkia tealli* Casey (HOLOTYPE); Mintlyn Beds; *stenomphalus* Zone; King's Lynn By-Pass, Galley Hill, Mintlyn Wood (BGS Ce4407)

2a–b *Peregrinoceras rosei* Casey (HOLOTYPE); Mintlyn Beds; *albidum* Zone; King's Lynn By-Pass, north of Church Farm, Bawsey (BGS GSM114730)

3a–b, 4a–b *Lynnia icenii* Casey; Mintlyn Beds; *icenii* Zone; No. 2 Gas Feeder Main trench, Manor Farm, North Runcton Fig. 3a–b (BGS Ce5311); Fig. 4a–b (HOLOTYPE; BGS Ce5298))

5a–b *Peregrinoceras* cf. *albidum* Casey; Glacial Drift (ex Hundleby Clay); King's Lynn By-Pass, Castle Rising (BGS GSM114748)

6, 7 *Hectoroceras kochi* Spath; Mintlyn Beds; *kochi* Zone; Fenland Flood Relief Channel, West Dereham (Fig. 6 (impression of BGS Ce3818); Fig. 7 (BGS Ce2428))

8a–b, 9a–b *Runctonia runctoni* Casey; basal nodule bed of Mintlyn Beds; *runctoni* Zone; No. 2 Gas Feeder Main trench, Manor Farm, North Runcton (Fig. 8a–b (BGS Ce5347); Fig. 9a–b (HOLOTYPE; BGS Ce5348))

10 *Subcraspedites* (*Subcraspedites*) *sowerbyi* Spath; Lower Spilsby Sandstone; *preplicomphalus* Zone; erratic block, Leziate Sand Pits, Bawsey (BGS Ce3188)

	Thickness m
Sand, weathered brown and green, barren up to	0.5
Clay-ironstone, tabular, rusty brown weathered; numerous rounded, bluish black, phosphatic pebbles in lower part	0.25
Sand, fine-grained and sandy clay, brownish and greenish grey, glauconitic	0.30
Basal nodule bed; complex concentration of large off-white, sandy, small bluish black and soft medium brown, phosphatic nodules set in clayey glauconitic sand; irregular base	0.35

RUNCTON BEDS

Barren glauconitic sands	seen 0.60

The basal nodule bed contains a rich phosphatised fauna derived from the Runcton Beds and possibly, in part, from the Roxham Beds. Dr S R A Kelly and Mr Morter have identified a terebratulid, *Anisocardia* sp. nov., *Camptonectes morini* (de Loriol), *Corbicellopsis claxbiensis* (Woods), *Cucullaea* (*Dicranodonta*) *vagans* Keeping, *Entolium orbiculare* (J Sowerby), *Hartwellia* sp., *Modiolus* sp., *Myophorella* sp., *Neocrassina* (*Lyapinella*) *asiatica* (Zakharov), *N.* (*Pressastarte*) sp. nov., *Pleuromya* sp., *Protocardia* (*P*) *concinna* (von Buch), *Sowerbya longior* Blake, *Thracia* sp., *Bathrotomaria* sp., *Neritopsis* sp., gastropods indet., *Subcraspedites* (*Volgidiscus*) spp. and *Acroteuthis* sp.

The base of the Mintlyn Beds is marked by an abundance of phosphatic nodule bed debris in the soil brash and in ditches between Hardwick and Fair Green. The basal beds were briefly exposed in the No. 4 Gas Feeder main trench near the windmill at West Winch [6325 1674] and near Fair Green [6480 1697 and 6573 1738]. No section was seen but the spoil at all three sites consisted of glauconitic, greenish grey, fine-grained, clayey sands with thin beds of dark green, glauconitic clay and thin tabular beds of clay ironstone. Phosphatic pebble/nodule beds occur at several levels in the basal part of the member including a pebbly, glauconitic clay ironstone (as seen at Fairstead Estate and Mintlyn Wood cutting) and, at the base,

1a

1c

1b

2a

2b

3a

3b

4a

4b

5a

5b

6

7

8a

8b

9a

9b

10

large off-white, glauconite-speckled, phosphatic nodules with derived *Subcraspedites*.

The basal beds of the Mintlyn Beds were exposed in three trenches [6370 1630] at few metres apart at Brook Farm, North Runcton (Figure 25), where the following composite section was recorded:

	Thickness m
QUATERNARY	
Soil and subsoil, sandy, with ironstone fragments	0.10
Ferruginous pan or 'chevick' of densely limonite-cemented sand and clay ironstone rubble (trenches 1 and 2), passing uphill (trench 3) into sandy subsoil with ironstone fragments; much disturbed by cryoturbation, very irregular base	0.40 to 0.65
MINTLYN BEDS	
Sand, fine-grained, yellowish brown, weathered	0.60
Clay ironstone, tabular, weathered orange-brown and brownish yellow; numerous small, mostly well-rounded, bluish black, phosphatic nodules in basal part of bed; base irregular	0.10
Sand, fine-grained, brownish, yellowish and greenish grey; glauconitic, slightly clayey, weathered	0.15
Nodule bed; widely spaced small, bluish black, phosphatic nodules scattered throughout bed of weathered, glauconitic sand; rare phosphatised small bivalves	0.10
Sand, fine-grained, brownish and greenish grey, weathered, as above; base of bed marks base of weathering in trench 2	0.50
Basal nodule bed, with widely spaced, dark brown, phosphatic nodules, mostly small round or irregular knobbly shapes, almost all enclosing small bivalves or nests of bivalves; mixed with glauconite-speckled, black and dark brown, phosphatic nodules with a creamy white weathering cortex	up to 0.05
RUNCTON BEDS (see p.63 for details)	

Phosphatic debris from the basal nodule bed and slabs of weathered ironstone are common as soil brash in that area. Several local walls and barns have been built from the ironstone and the secondary ferruginous pan derived from it. A group of shallow pits [639 157] in glauconitic sands and clays near Manor Farm, North Runcton, may have been dug for ironstone for this purpose.

The basal beds of the Mintlyn Beds were well exposed in the No. 2 Gas Feeder Main trench [6515 1555] near Manor Farm. The following composite section was measured (after Casey and Gallois, l973, pp.8–9):

	Thickness m
MINTLYN BEDS	
Clay, lilac-grey, with glauconitic 'rafts' and some iron-staining seen	1.00
Phosphatic nodules (up to l00 mm diameter), buff and grey in grey-green, brown-weathering, glauconitic, sandy clay; some glauconitic, loosely cemented, gritty nodules at base; ammonites including *Bojarkia* spp. nov. fairly common	0.05 to 0.10
Clay-ironstone, sandy, buff, glauconite-speckled, in lenses; ochreous ferruginous crust; *Bojarkia* spp. nov.	up to 0.10

Clay, sandy, grey, with 'rafts' of green, glauconitic, clayey sand	0.25
Clay-ironstone, sandy, glauconitic, buff-brown	0.20 to 0.30
Phosphatic nodules, small, dark inside, pale outside, in clusters in brown-weathering, glauconitic, clayey sand; *Hectoroceras* present	0.05 to 0.10
Sand, glauconitic, clayey, dark green	1.00
Basal nodule bed; black and dark brown, gritty, phosphatic nodules in matrix of dark green (with drab brown wisps) glauconitic, clayey sand; much iridescent phosphatised shell debris from broken ammonites, trigoniids and other shells; many nodules pyritic; ammonites tend to be concentrated in lower part of the bed including *Runctonia* with traces of iridescent test and *Subcraspedites* (s.l.) in black phosphorite; irregular base	0.10 to 0.15
RUNCTON BEDS (see p.63 for details)	

The basal nodule bed yielded a phosphatised derived fauna that includes a serpulid, *Camptonectes morini* (de Loriol), *Corbicella claxbiensis* (Woods), *Cucullaea* (*Dicranodonta*) *vagans* Keeping, *Entolium orbiculare* (J Sowerby), *Hartwellia* sp., *Myophorella keepingi* (Lycett), *Neocrassina* (*Lyapinella*) *asiatica* Zakharov, *Pleuromya* cf. *uniformis* (J Sowerby), *Protocardia* (P.) *concinna* (von Buch), *Sowerbya longior* Blake, *Thracia depressa* (J Sowerby), an indeterminate gastropod, *Praetollia* (*Runctonia*) *runctoni* Casey, P. (*Runctonia*) spp. nov., *Subcraspedites* (*Volgidiscus*) spp. nov., and *Acroteuthis* sp.

The higher beds yielded an indigenous fauna that includes *Hartwellia* sp. nov., *Pleuromya* cf. *uniformis* (J Sowerby), P. spp., *Lynnia icenii* Casey and *L.* spp. nov.

Nar Valley to Denver

The Mintlyn Beds can be traced southwards from North Runcton by means of a soil brash of green clayey sands, phosphatic nodules and ironstone slabs, to Blackborough End where they disappear beneath the thick drift deposits of the Nar Valley. They re-emerge on the south side of the valley (in the Wisbech district) to crop out as low, poorly drained ground around Wormegay, but are then again overlain by drift deposits southwards as far as Stow Bardolph. Between there and Denver, the Mintlyn Beds, together with the Roxham Beds, Runcton Beds and Leziate Beds, crop out on a low escarpment that forms the eastern limit of Fenland. The only exposures recorded in this area have been in trenches at Wormegay [615 005], near Wimbotsham [618 052], at Clackclose Estate, Downham Market [612 041] and at Denver [TL 645 997]. All were deeply weathered and disturbed by cryoturbation and showed the basal phosphatic nodule bed with derived *Subcraspedites* spp. overlain by brown and green, glauconitic, clayey sands with clay-ironstone bands at several levels. They were too deeply weathered to yield an indigenous fauna.

Between Denver and West Dereham, the Mintlyn Beds again crop out as low, poorly drained ground with much ironstone and secondary iron-pan occurring as debris in the deeper ditches.

The full thickness of the Mintlyn Beds, unconformably overlain by the Carstone, was exposed at the southern limit of this area of outcrop in the Fenland Flood Relief Channel excavations [TL 639 995 to 662 996] eastwards from Roxham Farm, West Dereham. The following composite section was measured by Casey (in Casey and Gallois, 1973, pp.11–12):

CARSTONE (see p.111 for details)

	Thickness m

MINTLYN BEDS

Clayey sand, grey-green, with pyritic nodules full of bits of black wood; lignite in logs up to 1 m long; a few phosphatic nodules, mostly moulds of bivalves — 1.75 to 2.00

Clay-ironstone, glauconitic, sandy, weathering reddish brown; crowded with moulds and casts of fossils, mostly *Neocrassina*; locally divided into two beds with parting of green, clayey sand; *Hectoroceras kochi* Spath, *Borealites* cf. *fedorovi* Klimova — 0.20 to 0.30

Sand, clayey, grey, with pyrite-encrusted lignite — 1.20

Clay ironstone; impersistent bed — 0.15 to 0.25

Sand, clayey, grey, with lignite — 0.50

Clay-ironstone, glauconitic; *H. kochi*, *H.* spp. nov. — 0.10 to 0.15

Sand, clayey, grey — 0.15

Clay-ironstone, glauconitic; *H. kochi*, *H.* spp. nov. — 0.10

Clay, glauconitic, sandy, grey-green — 0.25

Clay-ironstone, sandy, buff-grey green-speckled, glauconitic, with a few phosphatic nodules, pyrite crystals and bits of wood; moulds of fossils, especially *Neocrassina*, *Hectoroceras kochi*, *H.* spp. nov. — 0.20 to 0.25

Clay, sandy, glauconitic, blue-green, sulphurous weathering; some lignite and a few brown phosphatised aggregations of shell debris; *Hectoroceras* sp. — 1.30

Basal nodule bed; thickly clustered, rolled, black, phosphatic nodules in matrix of green sandy clay; abundant fossil wood and rolled moulds of bivalves and ammonites including *Subcraspedites sowerbyi* Spath, *S. preplicomphalus* Swinnerton, *S. lamplughi* Spath, *S. claxbiensis* Spath, *Craspedites* sp.; some pyrite; locally cemented into 'doggers'; line of brown, friable, phosphatic nodules sparsely distributed at top of bed; *Hectoroceras* spp. nov., cf. *Borealites*?; channelled into bed below — 0.15 to 0.30

ROXHAM BEDS (see p.57 for details)

To the south of the Relief Channel, the Sandringham Sands crop out only as two small outliers on the Fenland 'islands' on which the villages of Hilgay and Southery are built. Mintlyn Beds are preserved on the more northerly of these. The Southery area is close to their estimated southern limit, because of the overstep of the Carstone, and only the basal beds are likely to be present there. A deep ditch [TL 6204 9772] at Hilgay yielded off-white phosphatic nodules from the basal nodule bed, but the formation is otherwise poorly exposed on the outlier. An abundance of Mintlyn Beds debris, mostly the basal nodule bed and ironstones, in the till that partly caps the Southery outlier suggests that the lower part of the Mintlyn Beds formerly cropped out nearby.

Boreholes

The full thickness of the Mintlyn Beds was cored in the Geological Survey boreholes in The Wash (Borehole 72/77) and at Hunstanton, Gayton and Marham. The lithological and faunal sequences proved at Gayton and Marham are similar to those proved at outcrop, but the Wash and Hunstanton sequences, although lithologically similar to that at outcrop, show marked faunal differences. The Hunstanton and Borehole 72/77 sequences either correlate with the highest (poorly exposed)

part of the Mintlyn Beds at outcrop (Figure 26) or are entirely younger than them. In both boreholes, the basal Mintlyn Beds nodule bed must be a highly condensed deposit that represents much of the Mintlyn Beds sequence at outcrop.

The following sequence was recorded between 43.10 and 48.30 m in Geological Survey Borehole 72/77:

	Thickness m	Depth m
CLAXBY BEDS (see p.75 for details)		43.10

MINTLYN BEDS

Ironstone; pale brown, dense, burrow-mottled siderite mudstone with many small voids (weathered glauconite or 'chamosite' grains); locally plant-rich; disturbed by coring	0.05	43.15
Core lost; probably soft sand	0.50	43.65
Sand and silt, fine-grained with burrowfills of clay; glauconitic in part; ironstone bed as above at one level; sparsely shelly to shelly, with bivalves, gastropods, ammonite fragments, crinoid columnals, scaphopods and coalified wood fragments	2.40	46.05
Core lost	0.50	46.55
Silt and fine-grained sand, brownish grey, burrow-mottled, with clay and glauconitic or oolitic sand burrowfills at several levels; ironstone bed as above; locally very shelly, with rich bivalve and gastropod fauna, some in life position, naticid gastropods, polyptychitid ammonite fragments and common crinoid columnals; passing down into	1.60	48.15
Basal nodule bed: sand and clay, interburrowed; many rough coated, irregular shaped, phosphatic nodules and a few glauconite-speckled, well-rounded nodules; sparsely shelly, with bivalves and a belemnite fragment	0.15	48.30

ROXHAM BEDS (see p.62 for details)

The fauna is dominated by bivalves and includes *Astarte*, *Callicymbula*? *navicula* Harbort, *Corbula isocardiaeformis* Harbort, *Entolium demisssum* (Phillips), *E.* cf. *nummulare* (Fischer de Waldheim), *Modiolus*, *Myophorella* sp. nov., *Nicaniella claxbiensis* (Woods), *Nucula* (*Leionucula*) sp., *Oxytoma* cf. *articostata* Zakharov, *Panopea neocomiensis* Leymerie, *Pinna raricosta* Harbort, *Protocardia peregrinorsa* (d'Orbigny), *Resatrix neocomiensis* Weerth, *Spondylus* (*Hinnites*) sp. nov., *Tellina* (*Lavingnon*) *ovalis* Harbort, venerids and *Vnigriella*? *scapha* (d'Orbigny); *Ampullina laevis* Phillips, *Bathraspira* sp. nov., naticids, '*Trochus*' *quadricoronatus* Harbort, polytptychitid fragments, *Acroteuthis subquadratus* (Roemer), *Neocrinus tenellus* (Eichwald), *Antalis* cf. *valanginiense* (Pictet & Campiche) and coalified wood fragments are also present. It has many features in common with the faunas from Valanginian sequences in northern Germany.

The following sequence was proved between 85.42 and 96.83 m in the Hunstanton Borehole.

	Thickness m	Depth m
LEZIATE BEDS (see p.81 for details)		

MINTLYN BEDS

Interbedded and interlaminated greyish brown clay and soft, micaceous, fine-

	Thickness m	Depth m
grained sand with comminuted plant debris	0.28	85.70
Sand, fine- and very fine-grained, grey and brownish grey, soft; wisps and seams of brown and greenish brown clay becoming more common with depth; glauconitic in part; hard, dense, pale brown, siderite mudstone bands at several levels; bioturbated throughout and with cross lamination at several levels; plant debris and lignite fragments common; fossiliferous at some levels with fauna dominated by bivalves;	10.36	96.06
Basal nodule bed: soft, cream-coloured, phosphatic nodules, large rough-coated, brown, phosphatic nodules and rounded, bluish black, dense phosphatic pebbles enclosing shell fragments, all set in a green and brown glauconitic, bioturbated sand and sandy clay matrix	0.77	96.83

ROXHAM BEDS (see p.62 for details)

Mr Morter recorded the following fauna, mostly preserved in the ironstones, between 94.06 and 95.33m: *Camptonectes* (*Boreionectes*) sp., *Entolium* cf. *demissum*, *Lyapinella* cf. *laevis* (Phillips), *Myophorella* (*M.*) sp. nov. *Nucula* sp., abundant *Oxytoma articostata*, *Panopea neocomiensis*, *Procyprina* sp., *Protocardia* cf. *peregrinorsa*, *Thracia* cf. *neocomiensis*, and common *Neocrinus tenellus*.

This assemblage is similar to that proved in Borehole 72/77, but cannot be matched in detail with that recorded from the Mintlyn Beds at outcrop. Mr Morter considers the assemblage to be indicative of a late Ryazanian or early Valanginian age. In addition, he has identified the early Valanginian ammonite *Menjaites* sp. cf. *magnus* Sasonova in an ironstone at 92.56 m, which provides further evidence that the Ryazanian–Valanginian boundary falls within the Mintlyn Beds in the northern part of their subcrop.

When allowance is made for the loss of detail due to weathering at outcrop, the sequence proved in the Gayton Borehole is similar to that of the type section in the Mintlyn Wood cutting of the King's Lynn Bypass, and can probably be correlated with it at several levels (Figure 26). In the absence of faunal evidence the lateral persistence of the ironstones used for correlation in this figure cannot be established. The following sequence was proved between 62.17 and 75.39 m in the Gayton Borehole:

	Thickness m	Depth m
LEZIATE BEDS (see p.81 for details)		
MINTLYN BEDS		
Clay, dark green, glauconitic (3 cm), resting on an uneven surface of hard ferruginous sandstone that passes down into	0.10	62.27
Sand, glauconitic, fine-grained; clean white, green and grey, with many thin beds and laminae (up to 20 cm thick) of dark green, glauconitic clay forming small-scale rhythms; plant debris common; cross-bedding, lamination and bioturbation locally picked out by wisps of clay, plant debris or glauconite grains; densely cemented tubular sandy ironstones up to 30cm thick at several levels, locally with poorly preserved bivalves	10.86	73.13

	Thickness m	Depth m
Clay, dark green, sandy, highly glauconitic; lamination and common bioturbation picked out by sand grains; rare poorly preserved bivalves; pebbly in lowest 15 cm with cream-coloured, glauconitic, phosphatic nodules (up to 5 cm across), small black, glauconitic, phosphatic nodules and rare tiny, vein quartz pebbles set in sandy clay with many clay casts of bivalves	0.88	74.01
Ironstone, hard, medium brown, partially translucent, with numerous small bluish black, glauconitic, phosphatic pebbles	0.30	74.31
Sand, fine-grained, glauconitic, clayey, green; a few poorly preserved bivalves; pebbly in lowest part (Basal nodule bed) with large (up to 4 cm), cream-coloured, phosphatic nodules with included shell debris, small bluish black, phosphatic nodules and soft brown phosphate pebbles (presumably derived from Runcton Beds); resting on an irregular, burrowed surface	1.08	75.39

ROXHAM BEDS (see p.62 for details)

The Mintlyn Beds sequence proved between 54.36 and 68.33 m in the Marham Borehole, although similar in thickness (13.97 m) and made up of the same range of sand, clay and ironstone lithologies as that at Gayton, cannot be matched bed by bed with either the Gayton Borehole or the Mintlyn Wood sequences (Figure 26).

Leziate Beds

The Leziate Beds consist of loose, fine-grained, cross-bedded quartz sands, yellow, green, orange-brown or red in places, but mostly clean grey or white, with subordinate bands of silt and clay. Locally, the sands are sufficiently lithified to form a friable sandrock. Pyrite nodules are common throughout the formation, but above the water table these become oxidised to form limonitic concretions, geodes or merely ferruginous stains. Glauconite is abundant in some areas and at some stratigraphical levels in the Leziate Beds, but is rare in others. Phosphatic nodules or other signs of erosion and condensed deposition appear to be absent. At outcrop, the clay and ferromagnesian minerals are commonly weathered and leached to give a clean white sand with some patchy yellow staining. Cross-bedding occurs throughout the formation, mostly on a small scale. The sands are mostly fine grained, but thin medium-grained beds and scattered coarse grains are locally present.

The sands rise from beneath The Wash at Heacham and between there and King's Lynn crop out on the face of a steep feature that is capped by the basal ferruginous sandstones of the overlying Dersingham Beds. Southwards from King's Lynn they form a more subdued feature capped by Carstone. Much of the sandy heathland of west Norfolk that is regarded as typical Sandringham Sands scenery, including Dersingham Heath, Sandringham Warren, Roydon Common, Leziate Heath and Shouldham Warren, is underlain by the free-draining

Leziate Beds, basal Dersingham Beds and, at the foot of the Leziate Beds feature, by thick sandy wash derived from the Leziate Beds. The member underlies the whole of the eastern edge of the district, and was proved in the Hunstanton, Gayton and Marham boreholes (Figure 27). When traced north-westwards into The Wash the sands pass laterally into the argillaceous Claxby Beds.

The member is 20.4 m thick at Hunstanton but thickens steadily southwards from there to an estimated maximum of about 30 m in the Leziate–Middleton area. It is represented in Borehole 72/77 in The Wash by only 2.5 m of fine-grained sand in the top part of the Claxby Beds sequence. The absence of marker beds within the Leziate Beds makes it impossible to determine whether the variations in their thickness are due to overstep by the Dersingham Beds or to original differences in the depositional thickness. Their rapid attenuation beneath The Wash strongly suggests an original depositional variation. Southwards from Middleton [660 160], the Carstone rests on the Leziate Beds and, between there and their disappearance at Ryston Hall, Denver [628 012] in south-west Norfolk, they are progressively cut out in a southerly direction (Figures 20 and 27).

The type section of the Leziate Beds is the huge complex of pits dug for glass and foundry sands at Leziate Heath [675 193], near Brow of the Hill (Plate 6). The junction with the underlying Mintlyn Beds has been well exposed from time to time in the most westerly part [665 195] of the pits. There, and in the Gayton and Marham boreholes, the base of the Leziate Beds has been taken at the top of a thin bed of dark green glauconitic clay; the overlying lithologically uniform sands differ from the Mintlyn Beds in that they contain rare clay seams but no clay ironstone. This junction effectively marks the lower limit of the economically useful sands. It has proved to be relatively easy to map because it is marked by a spring line and a prominent feature. Throughout most of their outcrop the beds close to the junction are obscured by thick sandy downwash derived from the Leziate Beds, but the feature remains persistent.

The sand pits at Leziate are the most extensive workings of their kind in Britain. They were probably begun in the early part of the 19th century as small pits for glass sand. In recent years they have been worked by British Industrial Sands for glass and foundry moulding sands, and they now extend over about 2 sq. km. The full thickness of the Leziate Beds has been worked in the pits. In the past, mostly in the 19th century, the Leziate Beds were worked for glass sand in the Snettisham, Dersingham, Roydon, Castle Rising, North Wootton, Blackborough, Shouldham and Downham Market areas.

Although well exposed in comparison with the other Lower Cretaceous formations in the district, the Leziate Beds are palaeontologically the least rewarding part of the sequence. No body fossil has yet been recorded in situ in the sands but, in unweathered sections, pyritic concretions have yielded bits of coalified wood and, more rarely, poorly preserved bivalves and rare unidentifiable ammonite 'ghosts'. None of the determinable fossils is stratigraphically diagnostic. An ironstone cast of an ammonite body chamber found in an old sand pit at Roydon Common [678 220], and believed to have been derived from the Leziate Beds, was identified by Casey (in Casey and Gallois, 1973) as a ?*Polyptychites*. This supports the presumed Valanginian age of the member. Lithologically similar ironstones occur in the overlying Dersingham Beds, although much less commonly than in the Leziate Beds. However, the basal beds of the former formation contain a Hauterivian fauna (see p.84 for details) and could not have yielded a polyptychitid. The Roydon Common specimen is therefore believed to be the only fossil found to date that is indicative of the age of the Leziate Beds.

Bioturbation in the form of indeterminate mottling is common at all levels in the Leziate Beds in unweathered sections in boreholes. Vertical or subvertical tubes (cf. *Skolithos*), commonly cemented by pyrite or goethite, are very common; poorly preserved *Rhizocorallium* is also present, but rare. Possible burrows or casts of bits of rotted wood, lined with a carbonaceous film, are common at several horizons, especially in the lower part of the member. Tiny, hollow, pyritised tubes penetrate many of the clay bands within the sands; these are presumed to be burrows rather than the traces of rootlets. Almost all of these bioturbation features are destroyed by weathering. Plant stems and bits of wood, much of them cemented into pyritized masses, are common in the lower part of the member, where they appear to have formed waterlogged mats on the sea bed. These pyritised masses survive partial weathering and are common in excavations made below the water table in the Leziate Beds.

Rose (1835, p.178) noted that weathered sections in the sands contain many hollow spheres and cylinders of sandy ironstone, referred to by him as 'ferruginous geodes', that are infilled with clean white sand. Many of these geodes have been incorporated in the Pleistocene deposits in west Norfolk and their abundance led Fowler (1970) to suggest that they were derived from the 'Greensand' (Sandringham Sands and Dersingham Beds) or the Tertiary/Quaternary Crags, and that they had been formed by some, as yet, undescribed process. At Ling Common sand pits [652 238], Castle Rising, pyrite nodules, pyritised sand, geodes and irregular masses of limonite-cemented sand, and intermediate (part pyritic, part limonitic) forms all occur in situ. Many of the geodes are in the form of tapering, tube-shaped burrows. Elsewhere in the area, pyrite nodules are ubiquitous where the Leziate Beds are below the water table, and geodes are everywhere present where they are above it, irrespective of stratigraphical horizon. It seems clear, therefore, that the geodes are formed from pyritic concretions that have been oxidised and their ferruginous products reprecipitated in the adjacent sand. Many of the burrow-shaped geodes now have irregular shapes that reflect variations in the permeability of the adjacent sand, and they commonly preserve traces of sedimentary features such as cross-bedding.

At outcrop above the water table, the sands are extensively leached and are predominantly composed of fine-grained, angular, subangular and subrounded grains of clear quartz. Secondary iron-staining and cementation, ranging from pale yellow staining to a dense limonitic ce-

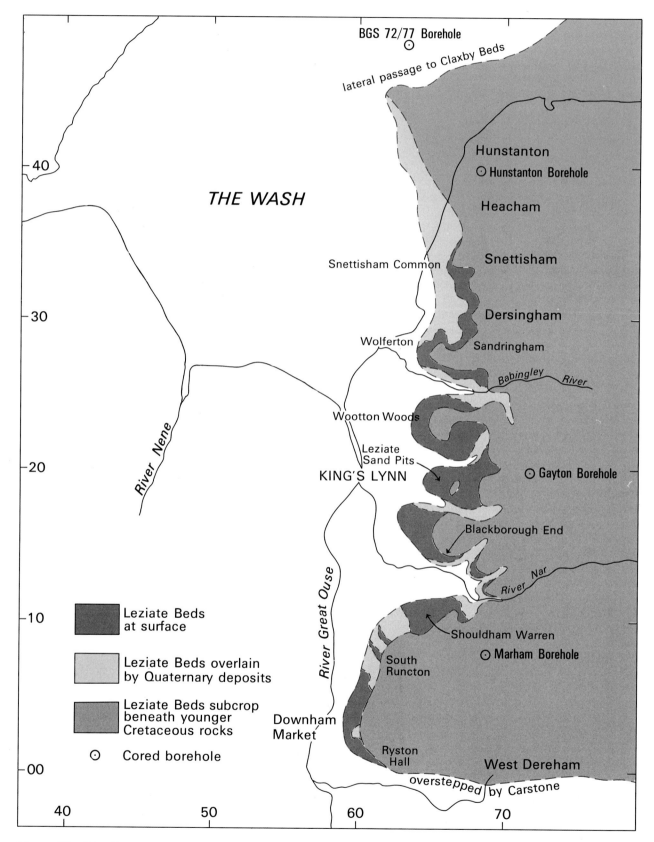

Figure 27 Distribution of the Leziate Beds in Norfolk.

ment, is patchily present at all localities and stratigraphical levels above the water table. In boreholes, and at outcrop where the water table has been artificially lowered, the sands are glauconitic and contain pyritised bands and concretions, and wisps and drapes of grey carbonaceous clay and green glauconitic clay. Leaching does not appear to affect sedimentary structures such as cross-bedding and burrowing, even though these features are commonly picked out by glauconite grains and concretions of clay minerals in unweathered sections, other than to make them less obvious.

An unusual feature of the Leziate Beds outcrop in the Castle Rising area, noted originally by Rose (1835), is the common occurrence of blocks of hard, pure quartzite up to 1 m across. Their field relationships and textures indicate that they are a form of silcrete, probably of Pleistocene age.

Although cross-bedding and other sedimentary features are present in even the smallest exposures of Leziate Beds, the sections are so widely spaced that they can only be qualitatively assessed. Lithological variations are also present whose significance, whether regional or local, are impossible to assess. For example, the upper part of the member at Snettisham Common contains up to 10 per cent of grains of soft white rotted limestone that were not recorded, except as a rarity, at any other locality. The same exposure contains a probable broad channel, another feature not seen elsewhere, and the unusual lithology and this sedimentary structure may be related. Despite these limitations, the following generalisations can be made:

1. the sands are predominantly fine grained throughout the district, but beds of medium-grained sand are more common in more northerly exposures and a few thin beds of coarse-grained sand, including tiny pebbles of ironstone and mudstone, are present in the Hunstanton Borehole.

2. the sands are finely micaceous at several levels at Hunstanton, and possibly Snettisham, but become glauconitic south of Snettisham, with the overall glauconite content, both as grains and as wisps and thin seams of glauconitic clay, increasing southwards. In the southern part of the district, at Blackborough, some thin bands of sand contain more than 40 per cent glauconite.

3. thin seams, wisps and clay drapes of grey carbonaceous clay occur throughout the district, but glauconitic clay is restricted to its southern part.

4. the commonest sedimentary feature is slightly curved planar cross-bedding. This occurs at all but the highest stratigraphical levels, mostly in units 0.15 to 0.45 m thick, whose boundaries are parallel over larger distances than those visible in the exposures. Current reversals ('herringbone' cross-bedding) are common in many of these units. Parallel-bedded, evenly spaced units of similar thickness are also common. Planar cross-bedded units up to 1.5 m thick were recorded at Leziate Sandpit, and a probable broad channel is present at Snettisham Common.

5. single lines of ripples are present, but rare, at several stratigraphical levels at most localities, and are present, usually accompanied by clay drapes, in the highest 0.5 m of the Leziate Beds at most of the localities where the formation is overlain by the Dersingham Beds.

Schwarzacher (1953) made a comparative study of grain size and cross-bedding in the Sandringham Sands of Norfolk and the Woburn Sands of Cambridgeshire. No precise localities are given, but the measurements in the Leziate Beds appear, from his site map (1953, fig. 2) to have been made in sand pits at Snettisham Common [6710 3362], Dersingham [6814 2937], Wolferton (widened railway cutting) [6622 2870], North Wootton [653 243], Leziate [666 190 to 688 195], Blackborough End [676 146] and South Runcton (not located). A preponderance of easterly dips led him to conclude that the sand had been derived from a north–south coastline that had lain between Lincolnshire and Norfolk. 'Herringbone' cross-bedding was noted to be a minor feature at most localities; it accounted for more than 5 per cent of the cross-bedding only at Snettisham Common, North Wootton, Leziate, Blackborough End and South Runcton.

Rastall (1919, pp.214–215), in a study of the heavy minerals of the Lower Cretaceous rocks of eastern England, described assemblages from the Leziate Beds samples from North Wootton [653 243]. A slide of the heavy minerals from this locality was figured by Hatch and Rastall (1913, fig. 5). The principal heavy minerals found were kyanite, tourmaline, staurolite, rutile, pyroxene, muscovite, ilmenite and garnet, with small amounts of zircon, green biotite, hornblende and apatite, and possible rare anatase, barytes and sillimanite. Many of the commoner grains are fresh, large and angular. Rastall (1919, p.272) drew attention to the abundance in some samples of kyanite, tourmaline, garnet and a bluish green amphibole (cf. arfvedsonite) and suggested (1919, p.272) that they were possibly derived from Scandinavia. He recorded similar suites of heavy minerals in the Woburn Sands of Cambridgeshire and the Carstone of Hunstanton; Ingham (1929) subsequently described a similar assemblage from the Spilsby Sandstone of Lincolnshire.

The occurrence of large, angular grains of minerals derived from igneous or high-grade metamorphic rocks remains an enigma at all three stratigraphical levels. No suitable nearby source rock is known that could have either cropped out in the early Cretaceous or have supplied these minerals to the Cretaceous via a sediment of intermediate age. The possibility that the younger Cretaceous strata have obtained their heavy minerals from the older Cretaceous rocks by erosion can be largely discounted because of the size and freshness of many of the mineral grains at all four stratigrapical levels.

In a subsequent study Rastall (1925) concluded that the overall petrography of the Woburn Sands suggested derivation from Millstone Grit or Coal Measures rather than from Lower Palaeozoic or Precambrian rocks. Boswell (1927) thought that the kyanites, in particular, were too fresh to have travelled far and suggested a nearby, but as yet unknown, source.

Sources as far distant as Scandinavia can be discounted now that the presence of continuous, thick sequences of predominantly argillaceous marine Lower Cretaceous deposits are known from the intervening North Sea. No Cretaceous rock has survived in the Midlands or Wales, the nearest areas in which igneous and metamorphic rocks crop out at the present day, so the likelihood of suitable source rocks being available there during the Cretaceous cannot be readily assessed. Large angular blocks of unweathered igneous and metamorphic rocks that occur in the basal bed of the Chalk in the Cambridge area were believed by Hawkes (1943) to have been derived largely from Wales and the Midlands.

The discovery in recent years of sandy and pebbly Carboniferous rocks at depth around the edges of the London Platform in Norfolk, Oxfordshire and Berkshire, has added weight to Rastall's (1925) suggestion that much of the quartz in the Lower Cretaceous sands in East Anglia could be derived from Carboniferous rocks. These Carboniferous rocks, and coarse Triassic sediments that also fringe the London Platform, contain relatively fresh clasts of Precambrian and Palaeozoic rocks, together with heavy minerals derived from them.

Details

Snettisham to the Babingley River

The most northerly outcrop of the Leziate Beds occurs on the lower slopes of Ken Hill, Snettisham, but the member there is obscured by thick downwash. It forms a prominent feature capped by Dersingham Beds sands that skirts the southern end of Lodge Hill in the area around Lodge Hill Farm, Locke Farm and Snettisham Common [673 338]. The following composite section was measured in the railway cutting [6680 3359] adjacent to Locke Farm:

	Thickness m
Soil and subsoil, sandy with ferruginous sandstone fragments	0.5
DERSINGHAM BEDS	
Sandstone, flaggy and fissile, rusty brown, densely ferruginously cemented, fine-grained; becoming leached to very soft, orange-brown, silty sandstone in upper part	1.4

	Thickness m
LEZIATE BEDS	
Sandstone, very soft, fine- to medium-grained; well-bedded, with trough cross-bedding mostly in 0.3 m-thick units; yellow-brown to orange-brown but with some bright carmine patches, mostly concentrated along cross-beds; clean white sand podsol developed where leached by vegetation	seen 1.5

The sands were formerly worked in a pit [6710 3362] at Snettisham Common, presumably for building purposes. Several degraded sections are still visible and the following composite section can be demonstrated:

	Thickness m
DERSINGHAM BEDS (see p.85 for details)	
LEZIATE BEDS	
Sands, fine- and medium-grained, mostly grey and white but with irregular patches of pale yellow,	

orange-brown and dark brown staining, locally forming a weak cement, mostly roughly aligned with bedding and jointing; some very irregular carmine staining in highest 1.5 m; quartzose, subangular and angular grains, but with common (up to 10%) grains of soft white rotted limestone; cross-bedded throughout in units mostly 0.15 to 0.45 m thick, with low-angle trough cross-bedding predominant and current reversals (herringbone structures) present at several levels; clay drapes and lenses and wisps of clay and clayey sand infill ripple-troughs in highest 0.5 m; these structures all truncated by overlying Dersingham Beds   seen 6.7

Because of the discontinuous nature of the exposures in the pit, the overall sedimentary structure is unclear, but it seems likely that a large part of the west face is occupied by a single broad channel. White sands have also been worked in a shallow pit [6818 3433] at Home Farm, Snettisham.

Between Snettisham and Dersingham, the Leziate Beds crop out on low ground at the foot of an escarpment capped by the Chalk; they are overlain by thick downwash, much of it probably of Pleistocene age. The highest beds of the member rise above this drift to crop out on the west side of Dersingham village, but there is no exposure. To the south of the village, the low regional dip causes the Leziate Beds to rise slowly southwards and westwards to form an impressive escarpment, capped by Dersingham Beds, at Sandringham Warren and Wolferton. Beds of clean white sand within the Leziate Beds were formerly the basis for an important local industry in that area. Small amounts of sand were won from pits [6857 2980; 6873 2970] at Dersingham Common, and much larger amounts from nearby pits [6807 2952; 6818 2947; 6814 2937] at Sandpit Cottages. The last of these is the largest and most recently worked and the only one in which the sands are still exposed. Up to 6 m of grey, white and greenish grey, fine-grained sand with patchy yellow- and orange-brown staining are visible in several small exposures. The sedimentary features, including cross-bedding, herringbone structures and clay-draped large ripples in the highest beds, are very similar to those at Snettisham Common. The junction with the Dersingham Beds (see p.88) is well exposed.

Sand was also worked in an adit that ran southwards from the south-east corner of this pit, beneath the King's Lynn to Hunstanton road, and beneath Jocelyn's Wood. This adit is now largely collapsed, but an entrance in the wood [6821 2914] shows a tunnel, about 2 m in diameter, in Leziate Beds sands supported by a roof of Dersingham Beds ferruginous sandstone. The age of the mine is uncertain: neither it nor the pit is mentioned by Whitaker and Jukes-Browne (1899) and thus it is likely to be early 20th century. There is no truth in the local legend that this tunnel was formerly connected to Castle Rising and the Red Mount at King's Lynn.

The same sequence as that at Sandpit Cottages is exposed, but more deeply weathered, in the deep railway cutting [6626 2870] adjacent to Wolferton Station. The cutting was widened to work the sand for ballast (Lamplugh in Whitaker and Jukes-Browne, 1899, p.17). It exposed about 10 m of cross-bedded, fine- and medium-grained, white and grey sands with yellow- and orange-brown patchy iron-staining. Thin, hard, ferruginously cemented beds occur at two levels and Whitaker and Jukes-Browne (1899, p.8) noted that the lower of these had, when fresh, a grey (? pyritic) cement. Lowe (1868, quoted in Whitaker and Jukes-Browne, p.8) drew attention to the large numbers of ferruginous 'geodes' present at this locality and suggested, correctly in the present author's opinion, that they were fossilised burrows. The lowest part of this sequence, over-

Plate 6   Type section of the Leziate Beds, Leziate Sand Pits.

Leziate Beds sands have been worked for glass, moulding and building sands in this area for over 100 years. The modern workings have cut back into the hillside and work the sands from beneath an overburden of Dersingham Beds and till. Below the water table (lower pit) the sands are grey and locally pyritic and glauconitic. In the higher workings (the bulk of the pit) these minerals have been oxidised and largely leached to give clean white sands. View south to Brow of the Hill.

lain by glacial deposits, is poorly exposed in a small cutting [6645 2878] on the north side of the railway.

Between Wolferton and the drift-filled valley of the Babingley River, the Leziate Beds crop out on the face and lower slopes of an escarpment that falls steadily eastwards. The outcrop becomes progressively more covered by sandy downwash derived from the Leziate Beds and the Dersingham Beds, and by Pleistocene drift when traced eastwards. Where the topographical features are best developed, in the Wolferton area, the Leziate Beds contain several harder, ferruginously cemented bands, which form subsidiary features on the main escarpment. None of these features persists for more than a few hundred metres, probably because the cemented beds are themselves impersistent.

Babingley River to the River Nar

The Leziate Beds have an extensive drift-free outcrop on the south side of the Babingley River between White Hills Wood, Roydon and Ling Common, North Wootton, but there is no exposure. The sands have been worked in small pits in White Hills Wood [6918 2449] and at Castle Rising [6620 2470], and in very extensive workings at Ling Common [653 243; 655 237]. More than 12 m of sands must once have been exposed in the Ling Common pits. A few small, deeply weathered exposures are still present and these show mostly clean white and grey, cross-bedded, fine- and medium-grained sands with some yellow and brown iron-staining. At one locality [6537 2398], an impersistent ferruginously cemented beds up to 20 cm thick, which has retained kernals of grey pyritic cement, is present. Ferruginous geodes, (many of them burrow shaped), nodules and boxstones filled with loose sand are common as loose debris and were probably derived by weathering from pyritic concretions and tabular beds.

An unusual feature of the Leziate Beds outcrop in that area, noted originally by Rose (1835), is the common occurrence of blocks of hard, pure quartzite up to 1m across. Whitaker and Jukes-Browne (1899, p.7) implied that this sandstone was visible in situ in the Leziate Beds at Ling Common. It was recorded only as loose debris in the drift in the present survey, and is here interpreted as a Pleistocene silcrete. Sand was also worked in small pits [6550 2265; 6580 2224] at South Wootton: the latter locality is probably the pit, at that time already overgrown, referred to by Whitaker and Jukes-Browne (1899, p.7) as Wootton Warren Wood.

The Leziate Beds have an extensive outcrop between the King's Lynn Bypass and Grimston Warren, and on the western and northern sides of Roydon Common, where they crop out on the face of a small, but locally steep, escarpment. There are many small- and medium-sized old sand pits at Warren Farm [6665 2180; 6687 2180], Grimston Warren [6823 2170; 6806 2135] and Roydon Common [6775 2248, 6793 2228, 6800 2220, 6802 2212, 6790 2193], and a very large pit at Grimston Warren [678 216]. Several of the Roydon Common pits show the basal strata of the Dersingham Beds overlying small scrapes in deeply weathered white, grey, yellow, brown or red fine-grained pure quartzose sands. The large pit at the Warren exposes 2.5 m of white and brown, fine-grained, cross-bedded sand with some weak, patchy, ferruginous cement, overlain by glacial deposits.

No determinable ammonite has been recorded from any locality in the Leziate Beds, but a ferruginous cast of a ?Polyptychites found loose in pebbly head at the top of one of the Roydon Common pits [6784 2194] is believed to have been derived from the member. Ferruginous geodes are common in all the old sand pits in the area, and in the drift deposits that overlie them.

Small outliers of Leziate Beds crop out at Mill Hill [695 220] and near Sugar Fen [693 213], but there is no exposure at either locality. Up to 6 m of Leziate Beds are preserved in a small fault-bounded outlier on the south side of the Gaywood River at Holding's Hill [652 665], Gaywood. Much of the core of the outlier has been dug away for sand in a large pit that exposes up to 1.5 m of fine-grained, cross-bedded, grey sand with some ferruginous cement and with a thin (1 cm) bed of glauconitic clay.

The present and former sand workings, usually referred to as Leziate Sand Pits, now occupy about 2km square of ground including Leziate Heath and Bawsey Heath on the north side of the Middleton to Brow of the Hill road, and the former golf course on the south side of the road. The full thickness of the formation is exposed in the workings, but in widely separated sections that are difficult to correlate. About 30 m of relatively uniform sands are estimated to be present. The availability of the exposures varies considerably with time due to the rapid rate of working, and the following sections have been chosen to summarise the sequence. The succession at any particular locality is to some degree repetitive, in that two of the main lithological components, glauconite and pyrite, have been oxidised and their oxidation products to a large degree removed, everywhere above the natural water table.

The lower part of the member was formerly well exposed in the more westerly workings in Hundred Acres [6672 1889 to 6654 1935] and Bawsey Warren [6676 1941 to 6662 1962], where the following composite section was measured:

Thickness
m

LEZIATE BEDS

Soil and subsoil; white fine-grained sand and sandy podsol with local concentrations of bleached angular flints — up to 0.6

Sand, fine-grained with a few interbeds of medium-grained sand; highest 0.5 to 1.0 m commonly structureless, grey and leached, with some black organic staining; clean white and greyish white, but with extensive patchy, yellow to deep orange-brown iron-staining; cross-bedded at most levels in 0.15 m to 0.3 m units, most planar and with some current-reversals; unit boundaries mostly planar over very large distances; rare ripples, some with clay drapes; thin grey carbonaceous clay lenses; scattered pyritic concretions and tabular beds in lower part, but mostly oxidised to ferruginously cemented

Thickness
m

sandstone patches, some with pyritic cores; becoming slightly greenish grey and glauconitic with depth and passing down into — 5.0

Sand, glauconitic, fine-grained, with some medium-grained beds; cross-bedded as above; grey, greenish grey and yellowish green, becoming darker grey and green below the natural water table; comminuted plant debris abundant at some levels; pyritic concretions, some enclosing a matted network of carbonised plant fragments and rare poorly preserved bivalves, common at several levels; ferruginously stained, greyish brown and green, glauconitic lenticular clay seams up to 5 cm thick in lower part — 3.0

Iron-rich seepages from the lowest, most clayey part of the sequence mark the lower practical limit of the workings.

The middle and upper part of the member is exposed in a series of workings [6765 1912 to 6882 1954] on the north side of the road at Brow of the Hill. The following composite section was measured in the more easterly of these [6843 1935] at Leziate Heath:

Thickness
m

SNETTISHAM CLAY (see p.90 for details)

LEZIATE BEDS

Sand, mostly fine-grained with some medium-grained lenses and beds; white and grey, with patchy pale yellow to dark orange-brown ferruginous staining; ferruginous sandstone patches (formerly pyritic), cross-bedding, rare ripples, thin clay seams etc. as elsewhere; passing down into — c. 17

Sand, dark brownish grey, greenish grey and yellowish green; common carbonaceous debris; thin plant-rich clay seams; pyritic concretions with much plant debris and some bivalves — c. 12

There is clearly some overlap between these two sections, but the lack of marker bands combined with the intensity of the weathering above the water table make correlation impossible. Where unweathered, the whole of the sequence appears to consist of glauconitic sands with rare thin clay seams and with pyritic concretions at numerous levels. There appears to be a broad change in the scale, but not the type, of cross-bedding upward within the Leziate Beds. Small-scale units, mostly 0.15 m to 0.3 m thick, pass up into larger-scale planar cross-bedded units that are commonly 0.5 to 1.0 m thick. The exposures in the lower part of the member are too widely separated from those in the upper part to determine whether or not this is a persistent stratigraphical feature or merely local lateral variation.

The Leziate Beds re-emerge from beneath the thick drift deposits that infill the valley of the Middleton Stop Drain, to crop out on the flanks of the ridge of high ground that runs from North Runcton to East Winch and Blackborough End. There are few exposures in the area. Sand was formerly worked in a pit [6676 1679] at Tower End and continues to be sporadically worked from beneath an unconformable Carstone capping at Blackborough End [676 146]. At this latter locality, the Carstone rests on an intensely burrowed surface of white, leached, fine-grained sands that pass down at about 2 m depth into relatively fresh greenish grey and yellowish green, moderately and highly glauconitic, fine-grained sands with some green glauconitic clay wisps. A total of about 7 m of Leziate Beds is exposed; the lower beds form a small Pleistocene cliff against which Glacial Sand and Gravel is banked. Sand has also been worked in two other pits [6676 1489; 6718 1438] at Blackborough End. The first of these shows Carstone resting on an intensely burrowed surface of white, fine-grained Leziate Beds sands with traces of cross-bedding. The second is entirely within the Leziate Beds and shows up to 2.5 m of grey and greenish grey, slightly clayey, fine-grained, glauconitic sand with patchy iron-staining and a ferruginously cemented band.

River Nar to Denver

The Leziate Beds are overlain by the thick Quaternary deposits of the Nar valley for about 3 km south of Blackborough End. They crop out on the south side of Shouldham Warren and The Sincks, where they again give rise to sandy heathland, now afforested. The highest 3 m of the sands have been worked from beneath the Carstone in a pit [6837 1132] at the Warren. The sands are leached, grey and white, fine grained, with the usual cross-bedding features. There is no other exposure in the area, although scrapes in white sand are common and, on the lower ground close to the water table, pyritic nodules are common in drainage ditches.

The Leziate Beds are overlain by glacial deposits everywhere between Shouldham Thorpe and Wimbotsham, except for a small inlier at South Runcton. Schwartzacher (1954) described the bedding features in a small, now degraded pit in this inlier.

Between Wimbotsham and Denver the Leziate Beds crop out on a prominent escarpment and have been worked in a large number of pits. All of these are now degraded but most contain scrapes of loose sand, numerous ferruginous geodes and slabs of ferruginous sandstone. Every pit is in the upper part of the member and well above the water table; pyrite and glauconite were not recorded. Two well-developed subsidiary features within the outcrop appear to mark the positions of ferruginously cemented beds that are locally persistent. Between Wimbotsham and Downham Market, there are several sand pits adjacent to the Downham Market–King's Lynn road [6194 0446; 6204 0431; 6185 0400; 6160 0374; 6145 0363], and other pits occur in Downham Market itself [6127 0324; 6124 0319; 6150 0313; 6168 0316; 6153 0340; 6148 0255].

Sand was also worked in a small pit [6181 0179] at Ryston Park, Denver, close to the southern limit of the Leziate Beds. White and brown sand with many ferruginous geodes form a weak feature in the northern part of the park, but these are rapidly overstepped by the Carstone near Home Farm, Ryston [622 008]. A narrow inlier of the basal strata of the Leziate Beds occurs between Mouse Hall Farm, Ryston and Coldham's Farm, Crimplesham, and exposes grey and pale green glauconitic sands, much disturbed by cryoturbation.

Boreholes

The Leziate Beds are relatively uniform in lithology in both their outcrop and subcrop in the district. The full thickness of the formation was cored in the Hunstanton, Gayton and Marham boreholes, although in the last-named it is incomplete because of the overstep of the Carstone. The following sequence was proved between 33.12 and 62.18 m in the Gayton Borehole: it is probably typical of the southern part of the district:

	Thickness m	Depth m
DERSINGHAM BEDS (see p.91 for details)		
LEZIATE BEDS		
Sand, fine-grained, greyish white, well-sorted; very poorly cemented; composed almost entirely (99% at most levels) of angular to subangular, clear quartz grains; traces of grey clay drapes and burrow linings at several levels; faintly glauconitic in part	2.36	35.48
Core lost in soft sand	0.31	35.79
Sand, as above but more glauconitic and clayey at several levels and with wisps of grey clay and green glauconitic clay	2.28	38.07
Core lost in soft sand	6.53	44.60
Sand, as above, becoming more clayey and less well-sorted with depth; traces of burrowing and small-scale cross-bedding; small patches of pyritic cement at several levels; becoming less clayey with depth but with widely spaced thin beds (up to 1 cm thick) of black or dark grey carbonaceous clay; in lower part of bed cross-bedding and burrows are picked out by concentrations of medium-grained sand; fining-upward, graded-bedding units present at several levels; carbonaceous plant debris (some pyritised) common throughout; weakly glauconitic throughout but with burrow-fill and cross-bedding concentrations of glauconite at several levels; planar base	17.58	62.18
MINTLYN BEDS (see p.74 for details)		

DERSINGHAM BEDS

The Dersingham Beds comprise a laterally variable, rhythmic sequence of thinly interbedded fine-grained sands, ferruginous sandstones, silts and clays. The formation has an almost continuous, largely drift-free outcrop between Heacham and the Babingley River. Southwards from there it crops out, mostly as a narrow strip from beneath extensive drift deposits, as outliers at Rising Lodge, Brow of the Hill and East Winch with the main outcrop on low ground between Roydon and Ashwicken. The Dersingham Beds are cut out by the southerly overstep of the Carstone in the vicinity of East Winch. The extent of their subcrop is poorly known. The formation was proved in the Gayton and Hunstanton boreholes and might be expected to occur beneath much of north-west Norfolk.

At outcrop, the formation undergoes a broad lithological change, in addition to minor local variations, from predominantly arenaceous in the south to a mixture of arenaceous and argillaceous in the north (Figure 28). Between the most northerly outcrop at Heacham and the Hunstanton Borehole, the formation undergoes a rapid change to become predominantly argillaceous and, in part, lithologically similar to the Tealby Beds of Lin-

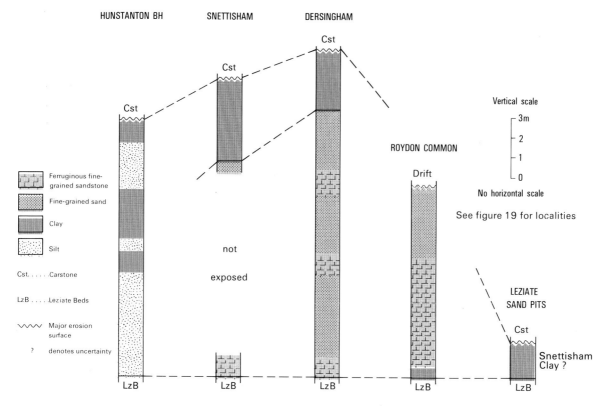

Figure 28   Correlations between the Dersingham Beds sequences in the district.

colnshire. The Dersingham Beds pass into the Tealby Beds beneath the Norfolk side of The Wash.

The formation takes its name from the Dersingham area where the harder beds within a predominantly sandy sequence about 16 m thick give rise to a series of drift-free terrace-like features that, despite the paucity of exposure, enable the lithological sequence to be demonstrated. There is no single section that exposes the full sequence, and the more argillaceous parts of it are known only as deeply weathered spoil from ditches. In that area, the Dersingham Beds consist of about 1 m of fine-grained ferruginous sandstone overlain by about 12 m of rhythmic alternations of fine-grained white sand and ferruginous sandstone, with thin beds of silt and silty clay, overlain by about 3 m of clay and silty clay (Figure 29).

Throughout the Dersingham Beds outcrop the highest bed of the formation is a thin but laterally persistent bed of clay, thickening from 1.5 m in the south to 6 m in the north, that Whitaker and Jukes-Browne (1899, pp.6 and 9) termed the Snettisham Beds [Clay]. The stratigraphical relationships of this bed to the remainder of the Dersingham Beds is still not entirely clear; the reasons for considering it to be part of the Dersingham Beds instead of a separate formation are discussed below.

In the Hunstanton Borehole, the Dersingham Beds consist of 11.6 m of silts, clays and silty, sandy and oolitic ('chamositic') clays that form a complex rhythmic sequence. Many of the rhythms have erosional bases that bound beds of sparsely fossiliferous strata, and few detailed correlations can be made with the succession at

outcrop. The Dersingham Beds are not exposed between the borehole and Heacham, and are poorly exposed between the latter locality and the type area. The lower part of the clayey sequence in the borehole appears, however, to pass southwards into a complex sequence of thinly interbedded sands, silts and clays. The Hunstanton sequence passes beneath The Wash into the even more argillaceous Tealby Beds. Both successions are made of fining-upward rhythms (Figure 30). The Snettisham Clay appears to be the correlative of one or possibly two rhythms in the upper part of the Hunstanton sequence; each of the lithologies in the generalised Hunstanton rhythm is present in Snettisham Clay spoil in the Heacham to Snettisham area, but the exposures are now too poor for their sequence to be determined. Even with good exposure, it would be difficult to make correlations in this type of sparsely fossiliferous sequence that contains numerous erosion surfaces.

The base of the Dersingham Beds, both at outcrop and in the Hunstanton Borehole, rests with sharp lithological contrast on the loose clean white sands of the Leziate Beds. The basal bed is usually a few centimetres of clay overlain by 3 to 5 m of fine-grained, ferruginous sandstone. The contrast in permeability and hardness between the Leziate Beds and the basal Dersingham Beds gives rise to a weak spring line and one of the most prominent features in west Norfolk, and the junction of the two formations is easy to survey.

The basal ferruginous sandstones have been well exposed in the past in quarries at Dersingham (Rose, 1862,

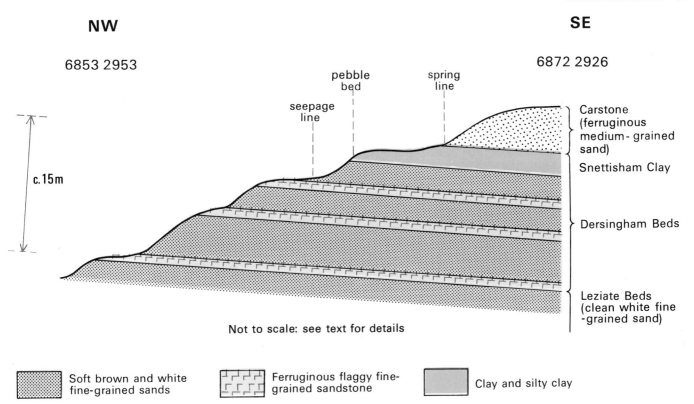

Figure 29 Dersingham Beds sequence at Dersingham.

p.31), Sandringham Warren and Wolferton (Lamplugh *in* Whitaker and Jukes-Browne, 1899, p.17), and the junction with the Leziate Beds has been exposed in sand pits at Roydon Common, Dersingham, Snettisham and in Wolferton railway cutting (Plate 7). The middle part of the formation has never been well exposed. The features in drift-free areas such as Sandringham Park and Lodge Hill [670 340] suggest that it includes several relatively thick clay seams; small, degraded brickpits are present in both areas.

At the top of the formation the Snettisham Clay can be traced almost continuously at outcrop from Heacham to Ashwicken. It was well exposed in Victorian times when it was dug for brickmaking at Heacham, Snettisham, Ingoldisthorpe, Dersingham and Brow of the Hill. Jackson (1911, p.63) recorded the full thickness of the Snettisham Clay at Heacham Brickworks [678 364] as 9 m of greyish brown clay with a line of fossiliferous clay ironstones about 3 m from the top, and with ironstone concretions scattered throughout. Loose spoil in the modern pit suggests that the clays rested on white sands, the junction being marked by a bed of ferruginously cemented sandstone with small quartz and other pebbles, and angular clay clasts. Evidence of a break in sedimentation at the base of the Snettisham Clay is also present at other localities and it is probable that it rests with minor unconformity on the underlying beds throughout its outcrop. In the Leziate, Ashwicken and East Winch areas the Snettisham Clay appears to cut out the lower and middle parts of the Dersingham Beds

and rests on the Leziate Beds. The critical sections, up to 2 m of soft mudstone at Leziate Sand Pits and in the Gayton Borehole, are sparsely fossiliferous and yielded insufficient faunal evidence for their ages to be determined. They probably belong with the Snettisham Clay, but they could also be a locally expanded equivalent of the persistent thin clay that marks the base of the Dersingham Beds.

Sedimentary features, other than lamination picked out by partings of sand, are rare in the Snettisham Clay at outcrop. In the Gayton Borehole, bioturbation in the form of tracks, trails, vertical and subhorizontal tube-shaped burrows and large poorly defined burrows, are common at several levels.

Many of the Dersingham Beds lithologies are sparsely fossiliferous, largely because of their unsuitability for preservation. The two notable exceptions are the ferruginous sandstones at the base of the formation and clay ironstone nodules in the Snettisham Clay, both of which have given rise to extensive museum collections. The three exposures in the basal sandstones, at Wolferton [662 287], Sandringham Warren [669 276] and Roydon Common [675 222], that were listed by Whitaker and Jukes-Browne (1899,p.27) as fossiliferous, are still open and they, together with a small number of additional exposures in the same area, have yielded a relatively rich fauna. This is dominated by bivalves (24 species), but includes common plant fragments, trace fossils, serpulids and gastropods and rare ammonites and echinoids. The ammonites include species of *Endemoceras* that indicate an early Hauterivian age. Casey (1967) recorded the late Hauterivian/early Barremian ammonite *Craspedodiscus* as a derived fossil in the Carstone at West Dereham, in south-west Norfolk, and presumed it to have been derived from a former outcrop of the Dersingham Beds. The non-ammonite fauna of the Dersingham Beds is long ranging and not stratigraphically diagnostic. Casey (1973) has noted, however, that the occurrence of large numbers of the Boreal bivalve *Buchia* in the basal part of the Dersingham Beds can be correlated with a similar influx of *Buchia* in the Hauterivian part of the Speeton Clay of Yorkshire.

The former brickworks in the Snettisham Clay at Heacham, Snettisham and Ingoldisthorpe yielded a rich bivalve fauna, common gastropods, ammonites, plant fragment and serpulids, and rare brachipods, belemnites, crinoids and fish debris. Many of the ammonites are well preserved as uncrushed red or brown, ironstone casts; they were much sought after when the pits were in work. Many of the finer specimens are now in museum collections. They all belong to crioceratid genera and were shown by Spath (1924) to be indicative of a Barremian age.

The sections are all now degraded and the published descriptions of several of them are either incomplete or in disagreement with one another. The provenance of much of the fauna can no longer be demonstrated but re-examination of museum collections by Mr Morter strongly suggests that the bulk of it belongs to one or other of two assemblages of differing ages. Contemporary descriptions of the brickpits suggest that these assemblages came from two separate horizons, one close to the base of the Snettisham Clay and one in its upper part. The older fauna is by far the more varied and includes plants, annelids, brachiopods (2 species), bivalves (22 species), gastropods (7 species), crioceratid ammonites including *Acrioceras* sp., species of the belemnites *Aulacoteuthis* and *Praeoxyteuthis*, and fish vertebrae, all preserved as nests in large, rusty brown, clay-ironstone concretions or in ferruginous oolitic clays. The ammonites and belemnites suggest correlation with the early Barremian zone of *Crioceratites (Hoplocrioceras) fissicostatum*. At Heacham and Snettisham brickworks this type of fossiliferous ironstone appears to have been present at the base of the Snettisham Clay and possibly at one or more levels in the lowest 2 m.

The younger fauna is more limited and consists of species of *Crioceratites (Hoplocrioceras)* and *C. (Paracrioceras)*, and a few bivalves and gastropods, beautifully preserved in buff, red or purplish red clay ironstone. The ammonites are indicative of the mid-Barremian zone of *Crioceratites (Paracrioceras) elegans*.

At Hunstanton, the Snettisham Clay is overlain conformably or with only minor sedimentary break by the Roach. Everywhere south of there, it is unconformably overlain by the Carstone, and is overstepped by that formation near East Winch. At outcrop between Snettisham and Leziate, the Snettisham Clay and Carstone appear to have a concordant relationship, in which their basal unconformities are almost parallel to one another. Correlation of the more important Dersingham Beds sequences in the district with one another is shown in Figure 28.

Details

Heacham to Snettisham

The Dersingham Beds outcrop occupies low ground that is entirely drift covered northwards from Heacham. The formation crops out on gently rising ground in the southern part of Heacham village and can be traced from there southwards by means of small features, pits for sand or clay and, at the top of the formation, by a well-developed spring line. Sands in the middle part of the formation have been worked at Heacham parish [6794 3677] and Draper's Hole pits [6754 3697], the former reputedly in 'silver' and red sands.

The former workings at Heacham Brickworks [678 364], at Mount Pleasant, probably penetrated the full thickness of the Snettisham Clay and about 3 m of the underlying sands. The section is now degraded and shows about 6 m of deeply weathered brown and grey clays with a few concretionary ironstone fragments, overlying about 2 m of fine- and medium-grained sands. Teall (1875, p.170) recorded the following section (metricated here) in the lower beds:

	Thickness m
Clay, passing up into sandy and then very sandy clay	2.4 to 3.0
Clay, stiff, dark blue, with ironstone nodules containing ammonites	0.6 to 0.9
Layer of indurated sandstone overlying sands of unknown thickness	

Lamplugh (in Whitaker and Jukes-Browne, 1899, p.12) measured what appears to be the full thickness of the Snettisham Clay, partly in the brick pit and partly in an excavation. The section can be summarised as follows:

	Thickness m
Head, brown loam with flints etc.	1.5
SNETTISHAM CLAY	
Dark blue pyritic clay with decomposed pyritic layer at base	0.5
Pale bluish brown clay with pale brown pyritic concretions containing crioceratid ammonites	0.9
Dark brown and blue clay passing down into	1.8
Dark pyritic silt	0.6
Clay with pebbly clay ironstone nodules at base, some bored and serpulid encrusted	2.3
DERSINGHAM BEDS (undifferentiated)	
Pale yellow sand	0.6

Jackson (1911, p.63) recorded the full thickness of the Snettisham Clay in the pit as 30 ft (9 m) of greyish brown clay with a line of fossiliferous clay ironstone with pyritic cores about 10 ft (3 m) from the top and with ironstone concretions scattered throughout. Loose spoil in the modern pit suggests that the junction of the clays and the sands was marked by a band of ferruginously cemented, medium- and coarse-grained sandstone, presumably Teall's 'indurated layer' and Lamplugh's 'pebbly clay ironstone', with small quartz and other pebbles, and angular clay clasts. Jackson (1911, p.63) recorded a bed or 'iron-grit' with many shells and wood fragments at about this level. The colour differences between the Teall and Lamplugh sections are probably due to weathering.

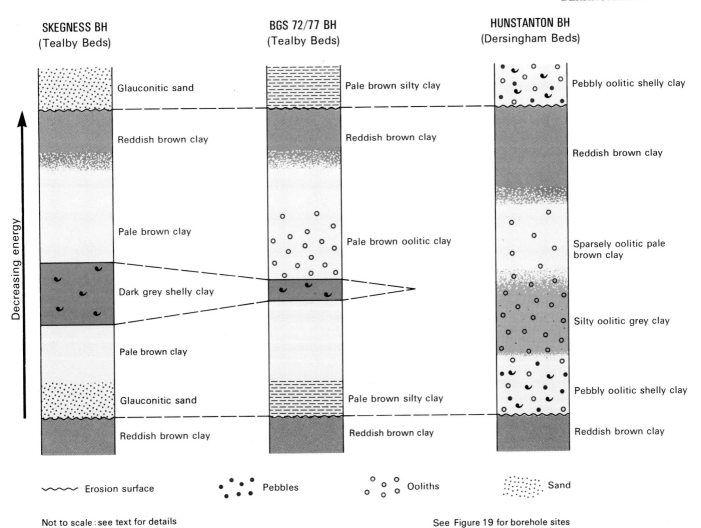

SKEGNESS BH
(Tealby Beds)

BGS 72/77 BH
(Tealby Beds)

HUNSTANTON BH
(Dersingham Beds)

Decreasing energy

Glauconitic sand

Reddish brown clay

Pale brown clay

Dark grey shelly clay

Pale brown clay

Glauconitic sand

Reddish brown clay

Pale brown silty clay

Reddish brown clay

Pale brown oolitic clay

Pale brown silty clay

Reddish brown clay

Pebbly oolitic shelly clay

Reddish brown clay

Sparsely oolitic pale brown clay

Silty oolitic grey clay

Pebbly oolitic shelly clay

Reddish brown clay

〜〜〜 Erosion surface   Pebbles   Ooliths   Sand

Not to scale: see text for details

See Figure 19 for borehole sites

Figure 30 Correlation of generalised rhythms in the early Barremian sediments of the Wash area.

The higher level of concretions have become justifiably famous because of their rich crioceratid ammonite fauna, usually preserved as casts in brick-red or purplish red oolitic ironstone, and fine specimens are included in many museum collections. The fauna includes *Acrioceras* cf. *tabarelli* (Astier), *Crioceratites* (*Paracrioceras*) cf. *elegans* (Koenen), *C.* (*P.*) *occultum* (Seeley), *C.* (*P.*) cf. *occultum* and *C.* (*Hoplocrioceras*) aff. *laevisculum* (Koenen). The remaining fauna from the pit, presumably derived from several stratigraphical levels, includes lignite fragments, *Corbula* sp., *C.* cf. *isocardiaeformis* Harbort, *Cucullaea cornueliana* (d'Orbigny), *Iotrigonia scapha* (Agassiz), *Nucula* (*Leionucula*) cf. *planata* (Deshayes), *Oxytoma* sp. indet., *Phelopteria* aff. *rostrata* (J de C Sowerby), *Pholadomya eberti* Wolleman, *Thetis minor* J de C Sowerby, *Vnigriella? scapha* (d'Orbigny) and *Nerineopsis* sp. nov. aff. *claxbiensis* Abbas.

The clay has also been worked in small pits [6800 3596; 6804 3587] between Heacham Brickworks and Heacham Bottom Farm; the underlying sands have been worked in a small pit [6798 3550] close to the farm.

The Dersingham Beds crop out as a series of steep features capped by the Carstone on the middle and upper slopes of Ken Hill and Lodge Hill. Most of the outcrop is obscured by

downwashed sands and there are a few exposures. The basal beds of the formation, including the junction with the Leziate Beds, are exposed in Snettisham Common sand pit [671 337] at the foot of Lodge Hill, where the following deeply weathered composite section was measured. The beds are laterally variable, although the relative proportions of the various lithologies remain roughly constant throughout the pit:

	Thickness m
Soil, subsoil and sandy wash	0.8
DERSINGHAM BEDS	
Sand, soft, silty and clayey, fine-grained, deeply weathered, pale greyish brown	0.23
Sand, clayey, fine-grained, bright orange-brown, with angular ironstone fragments	0.13
Clay, dark grey, with sand-filled burrows and wisps; irregular but persistent bed	up to 0.01
Sand, clayey, soft, greyish brown; seepages at base	0.13
Clay ironstone, rotted, rusty brown, porous, with many voids, some clay filled; irregular base	0.05 to 0.10

Plate 7   Junction of Leziate Beds and Dersingham Beds, Wolferton railway cutting [6626 2868].

Ferruginously cemented flaggy sandstones, here deeply weathered, at the base of the Dersingham Beds rest with marked lithological contrast on the soft white sands of the Leziate Beds. Ferruginous nodules close to the junction are probably a secondary feature produced by groundwater movement. (A 10856).

	Thickness m
Sand, fine-grained, very clayey, soft, with lenses of grey clay and ferruginous sandstone; hard dark brown ferruginous pan at base resting on Leziate Beds	0.08 to 0.13

Sand and clay have been worked in small pits [6682 3409; 6677 3391] on the side of Lodge Hill, but both pits are now degraded. The first appears to have been in interbedded fine-grained sands and silty clays. The second is the locality referred to by Lamplugh (*in* Whitaker and Jukes-Browne, 1899, p.19) as overgrown; it is probably that described by Teall (1875, p.18) as showing about 1.8 m of sandy clay and dark grey clay overlying 0.6 m of very fossiliferous, pebbly, sandy, highly oolitic (limonite) clay that rested on an unspecified thickness of fossiliferous sandy clay. Teall thought that Carstone sands might be present in situ at the top of the pit, but the field relationships suggest that the pit is either in the lower part of the Snettisham

Clay and the underlying beds, or that the whole section is stratigraphically lower than the Snettisham Clay.

The fauna of the sandy oolitic clay and of clay ironstone nodules associated with it includes *Lamellaerhynchia* sp., *Rouillieria* cf. *tilbyensis* (Davidson), *Camptonectes* (*Boreionectes*) *cinctus* (J Sowerby), cyprinid indet. *Entolium germanicus* (Wollemann), gervilliid, *Iotrigonia* sp., lucinid indet., *Mulletia mulleti* (Deshayes), *Mulletia* sp., *Oxytoma pectinatum* (J de C Sowerby), *Panopea neocomiensis* Leymerie, *Parmicorbula striatula* (J de C Sowerby), *Phelopteria* aff. *rostrata* (J de C Sowerby) *Resatrix* cf. *neocomiensis* (Weerth), *Thetis minor* (J de C Sowerby), *Dimorphosoma ancilochila* Gardner, naticid indet., *Nerineopsis aculeatum* (Sharman & Newton), *Semisolarium moniliferum* (Michelin), *Tessarolax* sp., *Praeoxyteuthis* cf. *brunsvicensiformis* (Stolley) and *P.* sp. indet.

Between Lodge Hill and Dersingham, the lower and middle parts of the Dersingham Beds crop out on low ground, much of it covered by sandy wash, and they are not exposed. The Snettisham Clay crops out on the face of a prominent feature and, although also largely covered by wash, its position is readily determinable by means of a spring line at its junction with the Carstone, the poorly drained nature of its outcrop and a number of small landslips. The clays have been worked for brick-making at Snettisham, Ingoldisthorpe and Dersingham, but all the pits are now completely degraded. Finds of Romano-British, Saxon, Anglo-Saxon and Medieval pottery at localities adjacent to the Snettisham Clay outcrop at Park House Farm, Snettisham, may indicate a long history of clay working in that area.

Snettisham Brickworks [6875 3380] was disused and overgrown at the time of the Old Series survey and no accurate measurement was possible. The bulk of the pit was in grey sandy clay which weathered brown, buff and red; a bed of very shelly brown ironstone formed the floor. A band of purplish red fossiliferous clay ironstone nodules occurred in the middle part of the section. In recent years, foundation excavations have exposed sufficient of the sequence for the following composite section to be determined. All the sections were small and deeply weathered and the thickness estimates are therefore unreliable.

	Thickness m
Head: sandy wash derived from Carstone	0.6

SNETTISHAM CLAY

Clay, mottled brown and grey, with some sand wisps, possibly burrow infillings; two (possibly three) horizons with widely spaced, small round (2 to 5 cm diameter) and elongate, flattened (up to 22 cm across) clay ironstone concretions with yellow centres and reddish brown limonitised weathering skins	c.3.0
Clay ironstone, hard, grey, with dark brown limonitised skin; continuous tabular bed; local nests of fossils, mostly moulds and casts of bivalves and wood fragments, weather to porous brown masses	0.07 to 0.15
Clays, silty, mottled red, brown and grey	c.1.5
Gap, not exposed	c.0.3

DERSINGHAM BEDS (undifferentiated)

Sands, dirty, bioturbated; brown and greyish white	seen c.1.0

Museum specimens of the basal ironstone, comprising deeply weathered rusty brown porous masses that are presumed to have originally been a clayey limonite oolite, contain a rich fauna including *Weichselia reticulata* (Stokes & Webb) Fontaine, *Mucroserpula* sp., *Cucullaea* cf. *cornueliana* (d'Orbigny), *Cucullaea* sp., *Entolium germanicus* (Wollemann), *Entolium* sp., *Iotrigonia scapha* (Agassiz) *Mimachlamys robinaldina* (d'Orbigny), *Nucula* (*Leionucula*) *planata* (Deshayes), *N.* (*L.*) sp., *Oxytoma pectinatum* (J de C Sowerby), *O.* sp. (smooth form) *pectinatum* gr., *Parmicorbula striatula* (J de C Sowerby), *Panopea* sp., *Pinna* sp., *Resatrix* cf. *neocomiensis* (Weerth), *Thetis minor* (J de C Sowerby), *Vnigriella* ? *scapha* (d'Orbigny), *Actaeon* cf. *dupiniana* (d'Orbigny), *A.* sp., *Anchura* (*Perissoptera*) sp., *Avellana* sp., *naticid, Nerineopsis aculeatum* (Sharman & Newton), *Semisolarium* cf. *moniliferum* (Michelin), *S.* sp., *Tessarolax bicarinata* (Deshayes), ammonite frag. indet., crioceratid ammonite, *Aulacoteuthis* cf. *ascendens* (Stolley), *Praeoxyteuthis* cf. *jasikowi* (Lahusen) and teleost vertebrae.

The purplish red ironstones contain a more limited fauna that includes *Iotrigonia scapha* (Agassiz), *Nucula* (*Leionucula*) *planata* (Deshayes), *Thetis minor* J de C Sowerby, *Vnigriella*? *scapha* (d'Orbigny), *Tessarolax bicarinata* (Deshayes), *Crioceratites* (*Paracrioceras*) cf. *elegans* (Koenen) and *C.* (*P.*) cf. *occultum* (Seeley).

The lithological and faunal similarities suggest that the basal ironstone at Snettisham may correlate with Lamplugh's 'pebbly clay ironstone' at Heacham Brickworks and with the fossiliferous ironstone at Lodge Hill. The purplish red ironstones at Snettisham correlate with the lithologically similar crioceratid-rich nodules at Heacham Brickworks.

Very old clay pits, presumably dug for pottery or brickmaking, occur at Park Farm [6888 3395] and Park House Farm [6900 3330], Snettisham, and at Manor House Farm [6880 3295], Ingoldisthorpe. Larger, more recent brickpits (probably Victorian) occur in the lower part of the Snettisham Clay and the upper part of the underlying sands at Ingoldisthorpe Hall [6873 3263], Brickley Wood [6882 3220], Ingoldisthorpe, and adjacent to Life Wood [6883 3150], Dersingham. The Ingoldisthorpe Hall pit is probably the locality 'near Mount Amelia' from which Fitton (1836) collected fossiliferous ferruginous concretions with *Entolium germanicus* (Wollemann), *Oxytoma pectinatum* (J de C Sowerby), *Parmicorbula striatula* (J de C Sowerby), *Resatrix* and *Nerineopsis* cf. *aculeatum* (Sharman & Newton) (= *Turritella granulata* Forbes non Sowerby). The fauna is not diagnostic, but both it and the lithology are similar to those of the basal ferruginous ironstone at Snettisham Brickworks.

Dersingham to Sandringham

The basal strata of the Dersingham Beds rise above the drift deposits at Dersingham and form a low plateau on which much of the village is built. The junction with the Leziate Beds is marked by a low but persistent feature capped by much ferruginous sandstone brash. A thin bed of sandy clay at the junction gives rise to an impersistent line of seepage. The higher parts of the formation, including the Snettisham Clay, crop out in the eastern part of the village. They give rise to several persistent features, but are poorly exposed. Small degraded sand pits occur at several horizons.

The base of the Snettisham Clay is exposed in a valley side [6935 2997] opposite the Feathers Hotel, Dersingham, where the following section was measured:

	Thickness m
Sand and sandy subsoil	0.6

SNETTISHAM CLAY

Clay and silty clay, brown and grey mottled, with a line of small nodular and flaggy clay-ironstone concretions in the middle part of the bed	0.6
Clay, silty, pale grey, with small well rounded pebbles of quartz, ironstone and rotted limestone, and larger phosphatic pebbles; irregular base resting on ferruginously cemented and channelled surface	0.1 to 0.2

DERSINGHAM BEDS (undifferentiated)

Sand, fine-grained, white	seen 0.3

Lamplugh (*in* Whitaker and Jukes-Browne, 1899, p.10) collected a fossiliferous phosphatic nodule from the pebbly clay; this yielded ostreids indet., *Panopea* cf. *neocomiensis* (Leymerie), *P.* sp., cf. *Parmicorbula striatula* (J de C Sowerby), *Pinna robinaldina* (d'Orbigny), *P.* sp. indet., *Resatrix neocomiensis* (Weerth) and *Crioceratites* cf. *hildesiensis* (Koenen) sensu Immel. The underlying sand contained a *Panopea* cf. *subhercynicus* Maas in life position. Lamplugh recorded a similar sequence in a nearby lane section [6939 3001], no longer visible, but with a less sharply defined junction of the clays and sands. The phosphatised fauna from the pebble bed is probably comparable in age to the in situ faunas of the ironstones at or close to the base of the Snettisham

Clay at Heacham Brickworks, Lodge Hill and Snettisham Brickworks. Its derived presence at Dersingham suggests that the basal beds of the Snettisham Clay are locally condensed.

The Dersingham Beds form a series of persistent, terrace-like features that can be traced southwards from Dersingham to the Babingley valley, where they disappear beneath thick drift deposits. The ferruginous sandstones at the base of the formation cap an impressive feature that runs from Dersingham Common to Wolferton, and they form an extensive plateau between Wolferton and Lynn Lodges, Sandringham. The middle and upper parts of the formation crop out on the face of a less impressive, but nevertheless large feature that is made up of small features that reflect individual changes in lithology within the Dersingham Beds. This large feature lies to the east of King's Lynn to Hunstanton road between Dersingham and Sandringham Warren, and then turns eastwards to run through Sandringham Park to West Newton.

The ferruginous sandstones at the base of the formation have been quarried at a large number of localities in the area for building walls (known locally as 'Small Carr') and for hardcore. Rose (1867, p.31) recorded a quarry at 'Dersingham Heath', probably a sand pit [6873 2970] in Leziate Beds, and basal Dersingham Beds at Dersingham Common, that yielded 'Thetis, Natica and Nautilus'

The basal Dersingham Beds and the junction with the Leziate Beds are also exposed in the upper part of a large sand pit [6814 2937] and a nearby adit [6821 2914] at Dersingham, and at the top of the southern face of the deep railway cutting [6626 2870] adjacent to Wolferton Station. All the sections are deeply weathered and show up to 3 m of rotted, permeable, rusty brown, ferruginous sandstone. Lenses and nests of poorly preserved bivalves and other fossils occur at several levels. Lamplugh (in Whitaker and Jukes-Browne, 1899, p.17) believed these fossiliferous patches to be deeply weathered ironstone concretions and graphically described them as having a slaggy or cindery texture. The following composite section was measured in the sand pit [6814 2937] at Dersingham, the least weathered of the three sections referred to above:

	Thickness m
DERSINGHAM BEDS	
Soil and sandy subsoil with angular fragments of ferruginous sandstone	0.6
Sandstone, fine-grained, ferruginous, rusty brown, deeply weathered; irregular flaggy bedding; local concentrations of secondary ferruginous cement	1.5
Clay, dark grey, carbonaceous, with sand wisps and burrow infillings; irregular base truncating sedimentary structures in the underlying sands	0.05 to 0.10
LEZIATE BEDS (see p.78 for details)	
Loose white sand	

Fossils collected by Lamplugh from the railway cutting include *Astarte subdentata* Roemer, *Glycymeris* cf. *marullensis* (Leymerie), *Parmicorbula* cf. *striatula* (J de C Sowerby), *Resatrix* sp., trigonid indet., *?Yaadia* sp. and indeterminate gastropods.

Shallow sandstone quarries are numerous along the northern edge of the sandstone plateau at Sandringham Warren [e.g. 677 288; 6705 2805; 6630 2846; 6622 2825]. At one locality [6734 2828] sandstone or, less probably, ironstone appears to have been worked underground by means of bell-pits. At least one of these quarries was open at the time of the original geological survey and yielded *Rotularia* sp., serpulids indet., *Entolium*, *Eodonax* cf. *claxbiensis* (Woods), *Nemocardium subhillanum*, (d'Orbigny), nuculid indet., trigoniid indet. and common wood fragments.

Shallow pits in the central part of this plateau [688 277], that reputedly supplied the stone for the northern (Norwich Gates) wall of Sandringham Park, were reopened in 1965 to supply hardcore for the King's Lynn Bypass. A total of about 3 m of deeply weathered, ferruginous, fine-grained sandstone were exposed, varying considerably in hardness and in colour from pale brown to very dark brown, depending upon the degree of leaching or secondary limonitic cementation. Lenses of densely limonite-cemented, but extremely porous, very fossiliferous sandstone occurred at several levels. These yielded *Weichselia reticulata* (Stokes and Webb) Fontaine, wood fragments and a rich fauna including *Ophiomorpha* sp., *?Mucrosepula* sp., serpulids indet., *Astarte* cf. *rocklumensis* Wollemann, abundant *A. subdentata* Roemer, *A.* sp., *Buchia lamplughi* (Pavlov), *B.* sp., Cucullaea cf. *corneuliana* (d'Orbigny), cyprinid indet., *Eodonax claxbiensis* (Woods), *Gervillella* sp?, *Glycymeris* cf. *marullensis* (Leymerie), *Limatula tombeckiana* d'Orbigny, *Mesolinga?* cf. *teutobergensis* (Wollemann), *Mimachlamys* cf. *robinaldina* (d'Orbigny), *Nemocardium subhillanum* (d'Orbigny) *Neocrassina* (*Lyapinella*) cf. *transversa* (Leymerie), *Nucula* (*Leionucula*) cf. *planata* Deshayes, *Panopea* cf. *neocomiensis* Leymerie, *Parmicorbula striata* (J de C Sowerby), *Pleuriocardia cottaldina* (d'Orbigny), *Pterotrigonia* cf. *caudata* (Agassiz), *P.* sp., *Resatrix* cf. *neocomiensis* (Weerth), *Rutitrigonia* cf. *longa* (Agassiz), trigoniid, venerids, *Vnigriella?* cf. *scapha* (d'Orbigny), *Yaadia* (*Quadratotrigonia*) *nodosa* (J de C Sowerby), *Actaeon* sp., gastropods indet., *Endemoceras* aff. *oxygonium* (Neumayr & Uhlig), *E.* spp. indet. and *Nucleolites* sp. indet.

A nearby pit [6737 2744], adjacent to the King's Lynn to Hunstanton road and described as a 'gravel pit' on the Ordnance Survey map, appears to have been in angular sandstone rubble.

Spoil from a shallow ditch [6853 2953 to 6872 2926] dug across the full thickness of the Dersingham Beds at Dersingham Common suggests that the following generalised composite sequence is present:

	Thickness m
CARSTONE	
Soft ferruginous sand; spring-line at base	>5
SNETTISHAM CLAY	
Grey clay and silty clay with lenses of fine-grained sand; clay ironstone concretions close to base; pebble bed at base or close above base, with small well-rounded pebbles of quartz, quartzite, rotted chert, rotted limestone and phosphate	c.3
DERSINGHAM BEDS (undifferentiated)	
Sands, fine-grained, soft, brown and white	c.3
Sandstone, fine- and medium-grained, highly ferruginous, and sandy ironstone; pebbly in part and with poorly preserved bivalves and plant debris	c.1
Sands, fine-grained, soft, brown and white	3
Sandstone, fine- and medium-grained, ferruginous; highly porous, with casts of plant fragments and other possible fossils; locally flaggy and densely limonite-cemented, and with sandy ironstone concretions	c.1
Sands, fine-grained, soft, brown and white	c.4
Sandstone, fine-grained, ferruginous, flaggy	c.1
LEZIATE BEDS	
Soft white fine-grained sands	>2

The relationships of these lithologies to the features is shown diagrammatically in Figure 29, together with two sections from nearby Woodcock Wood which illustrate minor variations in the shape of the features at this stratigraphical level. The first of the sections, at Brick Kiln Covert is adjacent to a small pit [6792 2786] that was worked for sharp white sand, which was used for cleaning the Carstone-floored terrace of Sandringham House. There is much brick debris in the vicinity and this may be the site of the brickyard, already derelict in 1882, noted by Whitaker and Jukes-Browne (1899, p.10) and by Teall (1875, p.19), in which white sands rested on grey clay. The field evidence indicates that the clay must have been a very thin seam within the middle part of the Dersingham Beds, and is unlikely to have been used for brickmaking. Nearby, two small pits [6759 2798; 6836 2786] in the Snettisham Clay may have supplied the clay for the works.

The Snettisham Clay and Carstone form a strong feature that can be traced from Woodcock Wood via Sandringham Park to West Newton, but the clay is largely covered by sandy wash and there is no exposure. At Appleton House, West Newton, the clay crops out on flatter ground that is largely free from wash. It has been worked in three pits [7051 2723; 7063 2727; 7072 2720] south of the house, presumably for brickmaking. A nearby ditch section [7061 2684] shows the basal beds of the clay to consist of about 1 m of brown and grey mottled clay with a band of clay ironstone concretions resting on 7 cm of pebbly clay with small rounded quartz, ironstone, chert and phosphate pebbles; this rests of weakly ferruginously cemented sandstone. The features in the West Newton area, although largely obscured by drift deposits, suggest that the Snettisham Clay has cut down to within 5 m of the base of the Dersingham Beds.

Castle Rising to Roydon

The Dersingham Beds form a large outlier on the high ground between Castle Rising, Grimston Warren and Roydon, but this is largely overlain by glacial deposits and only a few small areas of the basal ferruginous sandstones are exposed. Ferruginous sandstone debris, some with plant fragments and poorly preserved bivalves, was recorded at White Hills Wood [690 244], Ling Common [660 233], near Hall Farm [686 234], between the King's Lynn Bypass and Grimston Warren [663 223 to 676 218] and along the western edge of Roydon Common [677 234 to 678 219]. The basal beds of the formation are exposed in several pits [6796 2223 to 6786 2192] at the south-western end of the Common, that were dug for the underlying Leziate Beds. The following composite section is typical of the area:

	Thickness m
DERSINGHAM BEDS	
Sands, fine-grained, white and brown, passing down into	c.1.5
Sandstone, fine-grained, irregularly ferruginously cemented, varying from pale brown soft sandstone to dark brown sandy ironstone; rubbly and flaggy bedding; a few 'cindery' fossiliferous lenses; locally, coarse sand with tiny pebbles as base	c.1.5
Clay, medium grey, plant-rich, silty and sandy; ferruginously stained in part due to seepages; lenticular bed	0.01 to 0.10
LEZIATE BEDS	
Loose white and yellow sands	seen 2

Lamplugh collected *Rotularia* sp., annelids indet., ?*Cucullaea* sp., *Eodonax claxbiensis* (Woods), *Pterotrigonia caudata* (Agassiz), *Pterotrigonia?* and an indeterminate trigoniid from one of these pits.

The most complete section through the Dersingham Beds in the area is that exposed in an old stone pit [672 222] near War-ren Farm that has been extensively expanded in recent years to provide hardcore. The following composite section was measured in the south-east face [6737 2215] of the modern pit:

	Thickness m
QUATERNARY	
Glacial deposits; chalky till and flint gravel	1.5
DERSINGHAM BEDS	
Sand, very fine-grained, pale grey, with tiny ribs and kernels of brown ferruginous sand; traces of rotted pyritic concretions; flat-bedded; weathers to a white, running sand; darker grey in the lowest part; line of limonitised clay ironstone nodules at base	3.5
Clay, pale grey, soft	12 mm
Sandstone, fine-grained, soft, weakly ferruginously, cemented, honey brown; friable and rubbly in upper part, becoming flaggy and/or thickly bedded in lower part	3.7
Clay, pale grey; lenticular bed	up to 25 mm
Sandstone, fine-grained, weakly ferruginously cemented, massive, pale brown	0.41
Clay, pale grey, soft	25 to 50 mm
Sandstone, fine-grained, soft, weakly cemented, as above; passing down into more strongly ferruginously cemented sandstone with wisps and lenses of pale grey clay	0.32
Clay, pale grey; lenticular bed	up to 12 mm
Sandstone, fine-grained, weakly cemented, massive, honey-brown; as above but with poorly preserved casts of bivalves and, more rarely, ammonites including *Endemoceras*	0.46
Sand, fine-grained, with partings of grey clay	0.08
Sandstone, very fine-grained, hard, dark brown, with dense ferruginous cement and thickly bedded; line of limonitised clay ironstone concretions at base	0.48
Interlaminated fine-grained sand and grey clay	0.13
Clay, dark grey, plant-rich, with rare sand partings	0.15
Clay, interbedded pale and dark grey, plant-rich in part, with partings of clean, white, fine-grained sand	0.13
Sandstone, fine-grained, densely ferruginously cemented	3 mm
Clay, pale and dark grey, with wisps and partings of white sand up to 12 mm thick	0.08
LEZIATE BEDS	
Sands, fine-grained, loose, grey and white (up to 18 m reputedly proved in boreholes)	seen 6

The basal strata of the Dersingham Beds, about 0.6 m of grey sandy clay overlain by soft ferruginous sandstone with poorly preserved casts of bivalves and bits of wood, form a small outlier capping Leziate Beds sands at Sugar Fen [694 212]. This is the most southerly known occurrence of the lower part of the formation; between there and the Leziate area the Snettisham Clay oversteps the lower and middle parts of the formation and rests directly on the Leziate Beds.

Ashwicken to Brow of the Hill

The Snettisham Clay has a narrow, but continuous outcrop between East Farm and Closthorpe Manor, Ashwicken, and forms a small outlier, capped by Carstone, at Brow of the Hill. Ditch sections [7080 2032; 7104 2040] at East Farm show up to 1.5 m of sandy wash overlying deeply weathered and cryoturbated clays. The following composite section was estimated to be between 2.5 and 3 m thick:

Clay, thinly interbedded pale grey, dark grey and purplish red; smooth-textured; lenses of dense grey clay ironstone up to 0.7 m long and 30 mm thick, with brown limonitic weathered skins, occur at one level in middle part of bed.

Clay, dark grey, smooth-textured, with partings and thin lenses of fine-grained sand containing small crystalline pyrite concretions; widely spaced lenses of dense clay ironstone (lithology as above) in middle part of bed; closely spaced, rounded, limonitised clay ironstone concretions with purplish red weathered skins at base of bed.

Brown and grey clays with ironstone concretions have been worked in small pits at the farm [7110 1988] and westwards from there along the outcrop [eg 7042 1974; 7024 1970; 6959 2014], presumably for brickmaking.

The former Bawsey Brickworks at Brow of the Hill worked 'Snettisham Clay' and Boulder Clay (Whitaker et al., 1893, pp.18, 64) in several small pits, but the whole site has now been removed by the modern sand workings. The following section was recorded in the overburden face of the modern workings [6843 1935], close to the site of one of the original brick pits.

	Thickness m
CARSTONE	
Soft brownish green sand	seen 0.6
?SNETTISHAM CLAY	
Clay, medium and dark grey, smooth-textured, with many partings, wisps and burrow infillings of orange-brown-stained, very fine- and fine-grained sand; line of lenticular reddish brown clay ironstone concretions, up to 50 mm thick, in middle part of bed; irregular base, with sandy limonitic clay resting on deeply iron-stained surface of the underlying sands	2.0
LEZIATE BEDS (see p.80 for details)	
Sands, fine-grained, grey and white	>15

Snettisham Clay or another clay within the Dersingham Beds may also be present as transported masses within the till that caps the Brow of the Hill ridge. The following, somewhat puzzling section, was measured in a large erratic within the till in a nearby modern face [6777 1918] on the same side of the ridge:

	Thickness m
QUATERNARY DEPOSITS	
Chalky boulder clay	c.3
GAULT	
Clay and sandy clay	c.3.7
CARSTONE	
Greenish brown sands; pebble bed at base	5.8
?DERSINGHAM BEDS	
Sand, fine-grained, soft, white and yellowish brown, iron-stained, with partings of soft grey clay; traces of cross-bedding	1.37
Clay, finely laminated, greyish brown	75 mm
Sandstone, fine-grained, soft to hard, yellowish brown to dark brown, patchily ferruginously	

	Thickness m
cemented; flaggy-bedded	0.61
Clay, dark grey and pale greyish brown; finely laminated with lenses and partings of white sand picking out ripples and small washouts	0.57
Sand, fine-grained, soft, white; lenticular bed	0.10 to 0.15
Clay, dark grey, with sand partings, as above	1.73
QUATERNARY DEPOSITS	
Brown and grey, sandy, stony boulder clay	1.1
LEZIATE BEDS	
In-situ, fine-grained, white and grey sands	>10

Possible thrusts occur at the base of the Carstone and within the supposed Dersingham Beds, and parts of the sequence may be repeated. Nevertheless, the juxtaposition of the Dersingham Beds, Carstone and Gault is stratigraphically correct, although no other section has ever been recorded in which the Carstone rests on lithologies characteristic of the lower or middle part of the Dersingham Beds. If glacial tectonics are absent, then the sequence probably shows basal Dersingham Beds comparable to those exposed at Warren Farm, Grimston Warren, but with the basal clay expanded to more than 2.3 m in thickness. If this interpretation were correct, then all the 'Snettisham Clay', of the Ashwicken and Brow of the Hill areas might belong with the basal (Hauterivian) part of the Dersingham Beds and not with the true (Barremian) Snettisham Clay. The sands between the Hauterivian clay and the Carstone would then present an unresolved stratigraphical problem because the field evidence suggests that the Carstone rests directly on the true Snettisham Clay everywhere at outcrop in the area between Snettisham and Brow of the Hill. It seems likely that the erratic is of local origin and that beds as high as the Gault capped the Brow of the Hill ridge in preglacial times.

East Winch

A lenticular bed of pale grey, carbonaceous clay, less than 1 m thick, forms a spring line between the Leziate Beds and Carstone at Devil's Bottom, East Winch. This is the most southerly known occurrence of the Snettisham Clay.

Boreholes

Although the Hunstanton Borehole is located only 4 km north of the most northerly outcrop of the Dersingham Beds at Heacham, there is little lithological similarity between the two sequences, even when allowance is made for the considerable weathering effects that are present at outcrop. At Heacham and for a distance of about 17 km southwards along the outcrop, the formation can be divided, on gross lithology, into a lower part consisting predominantly of ferruginous sandstones and sands, a middle part consisting predominantly of sands, and an upper clay part (the Snettisham Clay). The following sequence, consisting almost entirely of clay and silt, was proved in the Hunstanton Borehole between 53.42 and 65.02 m:

	Thickness m	Depth m
ROACH (see p.93 for details)		
DERSINGHAM BEDS		
Clay, khaki, brown and red, oolitic (limonite) in part, silty in part; intensely burrow- mottled throughout; pebbly in part, with small-scale rhythms of pebbly clay passing up into silty clay and smooth		

	Thickness m	Depth m
textured clay; shelly and calcareously cemented at one level; passing down into	1.14	54.56
Silt, clayey, sparsely oolitic, brown and grey, burrow-mottled; rare small pebbles; ooids concentrated in burrows; small-scale rhythms with erosive bases of pebbly silt resting on burrowed surfaces with *Arenicolites* and *Skolithos* at several levels; sparse fauna includes bivalves and rare pyritised ammonite fragments; erosion surface at base	2.94	57.50
Clay, very silty and silty, pale and medium grey, shelly, with numerous tiny bivalves; becoming oolitic and pebbly with depth; passing down with burrowing into	0.38	57.88
Clay, pale and medium grey, very oolitic, pebbly, intensely burrow-mottled; poorly preserved bivalves and lignite fragments; burrowed surface in middle part of bed rests on sparsely oolitic silty clay with wisps and burrowfills of coarse silt; 1 to 3 cm-thick lenticular bed of pale brown clay ironstone at one level; bed of coarse sand grains and tiny pebbles at base resting on burrowed surface	0.76	58.64
Silt, fine- and coarse-grained, pale grey and brown, burrow-mottled, finely micaceous in coarser silts; burrows include *Ophiomorpha*; some burrows partially pyritised, others clay-lined; pebble bed at base with well-rounded pebbles of vein quartz, grey clay and soft chalky limestone set in plant-rich glauconitic silt; interburrowed junction with	1.12	59.76
Clay, medium grey, smooth-textured, with burrowfills of pale brown silt; sharp base	0.10	59.86
Silt, pale grey and pale brown, fine- and coarse-grained, burrow-mottled; micaceous in part; becoming clayey in lower part; faintly glauconitic in part; burrowfills of fine-grained quartz sand at several levels; burrows include *Chondrites* and *Ophiomorpha*; small bivalves and coalified wood fragments common at several levels; becoming increasingly sandy with depth and with tiny vein quartz pebbles; very clayey in lowest 0.2 m; interburrowed junction with	5.16	65.02

LEZIATE BEDS (see p.81 for details)
Medium- and coarse-grained grey sands

The full thickness of the Snettisham Clay was cored in the Gayton Borehole between 30.48 and about 32.6 m (core loss at base: boundary taken from geophysical logs). The following section, lithologically similar to that of the nearest outcrop in the Ashwicken area, is present:

	Thickness m	Depth m
CARSTONE (see p.110 for details) Pebbly yellowish green sand		
SNETTISHAM CLAY Clay, pale grey and brownish grey; *Skolithos*-type burrowfills of Carstone sand extend		

	Thickness m	Depth m
down for 0.5 m; lenses and burrow infillings of fine-grained micaceous silt occur throughout; small silt-filled washout at one level	0.61	31.09
Sandstone, fine-grained, dark brown, densely furruginously cemented; traces of oolitic grain coatings	0.02	31.11
Clay ironstone, dense, medium brown	0.04	31.15
Core loss; presumed clay with sand wisps and partings	0.30	31.45
Clay, pale grey and brownish grey, with partings, thin lenses, burrow- and trail-infillings and small washouts of fine- and very fine-grained white sand and micaceous silt; horizontal lamination where undisturbed by burrowing; burrows include many tiny horizontal and vertical tubes; parting of cream coloured phosphatised clay at base with pseudoalgal structure	0.86	32.31
Core loss; cavings of grey clay with small rounded pebbles of quartz and rotten chert suggest presence of pebble bed at base, similar to that at outcrop	0.76	33.07

LEZIATE BEDS (see p.81 for details)

ROACH

The presence of clay beneath the Carstone outcrop on Hunstanton beach appears to have been first recorded by Rose (1835) who presumed it to be the Kimmeridge Clay. Lamplugh (*in* Whitaker and Jukes-Browne, 1899) described the basal bed of the Carstone on the foreshore as a pebble bed set in a green clayey sand. The contact with the underlying clay was not at that time exposed, but he noted that 'hard grey clay', presumed to be Snettisham Clay, had previously been seen in the spoil from the pier foundations. The pebble bed yielded cobbles of fossiliferous ironstone and phosphatic pebbles with ammonites. Keeping (1883, pp.33, 57) described these ammonites as water-worn casts of early Aptian age and this was confirmed by Casey (1961b, p.571) who demonstrated their derivation from two zones of the Lower Aptian (see p.98 for discussion). The fossiliferous ironstones, described by Keeping (1883) as being 'as big as large cannonballs' contained numerous bivalves and rare ammonites; Casey (1961b) identified the latter as *Paracrioceras*, indicative of a Barremian age.

The present survey yielded little additonal data on the clay underlying the Carstone and no information on the sources of the derived faunas at the base of the Carstone. Furthermore, in the continuous cores of the BGS Hunstanton Borehole, the Carstone appeared to pass down without an erosion surface or phosphatic pebble bed from pebbly, oolitic, clayey sand into sandy clay and clayey sand, themselves locally pebbly and oolitic, that were lithologically unlike the Snettisham Clay. Deep trenches were therefore dug on Hunstanton beach to expose the base of the Carstone and to enable the nature of the underlying beds to be determined (Gallois, 1975a).

The composite section proved in these trenches is shown graphically in Figure 31. The ironstone nodules with the Barremian ammonite *Paracrioceras* were present in-situ in sands that were unconformably overlain by the Carstone. The sands were lithologically unlike the Snettisham Clay, but they are similar to parts of the Roach of Lincolnshire. This correlation was confirmed by Morter (1975) who concluded that the non-ammonite fauna of the in-situ ironstones could be matched with that of parts of the Roach, and that it was probably Middle Barremian in age.

The sequence in the trenches can be matched lithologically with that in the Hunstanton Borehole, with the exception that the borehole cores do not contain any macrofauna at that stratigraphical level, nor did they yield any phosphatic pebbles or reworked ironstone nodules. The trenches showed that the basal bed of the Carstone (Bed 2 of Figure 31) is laterally variable in lithology over short distances and that the distribution of the derived cobbles and pebbles is patchy. Correlation of the beach section and the sequence at outcrop in Lincolnshire has enabled 12.1 m of beds to be assigned to the Roach in the borehole. The sequence is more complete beneath The Wash and in Lincolnshire, due to the south-easterly overstep of the Carstone, and both the Skegness and IGS 72/78 boreholes proved 21.7 m of the formation.

The term Roach was first used by Jukes-Browne (1887, p.19) for a group of oolitic and pebbly sandy clays in the upper parts of the Lincolnshire Lower Cretaceous sequence. Swinnerton (1935, p.4) divided the formation into a Lower and an Upper member separated by a calcareously cemented, oolitic sandstone which he called the Roach Stone. Thurrell (in Owen and Thurrell, 1968, p.106) subsequently used Roach Stone for any hard bed that gave rise to a topographical feature at about the middle part of the Roach outcrop in Lincolnshire, irrespective of lithology. No 'Roach Stone' was recognised in

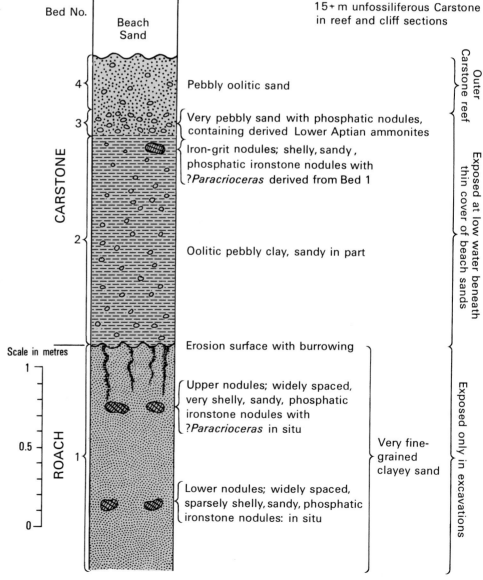

Figure 31 Junction of the Roach and Carstone, Hunstanton beach.

the Hunstanton Borehole, although thin calcareously or ferruginously cemented beds occur at several horizons.

The beds assigned to the Roach in the Hunstanton, Skegness and IGS 72/78 boreholes consist of a complex rhythmic sequence of interburrowed and interbedded lithologies made up of varying amounts of clay, 'chamosite' (berthierine) mud, 'chamosite' ooliths (mostly limonitised), quartz sand and small pebbles of quartz and ironstone. The base of the formation has been taken at a minor erosion surface that separates the predominantly oolitic and pebbly clays of the Roach from the smoother-textured, more fossiliferous clays of the Tealby Beds in Lincolnshire (Skegness and IGS 72/78 boreholes) and from the predominantly argillaceous, sparsely pebbly Dersingham Beds in Norfolk (Hunstanton Borehole). In all three boreholes the basal Roach erosion surface lies a little above the highest, well-developed, reddish brown, smooth-textured clay of the Tealby Beds and Dersingham Beds. Thin beds of grey and reddish brown clay occur in the Roach, but the presence of thick beds of this lithology appear to be a useful distinguishing characteristic of the Tealby Beds and Dersingham Beds.

The Roach sequence at Hunstanton continues the fining-upward rhythmic succession of the Dersingham Beds (Figure 30). In contrast to the Dersingham Bed rhythms, those of the Roach are commonly thin (less than 1 m) and end in silt or silty clay with the overall result that Roach is more silty, oolitic and pebbly than the Dersingham Beds. Most of the rhythm bases appear to be erosive and probably give rise to rapid lateral changes that make correlation of the Roach sequences across the mouth of The Wash (Figure 32) difficult and uncertain.

The outcrop of the Roach is almost entirely covered by drift deposits in Norfolk. It underlies much of the lowest part of the beach to the north of Hunstanton pier, but has been everywhere overlain by beach sand and gravel in recent years. Southwards from Hunstanton the formation is covered by till except for a small area near The Firs, on the northern side of Ringstead Downs valley. There, a deep trench [678 392] revealed pebbly sandy clay beneath a cover of sandy downwash. No evidence of the Roach was found to the south of the valley and it is presumed that the formation is overstepped by the Carstone close to that locality.

Although the ironstones obtained from the beach excavations at Hunstanton yielded a rich fauna, both in numbers and variety, the Roach in the boreholes is barren at many levels. It is probable that the fauna is concentrated, probably through selective preservation, in the more cemented lithologies and in the smoother-textured clays. The fauna of the beach concretions is dominated by small bivalves (18 species), but also contains fragments of ammonites (3 forms), gastropods (2 species and several small indeterminate forms), brachiopods (2 species), scaphopods and common wood fragments (Morter, 1975, pp.29–30). The faunas from the Roach in the Hunstanton, Skegness and IGS 72/78 boreholes are similarly dominated by small bivalves. The full faunal list is given in Table 11. Despite the length of this list the bulk of the Roach has a low diversity fauna and flora in which only the bivalves *Corbula isocardiaeformis* Harbort, *Entolium germanicus*

(Wollemann), *Paranomia laevigata* (J de C Sowerby) and *Parmicorbula striatula* (J de C Sowerby), and plant fragments are common. Most of the fauna and flora is long-ranging and none of the plants, scaphopods, echinoderms, bivalves or gastropods is diagnostic of a single stage. Many of the bivalves are characteristic 'Neocomian' forms and range from the Ryazanian to the Barremian. A few have Aptian affinities and suggest a Barremian or Aptian rather than an earlier age. *Hoplocrioceras* and ?*Paracrioceras* range from the Upper Hauterivian to the Middle Barremian and their record here adds little to the stratigraphical information already available for the age of the Roach. The brachiopod *Rhynchonella parkhillensis* Davison has only previously been recorded from the 'Roach Stone' of the Lincolnshire outcrop (Owen and Thurrell, 1968, p.119).

Details

Hunstanton

Up to 1.5 m of pale and medium grey, intensely bioturbated, fine- and very fine-grained sand and clayey sands were proved beneath the Carstone in excavations [6712 4152 to 6715 4177] made close to low-water mark on Hunstanton beach. Traces of cross-bedding were present, but mostly disturbed by burrowing. Fragments of coalified wood occurred scattered throughout, but no other fossil was recorded in the sands themselves. Lines of widely spaced concretions, 2 to 3 m apart, occurred at two levels, 0.5 and 1.1 m above the base of the section. Both sets of concretions were up to 0.2 m across, rounded, irregular-rounded and burrowfill shapes, composed of dark brown, phosphatic, sandy ironstone. Traces of cross-bedding and bioturbation were picked out by the sand grains within them. The lower concretions were only sparsely fossiliferous, but the upper ones were crowded with moulds of small bivalves and rarer brachiopods, gastropods and ammonites. Bits of coalified wood were also common. Their lithology and fauna are identical to that of the fossiliferous ironstones collected from the basal bed of the Carstone by earlier workers (Geological Survey and Natural History Museum collections), except that the derived concretions are generally more rounded and have a ferruginous weathering patina. The faunas of both types of in-situ concretion are identical (see Table 11 and Morter, 1975 for details). Small unfossiliferous pyritic concretions occurred in association with the lower concretions.

The following sequence was proved between 36.86 and 53.42 m in the Hunstanton Borehole:

	Thickness m	Depth m
CARSTONE (see p.108 for details) Very pebbly oolitic clay		
ROACH Sand, very fine-grained, very clayey, burrow-mottled, dark grey and greenish yellow-brown; oolitic (limonite) in part; barren except for plant fragments; passing down into	2.69	39.55
Clay, very sandy, burrow-mottled, sparsely oolitic, with ooids concentrated in bands; rare ironstone pebbles and calcite-shelled bivalves in lowest 10 cm; passing down into	0.79	40.34
Ironstone, densely calcite-cemented; muddy matrix with ferruginous ooliths and rare quartz grains; burrow-mottled; shelly, with thick-shelled bivalves;		

	Thickness m	Depth m
becoming softer in lowest part; interburrowed junction with	0.17	40.51
Sand, very clayey, burrow-mottled, medium and dark grey; sharp base with burrowed surface	0.77	41.28
Clay, smooth-textured, pale brown, khaki and reddish brown, burrow-mottled, with *Chondrites* and other burrows; line of small ?phosphatic concretions in top 5 cm; barren; small pyritised patches and pyritised trails; wisps of pale grey silt at several levels; interburrowed junction with	2.15	43.43
Clay, silty, oolitic and very oolitic, burrow-mottled, pale grey and brownish grey; rare bivalves and plant fragments; passing locally into clayey oolite or clayey silt as proportions of constituents vary; burrowfill concentrations of glauconitic silt at one level; becoming pebbly in lowest 35 cm with small quartz and ironstone pebbles, small oysters and rolled belemnite fragments; very pebbly in lowest 7 cm, with large pebbles of quartz, ironstone, chert, phosphate and rotted limestone resting on sharp, strikingly burrowed surface	1.81	45.24
Clay, silty, reddish brown, smooth-textured, with rare bivalves; becoming more silty with depth and passing down into	0.33	45.57
Clay, sandy and sparsely pebbly, burrow-mottled, medium grey with paler silt burrowfills; interbedded with oolitic burrow-mottled clays; sparsely pebbly in lower part; passing rapidly down into	1.14	46.71
Mudstone, tough, calcite- and phosphate-cemented; sparsely pebbly; scattered bivalves including oysters; sharp base on burrowed surface	0.15	46.86
Clay, very silty, intensely burrow-mottled, finely micaceous, sparsely oolitic and pebbly in part; burrowfills and thin interbeds of reddish brown clay; sharp base	1.68	48.54
Clay, pale slightly greenish grey, smooth-textured faintly burrow-mottled, with burrows picked out in pale grey silt; sharp base	0.46	49.00
Clay, silty, oolitic, intensely bioturbated; interbedded with clayey silts and sandy and oolitic clays in fining-upward, sharp-based rhythms; densely calcite-cemented in part; interburrowed junction with	4.42	53.42

DERSINGHAM BEDS (see p.90 for details)

The following complete Roach succession was proved between 24.60 and 46.30 m in Borehole IGS 72/78 in the central part of The Wash, the formation being conformably overlain in that area by the Skegness Clay:

	Thickness m	Depth m

SKEGNESS CLAY (see p.100 for details)

ROACH

	Thickness m	Depth m
Clay, dark and very dark clay, burrow-mottled, oolitic, becoming very oolitic in part; sparsely pebbly, but with some		
horizons made up of interburrowed clay, pebbles and ooids in roughly equal proportions; mottled pale and medium greys, with pinkish and yellowish browns in lower part; almost barren, with poorly preserved bivalves and *Weichselia* fragments	6.60	31.20
Clay, sparsely oolitic and oolitic; pale greenish brownish grey; intensely burrow-mottled in part, with some calcareously cemented burrowfills; sparsely fossiliferous, with oysters and other bivalves, *Oxyteuthis*, *Weichselia* and wood fragments	1.15	32.35
Core loss	0.10	32.45
Clay, pale greenish and brownish grey, 'chamositic' in part; oolitic and burrow-mottled; interbedded in lower part with pale brown, finely micaceous silt; clay locally weakly calcite-, pyrite-, or 'chamosite'-cemented; sparsely pebbly at one level; plant debris common; shelly at several levels, with a predominantly bivalve fauna	2.05	34.50
Core loss	0.90	35.40
Silt, pale brownish grey, finely micaceous, intensely burrow-mottled, with burrowfills of brownish grey clay; plant debris and bivalves common; fauna as in bed above; phosphatic pebble and interburrowing at base	0.42	35.82
Clay, pale and medium reddish brown; smooth-textured, with listric surfaces; a few pale greenish grey patches associated with pyritised trails and plant fragments; becoming dull brownish grey with silt wisps in lower part and oolitic in basal bed; almost barren throughout; passing down into	1.25	37.07
Ironstone, pale grey and brown, with ?siderite mudstone matrix; intensely burrow-mottled, with burrowfill concentrations of limonite ooids; barren; probably lenticular	0.23	37.30
Clay, pale grey and greyish brown, slightly silty, with thin beds and burrowfills of foraminifera-rich silt; faintly glauconite-speckled in part; sparsely to densely oolitic in part; becoming uniformly more silty in lower part; plant fragments including *Weichselia reticulata* (Stokes & Webb) Fontaine common at some levels; sparsely shelly, with fauna dominated by bivalves, some in growth position; complexly interburrowed junction with	2.00	39.30
Clay, pale and medium reddish brown with greenish grey mottled patches; smooth-textured, with listric surfaces; almost barren, with only a few plant scraps and pyritised pins and trails; passing down with interburrowing into	1.88	41.18
Clay, pale grey, slightly silty; passing down into burrow-mottled, medium and dark grey and greyish brown, oolitic and pebbly clays with burrowfills of foraminiferal silt; ooids and glauconitic silt at many levels in lower part; burrowfills of phosphatised clay at one level; sparsely shelly overall but		

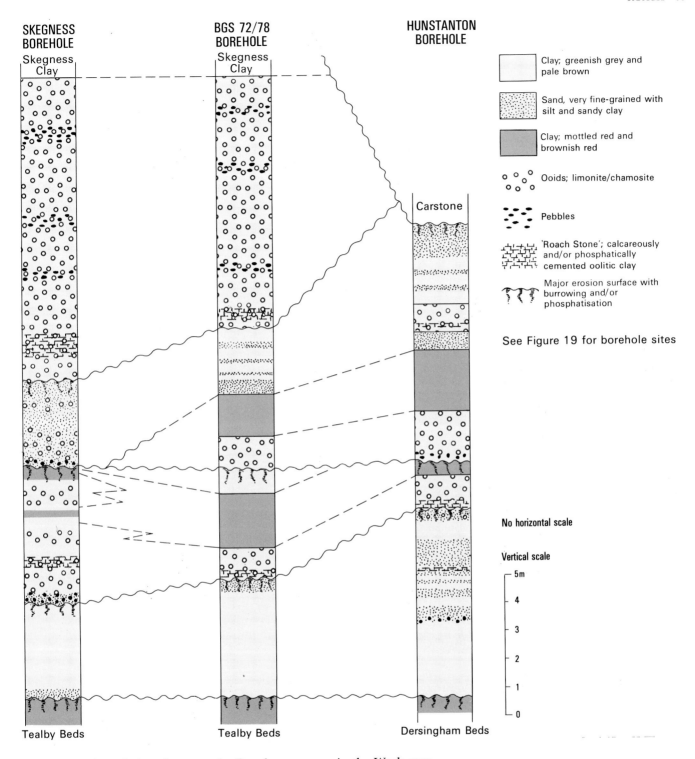

Figure 32   Correlations between the Roach sequences in the Wash area.

Table 11 Macrofauna of the Roach in the Hunstanton and Wash area.

✔ = present	Hunstanton Bh.	Hunstanton beach	IGS 72/78 Bh.	Skegness Bh.
Plantae				
Weichselia reticulata (Stokes and Webb)	✔		✔	✔
coniferous wood				✔
wood indet.	✔	✔	✔	✔
plant debris	✔	✔	✔	✔
Scaphopoda				
Antalis sp.		✔		✔
Brachiopoda				
Lamellaerhynchia sp.				✔
Rhynchonella parkhillensis Owen and Thurrell		✔		
Tamarella sp.	✔			
Terebratulina cf. *auriculata* d'Orbigny				✔
terebratelloid	✔			
Echinodermata				
'Cidaris' *phillipsii* Agassiz				✔
'Cidaris' *speetonensis* Desor	✔			
Nucleolites cf. *martini* (d'Orbigny)				✔
Bivalvia				
Acesta aff. *longa* (Roemer)		✔		
Aptolinter	✔			
Astarte cf. *cantabrigensis* Woods	✔			
Astarte sp.		✔		
Camptonectes cf. *cottaldinus* (d'Orbigny)	✔			
Ceratostreon minor (Coquand)	✔			
Ceratostreon tuberculifera (Koch and Dunker)	✔			✔
Chlamys aptiensis (d'Orbigny)		✔		
Clisocolus cf. *vectense* (Woods)	✔		✔	
Corbula isocardiaeformis Harbort			✔	✔
Crytochasma cf. *ovale* Casey	✔			
Entolium germanicus (Wollemann)	✔		✔	✔
Entolium germanicus lohmanni Wollemann			✔	✔
Entolium sp.	✔	✔		
Freiastarte cf. *subcostata* (d'Orbigny)	✔	✔	✔	
Gervillaria sp.	✔			
Limatula tombeckiana	✔	✔		
Linotrigonia (*Oistotrigonia*) sp.	✔			
Ludbrookia sp.				✔
Mesolinga teutobergensis (Wollemann)			✔	
Modiolus sp.	✔			
Mulletia mulleti (Deshayes)	✔	✔		✔
Nanonavis sp.	✔			

Table 11 continued

✓ = present	Hunstanton Bh.	Hunstanton beach	IGS 72/78 Bh.	Skegness Bh.
Nucula (*Leionucula*) *planata* Deshayes			✓	
Nuculana sp.	✓	✓		
Oxytoma (*Hyoxytoma*) *pectinatum* (J de C Sowerby)	✓	✓	✓	✓
Panopea neocomiensis (Leymerie)	✓			
Panopea plicata (J de C Sowerby)	✓		✓	✓
Paranomia? laevigata (J de C Sowerby)	✓	✓	✓	✓
Paramonia? striatula (J de C Sowerby)	✓			
Parmicorbula striatula (J de C Sowerby)	✓	✓	✓	
Pholadomya eberti (Wollemann)	✓		✓	
Pinna robinaldina (d'Orbigny)	✓		✓	
Pinna sp.	✓			✓
Pseudoptera cf. *subdepressa* (d'Orbigny)	✓		✓	✓
Resatrix sp.		✓		
Terebrimya sp. in wood		✓		
Teredo sp.	✓			
Teredolites sp. in wood		✓		
Thetis minor J de C Sowerby		✓		
Thracia neocomiensis (d'Orbigny)			✓	
Thracia phillipsii (Roemer)	✓		✓	
Thracia sp.	✓			✓
Vnigriella? scapha (d'Orbigny)	✓		✓	
Gastropoda				
Anchura (*Perissoptera*) *acuta* (d'Orbigny)			✓	✓
Cimolithium ascheri (Wollemann)			✓	
Semisolarium cf. *tabulatum* (Phillips)	✓			
Tessarolax bicarinata (Deshayes)	✓			
Tessarolax sp.		✓		
Turbo reticulatus Phillips	✓			
cf. *Trochus stillei* Wollemann		✓		
Cephalopoda				
Hoplocrioceras			✓	
?*Paracrioceras*		✓		
crioceratid fragments	✓	✓		
?desmoceratid fragments	✓	✓		
Oxyteuthis jasikowi (Lahusen)			✓	✓
Oxyteuthis sp.			✓	✓

	Thickness m	Depth m
with local concentrations of bivalves; *Oxyteuthis*, a crustacean chela and plant debris also present; interburrowed junction with	1.60	42.78
Clay, pale yellowish brown, with beds of hard, yellowish brown, phosphatic ironstone at two levels; becoming slightly silty, faintly reddish grey, burrow-mottled in lower part; almost barren, with scattered poorly preserved bivalves	0.62	43.40
Core loss	0.70	44.10
Clay, pale grey, slightly silty, sparsely shelly, with bivalves well-preserved in white and salmon-pink calcite; aporrhaid gastropods also common; fragment of *Hoplocrioceras*; plant fragments including common *Weichselia*; strikingly interburrowed junction with	2.20	46.30

TEALBY BEDS

The fauna of the Roach in this and the Hunstanton and Skegness boreholes is summarised in Table 11.

SKEGNESS CLAY AND SUTTERBY MARL

In the offshore area to the north-west of Hunstanton, the Roach is overlain by two marine argillaceous formations, the Skegness Clay and the Sutterby Marl, that are known only from boreholes in The Wash and at Skegness, and from a limited outcrop in Lincolnshire. These formations were deposited during the time interval represented by the unconformity between the the Roach and the Carstone at Hunstanton. Both formations were probably formerly widespread in the district, but were removed by erosion and are now represented in Norfolk only by derived faunas in the basal pebble bed of the Carstone.

Swinnerton (1935), in his description of the Lower Cretaceous rocks of Lincolnshire, noted that the Roach and the Carstone were separated throughout much of their outcrop by a thin bed of poorly exposed, pale grey, calcareous clay which he named the Sutterby Marl after a type section (1935, pp.18–19) near the village of Sutterby, Lincolnshire. The base of the formation was taken a little below a phosphatic nodule bed which yielded ammonites, including *Prodeshayesites*, indicative of the early Aptian, which Swinnerton presumed to be indigenous (1935, pp.24–25). However, Casey (1961b, pp.570–571) subsequently showed that this fauna was derived from three Lower Aptian zones, those of *Prodeshayesites fissicostatus*, *Deshayesites deshayesi* and *Tropaeum bowerbanki*.

The sequences proved between the Roach and the Carstone in the Skegness Borehole and in IGS Borehole 72/78 in The Wash were more complex than that described from Sutterby. Two clays were present, separated by an erosion surface. The upper clay could be matched in faunal and lithological detail with the Sutterby Marl, but the lower clay differed in lithology and age and was therefore given the new name Skegness Clay (Gallois,

1975b). The Skegness Clay has not been positively identified outside the two boreholes described above. The formation is absent in Norfolk, but probably underlies much of the Norfolk side of the Wash. The offshore seismic evidence (Wingfield et al., 1979) suggests that it may approach within 2 km of Hunstanton beach. It seems likely that it is also preserved to the north and west of Skegness and probably crops out in the southern part of the Lincolnshire Wolds.

The relationships of the Skegness Clay to the Roach, Sutterby Marl and Carstone in The Wash area, including the positions of the indigenous and derived faunas, are summarised in Figure 33.

The Skegness Clay in Borehole 72/78 consisted of 1.26 m of medium and dark grey and brownish grey, smooth-textured clays with a sparse fauna of small bivalves, common crushed iridescent ammonites and rare gastropods and corals. The junction with the underlying pebbly clays of the Roach was bioturbated but sharp, and apparently conformable.

The ammonites have been identified by Dr Casey as *Aconeceras nisoides* (Sarasin), *Prodeshayesites germanicus* Casey, *P. lestrangei* Casey and *P.* cf. *lestrangei*. These last three species are especially common as derived phosphatised specimens in the basal pebble bed of the Carstone at Hunstanton beach. In addition to the above, the Skegness Clay of the Skegness Borehole yielded *Prodeshayesites* cf. *bodei* (von Koenen), *P. fissicostatus*. (Phillips) and *Toxaceratoides royerianus* auctt. non d'Orbigny. This assemblage is indicative of the *P. bodei* Subzone of the *P. fissicostatus* Zone and is the earliest Aptian fauna yet recorded in situ in Britain. The junction of the Roach and the Skegness Clay is therefore presumed to mark the Barremian–Aptian boundary.

In the Southern Basin of Aptian deposition the earliest known indigenous cephalopod faunas have been assigned to the upper part of the *Prodeshayesites fissicostatus* Zone (*P. obsoletus* Subzone); the lower part of the Zone (*bodei* Subzone) is possibly represented by derived material in a nodule bed at the base of the Atherfield Clay (Casey, 1961b, p.506). In the Northern Basin, which includes the present district, *bodei* Subzone ammonites have been recorded as derived faunas in phosphatic pebble beds at the base of the Woburn Sands (Upper Aptian) at Potton and Upware, Cambridgeshire; at the base of the Carstone (Lower Albian) at Hunstanton beach, and at the base of the Sutterby Marl (Upper Aptian) in Lincolnshire. The Skegness Clay, or marine strata of equivalent age, were clearly deposited over a large area in eastern England.

The non-ammonite fauna of the Skegness Clay consists largely of small bivalves, most of which are long-ranging forms of *Camptonectes*, *Corbula*, *Grammatodon*, *Paranomia* and *Vnigriella* that have also been recorded from the Barremian Tealby Beds and Roach. A small ahermaphytic coral, '*Trochocyathus*', and a small ?fusinid gastropod were recorded in IGS Borehole 72/78, together with the belemnite *Hibolites* cf. *minutus* Swinnerton.

The Sutterby Marl consists of very pale and pale yellowish grey, bioturbated, highly calcareous clays with a rich and diverse fauna of ammonites, belemnites, bivalves and brachiopods, and rarer echinoids and corals. Much of the calcareous content is made up of coccolith debris (Black, 1972–75). The formation rests with minor unconformity on the Skegness Clay and has at its base a pebble bed composed largely of phosphatised body chambers of ammonites derived from the Skegness Clay and water-worn belemnite fragments, set in a matrix of silt and comminuted shell debris. The highest part of the formation consists of pale green, calcareous, silty clay that is separated from the underlying beds by a minor erosion surface. Another, apparently minor erosion surface separates this bed from typical Carstone lithologies and it is not clear, in the absence of faunal evidence, which erosion surface represents the major stratigraphical break that marks the base of the Carstone (see p.100 for discussion).

The Sutterby Marl was at least 3.14 m thick in Borehole 72/78, where the junction with the Carstone was lost during coring. The full thickness of the Sutterby Marl was 1.9 m in the Skegness Borehole, and Swinnerton (1935) recorded only 1 to 2 m of the formation in boreholes and at outcrop in Lincolnshire. The contact with the Carstone is therefore probably irregular on a regional scale and any variations there might have been in the depositional thickness of the Sutterby Marl can no longer be determined. Between Borehole 72/78 and the Norfolk coast, a distance of 8 km, the Carstone cuts out about 10 m of Roach, 1.3 m of Skegness Clay and 3.1 m of Sutterby Marl. It seems likely, therefore, that the Sutterby

Marl is overstepped by the Carstone about 5 km northwest of Hunstanton. The Skegness Clay and the Sutterby Marl could not be differentiated in offshore seismic traverses (Wingfield et al., 1979); it is assumed here that the older formation extends farther south than the younger.

The fauna of the Sutterby Marl in Borehole 72/78 included delicately preserved fossils, such as common thin-shelled terebratulid and terebratellid brachiopods and finely tapered forms of *Neohibolites*, which were not recorded by Swinnerton (1935) at outcrop and which were presumably destroyed by weathering there. The borehole yielded the echinoid *?Toxaster*, the brachiopods *Platythyris* sp. ('*Terebratula*' *montoniana* auctt.) and cf. *Tamarella tamarindus* (J de C Sowerby), the bivalves *Astarte*, *Aucellina aptiensis* (d'Orbigny), *?Ctenoides rapa* (d'Orbigny), *Inoceramus* cf. *ewaldi* Schlüter, *Nanonavis carinata* (J Sowerby), indeterminate ostreids and *Plicatula*, the ammonites *Aconeceras nisoides* (Sarasin), *Cheloniceras* sp., *Dufrenoyia* sp. and *Tonohamites?*, the belemnite *Neohibolites ewaldi* (von Strombeck) and indeterminate aporrhaid and cerithiid gastropods. The pebble bed at the base of the formation contained fragments of water-worn and bored phosphatised *Prodeshayesites bodei* (Koenen), *P. fissicostatus* (Phillips), *P.* cf. *fissicostatus* and *P.* spp., all of which were presumably derived from the Skegness Clay.

The belemnites from the Sutterby Marl at outcrop were described by Swinnerton (1935; 1936–55) and the

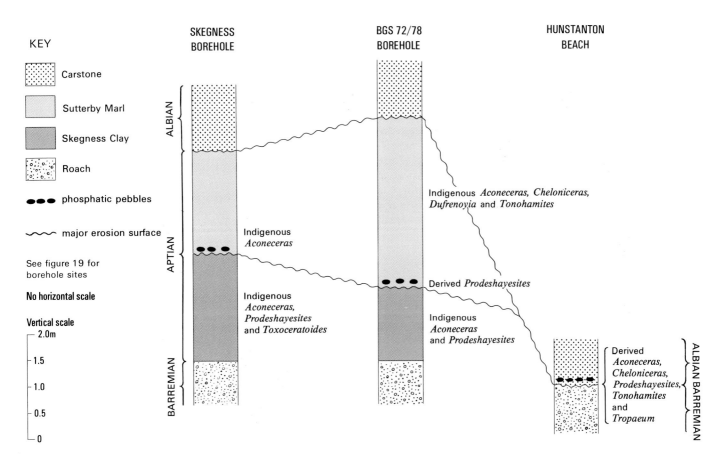

Figure 33 Skegness Clay and Sutterby Marl sequences in the Wash area.

ammonites, both indigenous and derived, have been discussed by Casey (1961b, pp.570–571). Both authors assigned the formation to the Upper Aptian; the first on the basis of *Neohibolites ewaldi* Strombeck) and the second on the ammonite *Colombiceras*. The presence of indigenous *Cheloniceras* and *Dufrenoyia* in the lowest part of the Sutterby Marl in the 72/78 and Skegness boreholes suggests that the beds at that level are late Lower Aptian or early Upper Aptian in age (Casey, personal communication). They are probably, in part, contemporaneous with the Woburn Sands of Cambridgeshire.

The bivalve fauna of the Sutterby Marl is distinctly Aptian and contrasts strongly with that of the Skegness Clay. The Sutterby Marl fauna contains numerous small oysters, including forms of *Aucellina* and *Inoceramus* characteristic of the Lower Greensand of southern England and the Ewaldi Marls of North Germany. Considering the fauna as a whole, it seems likely that the bulk of the Sutterby Marl is Upper Aptian in age.

The upper limit of the formation in the type section at Sutterby was taken by Swinnerton (1935) at an erosion surface at the base of the Carstone. In the Skegness and 72/78 boreholes, the lowest beds of the Carstone consisted of greyish green sandy clay and green silty clay. Two erosion surfaces occurred at the approximate level of the Sutterby Marl–Carstone junction in each borehole. In the Skegness Borehole these surfaces were separated by only 8 cm of burrow-mottled green silty clay and it was impossible to determine which surface correlated with the widespread unconformity that elsewhere marks the base of the Carstone. A specimen of *Neohibolites strombecki* Miller (indicative of the *Leymeriella tardefurcata* Zone of the Lower Albian) was recorded in this clay and suggests that these beds may be a remnant of a hitherto unrecorded formation which separates the Sutterby Marl and the Carstone. The comparable beds in Borehole 72/78 were mostly lost during coring. Burrow-mottled, pale and dark green, silty clays (5 cm), lithologically similar to the green silty clay (Albian) at Skegness, occurred immediately above the core loss. The total-gamma-ray log of the lost interval suggests that it consists of about 2.6 m of loose sand which rests on an erosion surface cut into the Sutterby Marl. It is not clear whether these sands and the overlying thin clay bed belong with the Sutterby Marl, the Carstone, or an as yet undescribed formation.

Details

Boreholes

The following section was proved between 20.20 and 24.60 m in IGS Borehole 72/78:

	Thickness m	Depth m
CARSTONE (see p.108)		
Core loss at junction		
SUTTERBY MARL		
Clay, very pale grey, very calcareous, slightly silty; burrow-mottled, with chondritic and other mottling; locally shelly, with thin-shelled terebratulids and the bivalves	0.12	24.60

	Thickness m	Depth m
Aucellina and *Inoceramus* cf. *ewaldi* Schlüter; belemnites including *Neohibolites ewaldi* (von Strombeck) common; echinoid debris and pyritised trails also present; fragments of *Cheloniceras* and *Dufrenoyia* occur at 21.85 m	1.80	22.00
Core lost	1.20	3.20
Clay, pale and very pale grey, very calcareous as above	0.12	23.32
Basal Nodule Bed: black phosphatic nodules up to 3 cm across enclosing fragments of *Prodeshayesites* including *P.* aff. *bodei* (von Koenen) and *P. fissicostatus* (Phillips), set in very pale grey clay with crushed and distorted indigenous *Aconeceras* and *Cheloniceras*; interburrowed junction with	0.02	23.34
SKEGNESS CLAY		
Clay, medium and dark grey, smooth-textured; burrowfills of the overlying clay extend down to 23.45 m; sparsely shelly but with crushed iridescent *Aconeceras nisoides* (Sarasin) and *Prodeshayesites*, including *P. fissicostatus* and *P.* sp.nov.; becoming brownish grey below 23.50 m, with scattered cream-coloured phosphatic nodules; burrowfills of limonite ooliths occur below 24.03 m and become more common with depth; interburrowed junction with	1.26	24.60
ROACH (see p.94 for details)		

CARSTONE, RED CHALK AND GAULT

The Albian Stage is represented in Norfolk by three interrelated formations of markedly different lithology, the Carstone, Red Chalk and Gault. In the northern part of their outcrop and in their subcrop beneath The Wash, the Albian sediments consist of up to 19 m of rusty brown and green, pebbly, limonite and 'chamosite' oolith-rich sandstone (Carstone) overlain by 1 to 5 m of thinly interbedded pink limestone and brick red marl (Red Chalk). When traced southwards along its outcrop from Hunstanton Cliffs (Front cover), the Red Chalk thickens from about 1.5 m of pink and red limestone to about 3m of interbedded creamy white limestone and pink marl at Dersingham. Between Dersingham and the Babingley River, the Red Chalk undergoes a rapid lithological change at outcrop and passes southwards into about 5 m of pale grey calcareous clay with bands of white chalky limestone and then, in the Gayton area, into about 8 m of clay and calcareous clay (Gault). The Carstone thins southwards from Hunstanton to Dersingham in almost direct proportion to the thickening of the Red Chalk in the same direction, and from Dersingham southwards in proportion to the thickening of the Gault. In south-west Norfolk, the Gault is over 15 m thick at outcrop and the Carstone consists of only 2 to 3 m of clayey sand at its base

When Rose (1862) subdivided the beds between the Kimmeridge Clay and the Gault in Norfolk, he adopted the local name Carstone (Carr Stone = Fen Stone) for his youngest division, as typified by the ferruginous sandstones in the cliffs at Hunstanton. The outcrop of the Carstone was traced southwards from Hunstanton on Old Series Geological Sheet 69 (Whitaker et al., 1893) as far as Ashwicken by means of a topographical feature and spring line which marks the junction of the Carstone and the Snettisham Clay. Southwards from Ashwicken, in the area where the Carstone rests on the Dersingham Beds or the Sandringham Sands, all the beds between the Kimmeridge Clay and the Gault (southern part of Old Series Sheet 69 and Sheet 65, 1884) were mapped as 'Lower Greensand'. It was found in the present survey that the Carstone is lithologically sufficiently distinctive from the Dersingham Beds and the Sandringham Sands for its outcrop to be mapped in both areas.

The striking colour contrast formed in Hunstanton Cliffs by the rusty brown Carstone, the Red Chalk and the white Chalk, together with the richly fossiliferous nature of the Red Chalk, has attracted the attention of large numbers of geologists, both amateur and professional. In addition, the complex sedimentology of the Red Chalk and the difficulties of demonstrating its stratigraphical relationships to the Carstone and the Gault have generated a wealth of geological literature.

William Smith appears to have been the first geologist to have recorded the Red Chalk at Hunstanton in his Table of Strata (1815b) and geological map of Norfolk (1819). Early descriptions of the cliff sections include those of Taylor (1823), Woodward (1833), Rose (1835) and Fitton (1836). More detailed descriptions of the Red Chalk were provided by Wiltshire (1859, 1869), Seeley (1861b, 1864a, 1864b, 1866), Whitaker (1883), Jukes-Browne and Hill (1887) and Whitaker and Jukes-Browne (1899), and they together with Barrois (1876), Jukes-Browne (1900), Kitchin and Pringle (1922, 1932) and Rastall (1930) discussed its stratigraphical relationships with the Carstone, Gault and Chalk. The sedimentology of the Red Chalk of Lincolnshire and at Hunstanton has been described by Jeans (1973, 1980), and Clarke (1964) has examined belemnite orientations in it at Hunstanton. Casey (in Larwood, 1961, pp.290–291) has re-examined the bivalve, gastropod and cephalopod faunas from Hunstanton, and Owen (1979, p.580) has reviewed the significance of the ammonites in relation to the modern zonal and subzonal scheme for the Albian.

The stratigraphical and sedimentological relationships of the Carstone, Gault and Red Chalk in Norfolk have been described by Gallois and Morter (1982), and Wilkinson and Morter (1982), in a complementary study, have described the ostracod sequences. Andrews (1983) has described the sequence of micro- and macrofaunas in the Red Chalk.

The upper limit of the Carstone, its relationship to the overlying Red Chalk, and the age of the two formations was for some time a matter of controversy. All the earlier workers noted that the Carstone passes up, by the rapid addition of calcareous content and diminution of sand and oolith content, into the Red Chalk in the cliffs at Hunstanton, and that the two formations appear to be conformably bedded throughout the 1.2 km of continuous cliff exposure. In additon, Rastall (1930, p.445)

showed that the insoluble sandy residue of the Red Chalk is identical in lithology and heavy-mineral assemblage to that of the Carstone. Difficulties arose, however, when Spath (1930) assigned the Carstone to the early Aptian on the basis of phosphatised ammonites that occur in its basal bed at Hunstanton. As a result of Spath's observations, Kitchin and Pringle (1932) reiterated their own earlier (1922) conclusion that the Red Chalk rests unconformably on the Carstone, and that a time break (late Aptian to mid Albian) is present at the junction between the two formations. They threw doubt upon, but admitted they could not disprove, the apparent parallellism of the bedding in the Carstone and the Red Chalk of Hunstanton Cliffs, and explained Rastall's petrological results by suggesting that the heavy minerals in the Red Chalk were derived from Carstone.

No field evidence has yet been reported to support the presence of an unconformity between the Carstone and the Red Chalk of Hunstanton. With the recognition that the fauna from the basal nodule bed of the Carstone is derived (Casey, 1961b, p.571) from the Skegness Clay and Sutterby Marl, no unconformity is required. In recent years, palaeontological evidence has accumulated which suggests that the Carstone is early to mid Albian in age (see p.104 for discussion).

Despite the early suggestion by Sedgwick (1846) that the Red Chalk at Hunstanton was the lateral equivalent of the Gault of Cambridgeshire, there was much subsequent discussion as to its precise age. Whitaker (1883), on the basis of the diverse fauna which had been collected from the Red Chalk, thought that it was probably the equivalent of the Chalk Marl, or possibly partly of the Gault and the Chalk Marl. However, Jukes-Browne and Hill (on Old Series Sheet 65, 1887) claimed to be able to demonstrate that the Chalk Marl thinned northwards at outcrop in Norfolk and that its basal bed passed laterally into the basal bed of the Chalk at Hunstanton. Whitaker's main objection to the correlation of the Red Chalk with the Gault was the occurrence in the Red Chalk of certain non-ammonite species that were at that time known only from the Lower Chalk. Most of these species are now also known from the Gault; others are known to be facies-dependant and associated with chalky limestones. Over 30 species of ammonite have been collected from the Red Chalk at Hunstanton (Casey in Larwood, 1961, p.15), and these have been shown by Spath (1923–43) and Owen (1979) to be indicative of the same mid to late Albian time span in which the Gault was deposited (see p.104 for discussion).

The Gault was first shown to be a discrete formation in East Anglia by William Smith, who referred to it as the 'Golt Brick Earth' on his geological map of England and Wales (1815a) and subsequently, in more detail, on his geological maps of the counties of Norfolk (1819) and Cambridgeshire (1819). The formation was extensively worked in the region for brickmaking in the 19th century and, additionally in Cambridgeshire and Norfolk, for agricultural phosphate. The Gault is richly fossiliferous at many stratigraphical levels and extensive collections, especially of ammonites and bivalves, were made from the exposures available to collectors at that time. These collections are commonly stratigraphically poorly documented because the exposures were separated from one another by long stretches of outcrop where the Gault is deeply weathered and rarely, if ever, exposed. There are no longer any permanent or semipermanent sections in the Gault in the region.

Most modern stratigraphical studies of the Gault have been concentrated on working brickpits in the Weald and on the type section at Folkestone, Kent, where the whole of the formation has been available for study at some time or another, albeit in separated sections. De Rance (1868, 1875), Price (1874, 1876), Jukes-Browne (1900), Spath (1923–43), Casey (1961b) and Owen (1971, 1976) have all contributed to the understanding of the stratigraphy of the Gault and, as a result, the succession of ammonite assemblages on which the formation is zoned is known in considerable detail.

Attempts have been made to correlate the Gault in East Anglia with the Folkestone sequence, but these have mostly been limited to the correlation of individual, usually small sections with the type section. Little has been published about the Gault at outcrop in Norfolk since Jukes-Browne's review in the Cretaceous Rocks of Britain (Vol. 1, 1900) in which the local details are based largely on those given in the memoirs for Geological Sheets 65 and 69. A number of critical sections in Norfolk were described by Jukes-Browne and Hill (1887) and Kitchin and Pringle (1922). Brighton (1938) reviewed the data from boreholes and temporary sections in East Anglia. He noted that large parts of the Folkestone sequence had not been recorded in the region and suggested that this was probably because of lack of suitable exposure rather than due to non-sequences within the Gault. Detailed work on the palaeontology of the Gault successions proved in the Gayton and Marham boreholes as part of the present survey have confirmed that a condensed but complete Gault sequence is present in the south-eastern part of the district (Gallois and Morter, 1982).

CARSTONE

The Carstone has an almost continuous outcrop in Norfolk from Hunstanton to West Dereham that is broken only by the wide drift-filled valleys of the rivers Babingley and Nar, and by patches of glacial deposits. Between Hunstanton and the Babingley River, this outcrop is largely free from drift deposits; southwards from the river the outcrop becomes increasingly obscured by drift until, between the River Nar and West Dereham, large parts of it are wholly obscured. The Carstone forms small outliers that cap broad ridges at Brow of the Hill and Middleton, but these too are largely obscured by drift.

Between Hunstanton and Sandringham, the Carstone mostly crops out on a steep feature that is capped by Red Chalk; it has its most extensive outcrops at Lodge Hill, Snettisham and Sandringham, where it forms small plateaux. The Carstone becomes thinner and more argillaceous southwards from Sandringham and the outcrop, in consequence, mostly occupies low ground. The formation is poorly exposed except for its type section in Hunstanton cliffs and a small number of quarries dug for

building stones and hard core between there and the River Nar. In drift-free areas north of the River Nar, it gives rise to a well-drained, rusty brown, sandy soil with common fragments of ferruginous sandstone, and its outcrop can be readily mapped. The formation is more argillaceous and less massive south of the River Nar, and has not been worked. It locally caps a weak feature, in which it is underlain by the soft sands of the Leziate Beds, that is separated from the Chalk escarpment by the low ground of the Gault outcrop. Parts of the Carstone in this area weather to the rusty brown sandstone typical of more northerly areas but, where close to the water table, much of it consists of relatively unaltered green and grey clayey sands that can readily be distinguished by augering from both the green and white sands of the underlying Leziate Beds and the grey clays of the overlying Gault.

The maximum recorded thickness of the Carstone is 18.9 m in the Geological Survey's Hunstanton Borehole; a similar thickness is probably present in Hunstanton Cliffs and foreshore, but it is difficult to measure accurately. The formation thins north-westwards beneath The Wash and is only 6.4 m thick in Borehole 72/78. It thins southwards at outcrop and is 8.5 m thick in the Gayton Borehole, about 5.6 m thick in the Marham Borehole and less than 2m thick beneath Methwold Fens.

The Carstone thins steadily southwards from West Dereham to the county boundary. Peat wastage has revealed a small deeply weathered outcrop at Stubb's Hill, Southery, but apart from this the formation is known only from boreholes in the most southern part of the county. The Carstone is not a discrete mappable formation in Suffolk and Cambridgeshire, but is probably present as a thin pebbly sand at the base of the Gault.

The extent of the subcrop of the Carstone in Norfolk is poorly known. About 17.7 m of the formation was proved in the North Creake Borehole (Kent, 1947), a similar amount in the South Creake Borehole, and about 6.8 m in the Great Ellingham Borehole (Cox et al., 1989). The formation was not recorded in the deep hydrocarbon-exploration boreholes in the eastern part of the county at Saxthorpe, Somerton and East Ruston, but this could be due to the difficulty of distinguishing it from other sandy Jurassic and Cretaceous lithologies in rock cuttings and geophysical logs. The Carstone was represented by only a few small pebbles and sand grains in the Geological Survey's Trunch Borehole (Gallois and Morter, 1976) in north-east Norfolk.

The best exposed development of Carstone in England is that at Hunstanton (Plate 8). There, about 19 m of rusty brown, massive, ferruginous sandstone is exposed in the cliffs and foreshore. Only the lower part of the formation is poorly exposed, but all the sections are deeply weathered and the lithologies and sedimentary structures are largely obscured by secondary limonitic weathering products. Cliff falls reveal fresher material from time to time and these usually show greenish brown, oolitic, cross-bedded sandstones with common *Arenicolites* and *Skolithos* burrows. Such sections weather within a few months to produce the typical limonite-crust and box-stone weathering. Unweathered Carstone has been seen only in boreholes and even there limonitisation has occurred at a number of levels due to oxidation within the zone (and former zone) of water-table fluctuation.

The junction with the Roach was exposed in trenches dug at low tide on Hunstanton beach [6712 4152 to 6715 4177]. It was taken at a prominent erosion surface (Figure 31) 1.0 to 1.5 m below the horizon of phosphatic nodules (well known as beach pebbles) which contain abundant early Aptian ammonites, including *Cheloniceras, Dufrenoyia, Prodeshayesites* and *Tropaeum*, derived largely from the Skegness Clay and Sutterby Marl. Rarer ferruginous concretions containing bivalves and the Barremian ammonite *Paracrioceras*, derived from the Roach, are also present.

The best-preserved section recorded to date is that in the cores of the Hunstanton Borehole between 17.96 and 36.86 m (see p.108). The base of the Carstone is not marked by a major lithological change in the borehole, probably because much of the content of the basal bed has been derived from the underlying Roach. Comparison of the basal beds of the Carstone in the borehole with those recorded in the trenches dug on Hunstanton beach has enabled the boundary to be accurately placed in the borehole. The patchy distribution of the ammonite-rich phosphatic pebble bed close to the base of the Carstone has made identification of the precise junction with the underlying beds difficult at many localities. Elsewhere at outcrop in Norfolk, sections were recorded in the present survey in which the Carstone and the underlying beds are lithologically similar and the junction

is marked only by an erosion surface with burrowing. This would be difficult to identify in borehole cores.

When fresh, the bulk of the Carstone appears in hand specimen to be a relatively pure 'chamosite' (berthierine) oolite, with a partial cement of 'chamositic' and/or sideritic mudstone (Plate 9). This is deceptive, however, and in thin section many of the ooliths can be seen to consist of 'chamosite' overgrowths on quartz grains. In all but a few samples, the total quartz content of the rock is greater than 50 per cent by volume and, in the terminology developed by Taylor (1949) for the lithologically similar Northampton Sand (Jurassic) ironstones, they are chamosite-oolite sandstones. Most of the ooliths are partly or wholly oxidised; so too are the pebbles of clay ironstone and oolitic ironstone that are abundant in the more pebbly parts of the rock. Typical ooliths from the Hunstanton Borehole were examined by X-ray diffraction by Mr R J Merriman and shown to consist largely of goethite and quartz with minor amounts of siderite, 'chamosite' and a montmorillonoid mineral (?nontronite). By contrast, the chamosite mud in the matrix is mostly fresh; in places a small part of it is replaced by siderite. This suggests either that some of the oxidation of the ooliths was penecontemporaneous with their deposition, or that they have been preferentially oxidised at some later date.

None of the pebbles in the Carstone can be identified with confidence, but most of the ironstones and mudstones can be matched with lithologies in the Dersing-

Plate 8
Rectilinear jointing in the Carstone, Hunstanton foreshore.

The high iron content of the Carstone makes it susceptible to deep chemical weathering at outcrop, where it gives rise to spheroidally weathered blocks bounded by vertical joints. Whitaker (1893) likened this foreshore outcrop to the paddle of a giant plesiosaur. (A 10850).

ham Beds or Roach of Norfolk, or with the Claxby Beds, Tealby Beds or Roach of Lincolnshire. The chalky limestones may have been derived from the Tealby Beds. Clastic materials other than quartz and the pebbles noted above make up less than 2 per cent by volume of the rock. Some unweathered Carstone samples from the borehole were shown by Dr J R Hawkes to contain small fragments of metaquartzite, quartz-schist, acid-igneous rock of possible volcanic origin, and grains of plagioclase and perthite: these materials are presumed to have been derived from distant sources.

Burrows and cross-bedding are present at many levels in the Carstone, and may be ubiquitous, but their presence is almost always masked at outcrop by secondary ferruginous weathering products. Where seen in less unweathered faces after fresh cliff falls at Hunstanton, in temporary excavations and in boreholes, the burrows are mostly long (commonly more than 1 m), thin, vertical tubes, either single (cf. *Skolithos*) or double (cf. *Arenicolites*). They are especially abundant at the base of the formation and large numbers commonly penetrate the subjacent formation irrespective of its lithology. Where the underlying formation is a soft sand, a variety of larger, more irregular burrows are present (Plate 10). Planar cross-bedding, in units 0.3 to 0.5 m thick, is present in the Carstone at Hunstanton, Gayton and Marham boreholes, but was too rarely observed at outcrop for its orientation to be determined. Schwarzacher (1953, p.322) did not record any cross-bedding in the Carstone. At Hunstanton, cross-bedding is present at three localities and dips northwards at each. The adjacent beds at each locality are parallel-bedded units up to 0.5 m thick, which are themselves within parallel-bedded units up to several metres thick. The thicker units are commonly bounded by possible erosion or still-stand surfaces that have large numbers of burrows associated with them.

The distribution of heavy minerals in the Carstone and Red Chalk has been described by Rastall (1919, 1930). Samples from several horizons in the Carstone at Hunstanton yielded kyanite, staurolite, rutile, zircon, tourmaline, garnet and limonite, and small quantities of hornblende, diopside and muscovite (1919, p. 213). Many grains of the commoner minerals are large and angular, and Rastall thought they had not travelled far from their source.

The upper limit of the Carstone is well exposed in the cliffs at Hunstanton. The junction with the Red Chalk is gradational but rapid. In the present work it has been taken at a weakly developed bedding plane that separates a 23 cm-thick bed of tough red calcareously cemented clayey sandstone that becomes less calcareous, less clayey and paler red with depth (Red Chalk) from a 28 cm-thick bed of soft yellowish brown, fine-grained sandstone (Carstone) containing numerous burrow-infillings of the overlying pink, calcareous sandstone, some including wisps of red clay. A line of widely spaced, pale brown, phosphatised sandstone burrowfills and nodules, mostly 5 to 10 cm in length, forms a useful marker band 28 to 60 cm below the top of the Carstone (Figure 34).

Fossils, other than plant debris, and burrows are rare in the Carstone and its age was in doubt until relatively recently. At Hunstanton, Casey (1961b, p.571) showed

Plate 9 Carstone mineralogy and textures. All photographs in transmitted light. Scale bar 1.0 mm. Magnification approx 34×.

A. Dark brown, limonitised ooids and angular quartz grains cemented by crystalline siderite; the siderite is probably a secondary replacement of a 'chamosite' mud matrix. Hunstanton Borehole; 21.49 m (70 ft 6 in). E 40049.

B. Opaque, limonitised ooids and well-rounded quartz grains with green overgrowth skins of 'chamosite'. Hunstanton Borehole; 22.86 m (75 ft). E 40050.

C. Limonitised ooids, some broken and some with concentric growth laminae, sand- and silt-grade quartz grains and green 'chamosite' pellets, cemented by 'chamosite' overgrowths. Hunstanton Borehole, 33.53 m (110 ft); E 40057.

D. Limonitised pellet enclosing silt-grade angular quartz grains. Hunstanton Borehole, 33.53 m (110 ft); E 40057.

E. Limonitised ooids and angular quartz grains set in a matrix of finely crystalline siderite rhombs, probably a replacement of an original 'chamosite' mud. Hunstanton Borehole, 35.05 m (115 ft); E 40058.

F. Part of a large (4 mm long) limonitised pellet which encloses limonitised ooids and quartz grains. Hunstanton Borehole, 19.96 m (65 ft 6 in); E 40048.

that the phosphatised ammonites from the basal beds of the formation were derived from two zones of the Lower Aptian. The first undoubtedly indigenous shelly fossils to be recorded from the formation in Norfolk were terebratulid brachiopods collected by Casey (1967, p.91) from the Cut-Off Channel near West Dereham [TL 662 996]. These were considered by him to be conspecific with forms from the Lower Albian Shenley Limestone of Bedfordshire; other brachiopods collected by Casey from the same locality include specimens of *Burrirhynchia leightonensis* (Walker) (Owen et al., 1968, p.518), a species known elsewhere only from the Shenley Limestone and the Carstone in Lincolnshire and Humberside.

Casey (in Casey and Gallois, 1973, p.11) also recorded the ammonites *Beudanticeras newtoni* Casey, *Douvilleiceras mamillatum* (Schlotheim) and *Leymeriella* sp., indicative of the *Leymeriella tardefurcata* and *Douvilleiceras mammillatum* zones of the Lower Albian, in phosphatic nodules in the top part of the Carstone in the Cut-Off Channel. Teall (1875) had recorded a similar fauna, but including *Sonneratia kitchini* Spath (indicative of the *mammillatum* Zone), from nodule beds at the same stratigraphical level in an agricultural-phosphate working in the West Dereham area. It is difficult to determine how closely these ammonites reflect the age of the deposit because, although they are completely phosphatised, they appear to be little worn even though they are preserved in an abrasive, pebbly sand matrix. It seems reasonable therefore to assume that this part of the Carstone is of *mammillatum* or post-*mammillatum* Zone age.

The Gault at West Dereham, and throughout its Norfolk outcrop at least as far north as Gayton, overlies the Carstone without any obvious major break in sedimentation. The presence of abundant *Hoplites* spp., including *Hoplites* (*H.*) *dentatus* Spath, in the lowest part of the

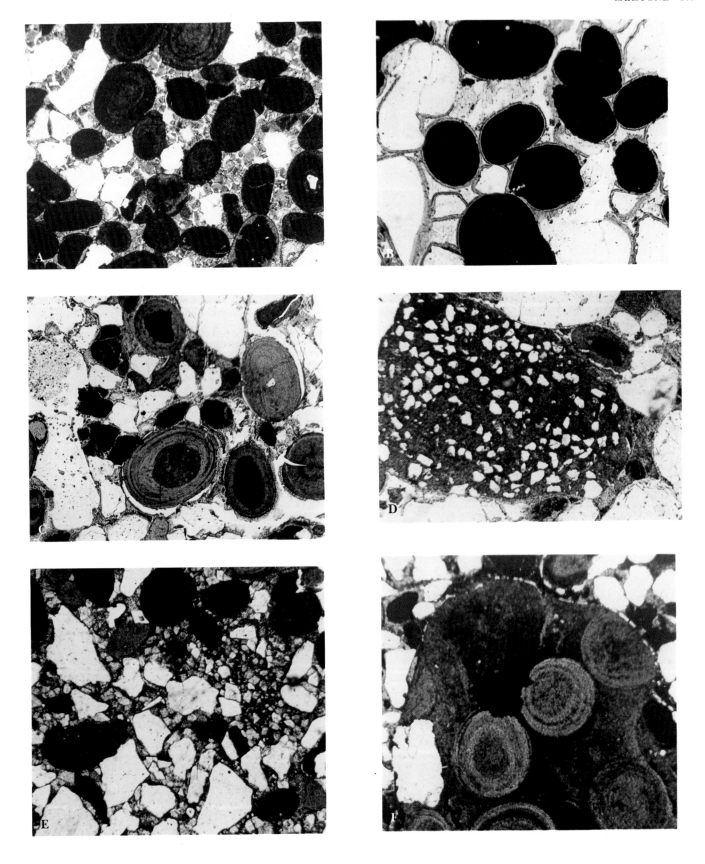

Gault suggests a *mammillatum* or early *Hoplites dentatus* Zone age for the Carstone there. At Hunstanton, the base of the Carstone can be dated as post-Lower Aptian on the basis of the youngest (*bowerbanki* Zone) derived ammonites in its basal bed. There too, the transitional contact with the overlying Red Chalk, which contains indigenous Middle Albian ammonites, suggests that the Carstone is late Lower or early Middle Albian in age.

A loose *Hoplites* (*H.*) *dentatus* found at Hunstanton cliffs, in the same preservation as the phosphatic nodules that occur close to the top of the Carstone there, may be the first direct evidence of *dentatus* Zone age for the top of the formation. In-situ evidence will be needed before this can be confirmed.

Details

Hunstanton to Heacham

The full thickness of the Carstone, totalling about 19 m, crops out on the foreshore and in the cliffs at Hunstanton. The whole of the formation is exposed with the exception of the basal beds and a thin, presumed softer, bed in the middle part of the formation, but even these are overlain by only a thin covering of beach sand.

The basal beds and the junction with the Roach were exposed by trenching on the foreshore (Gallois, 1975a). The following composite section, shown graphically in Figure 31, is based on those proved in three trenches [6712 4152; 6705 4168; 6715 4177] dug close to the outer Carstone reef:

	Thickness m
Beach sand, gravelly in places	0.3 to 0.5

CARSTONE

Sandstone, soft, ferruginous, fine-grained, orange-brown pebbly, oxidised and rotted; ghosts of limonite ooliths throughout; very rare phosphatic nodules as below; passage at base into — 0.3

Sandstone, as above but very pebbly and including dark brown, generally well-rounded, phosphatic nodules (up to 0.1 m across but mostly 0.02 to 0.03 m), some with worn specimens of Lower Aptian ammonites, including *Cheloniceras*, *Dufrenoyia*, *Prodeshayesites* and *Tropaeum*; larger unfossiliferous phosphatic nodules (up to 0.12 m across) also occur, mostly of burrow-fill shape; phosphatic nodules are extremely rare in some sections, in others they occur in clusters; locally less pebbly and indistinguishable from bed above; ferruginous seepage at base — 0.15 to 0.20

Pebbly clay, sandy in part; dark grey and greenish grey oolitic ('chamosite', largely oxidised to limonite); intensely burrow-mottled and varying both laterally and vertically in composition with pebbles and ooliths concentrated in bands generally 0.15 to 0.20 m thick; pebbles (mostly iron-stained quartz) lithologically identical to those of overlying bed. Large well-rounded nodules of very fossiliferous, sandy, phosphatic ironstone (the 'iron-grit' nodules of Keeping 1883, p.33) were collected from near the top of

Plate 10
Junction of Leziate Beds and Carstone, Blackborough Carstone Pit [6767 1475].

Loose brown pebbly sand (Carstone), up to 0.3 m thick, rests on an irregular eroded surface of soft white sands (Leziate Beds). Large irregular burrows occur close to the erosion surface, and numerous double-barrelled, ferruginously cemented worm burrows extend down to about 1 m below the surface. Wind erosion has revealed their 3-D shapes. (A 11796).

Figure 34 Lithological comparison of Carstone and Red Chalk sequences at Hunstanton.

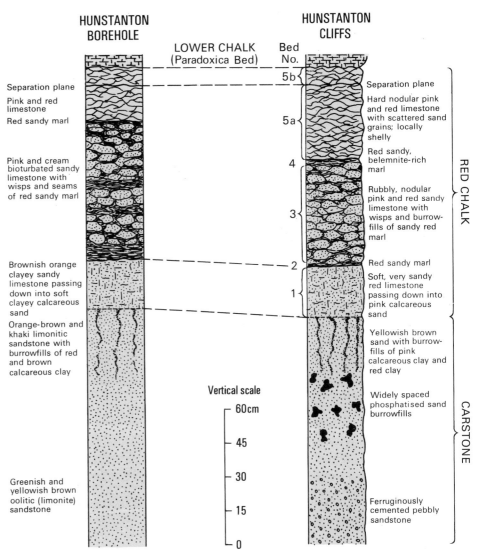

HUNSTANTON BOREHOLE

HUNSTANTON CLIFFS

LOWER CHALK (Paradoxica Bed)

Bed No.

Separation plane
Pink and red limestone
Red sandy marl

Pink and cream bioturbated sandy limestone with wisps and seams of red sandy marl

Brownish orange clayey sandy limestone passing down into soft clayey calcareous sand

Orange-brown and khaki limonitic sandstone with burrowfills of red and brown calcareous clay

Greenish and yellowish brown oolitic (limonite) sandstone

5b

5a

4

3

2

1

Separation plane
Hard nodular pink and red limestone with scattered sand grains; locally shelly

Red sandy, belemnite-rich marl

Rubbly, nodular pink and red sandy limestone with wisps and burrow-fills of sandy red marl

Red sandy marl

Soft, very sandy red limestone passing down into pink calcareous sand

Yellowish brown sand with burrow-fills of pink calcareous clay and red clay

Widely spaced phosphatised sand burrowfills

Ferruginously cemented pebbly sandstone

RED CHALK

CARSTONE

Vertical scale

— 60cm

— 45

— 30

— 15

— 0

Bed numbers are referred to in text

Thickness m

this bed in the 1930s (e.g. Natural History Museum specimen nos. C35512 and C35238); thickness variable; very irregular sharp base with burrowfills of fine- and medium-grained greenish yellow sand (Carstone lithology) extending down to 0.5 m into Roach sands; irregularities of up to 0.15 m observed in 2 m length of junction   1.0 to 1.5

ROACH (see p.93 for details)

Carstone forms two prominent reefs on the southern part of the beach, separated by a sand-covered slack that may conceal softer, possibly finer-grained or more clayey Carstone. A line at springs at the base of the higher reef adds weight to this suggestion. The outer reef runs north for about 700 m from near the Esplanade Gardens [6714 4115] and consists of dark greyish green, fine-grained, sparsely pebbly, oolitic sandstone weathering to the usual rusty brown. Its outer edge is close to the base of the formation and phosphatised fragments of ammonites derived from the basal beds are common as beach pebbles. The inner

reef occupies much of the higher part of the intertidal area northwards from the northern end of the Esplanade [6724 4131] and its width expands northwards as progressively more of the formation is brought down to the foreshore by the northerly component of the dip. Most of this reef is composed of sparsely pebbly and pebbly, fine- and medium-grained, dark greyish green Carstone that weathers to the typical rusty brown. Its most striking feature is the manner in which it has been eroded; at beach level it appears to form an irregular collection of rounded lumps of sandstone, but when viewed from the adjacent cliff-top these can be seen to be separated by a remarkably uniform recti-linear set of joints that give rise to an outcrop pattern that was aptly likened by Whitaker (*in* Whitaker and Jukes- Browne, 1899, p.16) to an enormously magnified ichthyosaur paddle (Plate 8).

When traced northwards from the Esplanade, the two reefs appear to merge north of the latitude of St Edmund's Chapel [6758 4195], but the outcrop may be complicated there by a small fault. At very low tides, a small reef composed of sparsely pebbly, medium-grained Carstone is exposed offshore of the outer reef [6708 4152 to 6709 4180] and therefore appears, from the gentle north-easterly dip of the reefs, to be at a lower stratigraphical level than the basal pebble bed. A small fault, downthrowing to the west, has been postulated to explain this apparent anomaly. The possibility cannot be excluded that this bed has been downthrown by an ancient landslip that failed in the argillaceous Dersingham Beds and Roach at some time in the Pleistocene when they, the Carstone and the Lower Chalk formed a single high cliff.

The upper 10 m of the Carstone, including its junction with the Red Chalk, are well exposed in the cliffs between the Esplanade [6723 4124] and St Edmund's Point [6786 4239]. The Carstone there is composed of fine-, medium- and coarse-grained sands and pebbly sands; some levels in the upper part of the sequence, notably a 4.5 m-thick band 3 m from the top of the formation, are sufficiently pebbly to be termed pebble beds. The top part of the sequence is summarised in Figure 34. The section is deeply weathered rusty brown with much secondary veining, cementation and Liesegang precipitation of limonite and other iron minerals. Planar cross-bedding, in units 0.3 to 0.5 m thick, is discernible at a number of levels, but other sedimentary features are usually obscured by the limonitic weathering. However, fresh cliff falls show the rock to be dark greyish green, oolitic ('chamosite', limonite and limonitised 'chamosite'), cross-bedded sandstone with large numbers of long, thin vertical burrows of the *Arenicolites* and *Skolithos* types. No body-fossil has been recorded from the Carstone at Hunstanton, probably partly because of their rarity at the time of deposition, and partly because of their susceptibility to weathering in such a porous medium.

The thickest development of Carstone recorded to date is that proved in the Hunstanton Borehole [6857 4078] between 17.96 and 36.86 m. At first sight, the lithologies there are markedly different from those at outcrop, the unweathered materials of the borehole being dominated by green and blackish brown 'chamosite' oolites:

	Thickness m	Depth m
RED CHALK (see p.113 for details)		
CARSTONE		
Sandstone, fine- and medium-grained, orange-brown and khaki; earthy-textured with limonitic weathering products in matrix; burrowfills of red and brown calcareous clay extend down from the Red Chalk for 0.3 m; becoming tougher, greenish and yellowish brown with depth with many limonitised ooliths and patchy 'chamosite'-mud cement; burrow-mottled in part; small pebbles common in lower part; passing down into	1.75	19.71
Oolite, 'chamosite' and limonitised-'chamosite' coatings of fine-grained quartz sand; very sandy in part; glistening blackish brown; burrow-mottled in part; pebbly in part with small quartz and ironstone pebbles; very pebbly from 20.5 to 23.5 m with ironstone pebbles up to 3 cm across; traces of cross-bedding; graded bedding in part with pebbly oolite passing up into sandy oolite; *Arenicolites* and *Skolithos*-type burrows common at several levels; pale greyish green and dark green 'chamosite'-mud cement at some levels; passing down into	13.82	33.53
Sandstone, fine- and medium-grained; very oolitic as above; greyish green with patchy pale green 'chamositic' cement; burrow-		

mottled in part; becoming uniformly coarser and pebbly below 35 m; pebbles up to 3 cm across consisting of milky and brown vein quartz, brown to brick red ironstones, chert, soft white chalky limestone and soft green and brown clays; passing down with interburrowing into   2.26   35.79

Clay, dull khaki and greenish yellow-brown, very sandy and pebbly; burrow-mottled; becoming predominantly sandy with depth, with dark grey clay burrow-fills; irregular contact with apparent channelling into underlying Roach   1.07   36.86

ROACH (see p.93 for details)

The Carstone thins rapidly north-westwards along its submarine outcrop beneath The Wash; this change appears to be accompanied by a decrease in the proportion of ooliths. The following section was proved between 11.00 and 17.35 m in Borehole 72/78 [6494 4972] drilled in the mouth of The Wash:

	Thickness m	Depth m
RED CHALK		
CARSTONE		
Clay, pink and deep red, silty and gritty, interburrowed with yellowish brown and khaki oolitic and sandy limonitic clay (decomposed products of Carstone); a few belemnites and terebratulids; passing down into predominantly rotted Carstone sand with pink clay burrowfills (0.20 m) with phosphatic nodules just above base; resting on soft fine-grained sandstone with dark grey clay wisps; tougher in part due to 'chamositic' cement; pebbly in part; rare wood fragments	3.05	14.05
Sandstone, yellowish green, fine-grained, very oolitic ('chamosite' and limonite) with small pebbles of vein quartz; ironstone and oolitic ironstone, the whole set in a pale green amorphous, 'chamositic', weak cement; barren; darker green (less oxidised) and with ooliths more shiny at some levels; dark grey-green and grey clay wisps and burrow linings at several levels	2.75	16.80
Core lost, probably loose sand	0.30	17.10
Sandstone, fine- and medium-grained, weakly calcareously cemented, cavernous in part (?washed out by drilling), with scattered pale brown ooliths and burrowfills of dark greyish green oolitic clayey sand; densely pyrite-cemented bed (0.02 m) at base containing burrowfills of pale green clay from underlying bed	0.25	17.35
?CARSTONE		
Clay, mottled dark green, brownish green and very pale green	0.05	17.40
Core lost; presumed from gamma-ray log to be sand	2.80	20.20

SUTTERBY MARL (see p.100 for details)

Southwards from Hunstanton Cliffs, the Carstone forms a steep feature through the town and through farmland to the Heacham River at Heacham. When ploughed, it reveals a dark brown sandy soil with common soft sandstone fragments and the outcrop can be readily traced. Up to 1 m of coarse-grained Carstone, probably a Pleistocene wave-cut platform, was formerly exposed beneath about 8 m of drift deposits in a gravel pit [6720 4032] near Hunstanton station (Lamplugh *in* Whitaker and Jukes-Browne, 1899, p.91).

Carstone has been worked in a small pit [6760 3964] near the Firs, Hunstanton, and in a slightly larger pit [6852 3780] near Church Farm, Heacham. The latter still exposes 0.5 m of deeply weathered, sparsely pebbly, medium-grained sandstone. Jackson's (1911, p.63) record of extensive reefs of hard, coarse-grained pebbly Carstone on Heacham foreshore [662 375] remains a mystery. The reefs are no longer exposed and both the onshore and offshore evidence from the present survey suggest that Carstone could only occur in that position if transposed there by faulting or transported there by glacial action.

Heacham to Dersingham

Between the Heacham River and Snettisham, the Carstone occupies ground of relatively low relief between the features of the Dersingham Beds and the Chalk. The outcrop rises slowly southwards from the valley of the Heacham River until it caps the eminences of Ken Hill and Lodge Hill, Snettisham. The lower boundary of the formation, resting with marked lithological contrast on the Snettisham Clay, can be traced by means of a spring line and feature, and the upper boundary by the prominent soil brash of the overlying Red Chalk.

Small quarries [6732 3481 and 6691 3423] near Ken Hill show up to 2 m of deeply weathered, sparsely pebbly Carstone close to the base of the formation. Carstone has been dug at Snettisham almost continuously for well over 100 years from a large quarry [685 349] adjacent to the Hunstanton road. This and adjacent, now defunct quarries [687 347 and 6853 3469] at Norton Hill provided much of the stone for New Hunstanton and for many of the grander Victorian buildings in west Norfolk. The two larger quarries are in the upper part of the formation and expose the junction with the Red Chalk. The working quarry exposes about 5 m of massive, well-jointed sparsely pebbly brown sandstone with traces of cross-bedding, burrows and rare plant fragments, overlain by up to 2 m of rubbly, fine-grained, sparsely and pebbly brown and khaki sandstone with much secondary iron staining. The contact between the rubbly and massive types is irregular and presumably a weathering feature. The use to which the stone has been put has varied over the years; most of the rubbly material has been used as hardcore and the massive material for building stone, either as roughly trimmed or dressed blocks.

Between Snettisham and Dersingham, the Carstone again crops out as a low-relief area capping the Snettisham Clay feature and can readily be traced by means of features, a spring line and soil fragments. Fine-grained sandstone at the the base of the formation appears to have been worked on a small scale in pits [6874 3626; 6879 3220] at Ingoldisthorpe that were dug primarily for the underlying Snettisham Clay.

About 1.5 m of strongly cross-bedded sandstone crops out in an old quarry [6984 7568], dug in the middle part of the Carstone near Doddshill, Heacham. Several small pits [e.g. 6979 3045; 6981 3033; 6951 3031] have been dug into the lower part of the formation in the same area; all are shallow and must have been in deeply weathered material.

Dersingham to the Babingley River

The Carstone has an extensive outcrop in Dersingham Wood and Sandringham Park where it forms the capping for a prominent feature. The lower boundary of the formation is marked by an almost continuous spring line, and ferruginous sandstone brash is common in the sandy, well-drained soil that characterises this and much of the remaining Carstone outcrop. The sandstone appears to have been worked in a few small, shallow pits in this woodland area, presumably for roads and paths; all are dug in the weathered zone.

The formation thins rapidly southwards between Sandringham House [695 288] and the Babingley River, from an estimated 15 to 10 m, in response to the appearance and subsequent thickening of the Gault in the same direction. It is poorly exposed in this area, but has been worked in shallow pits [709 521; 7107 2704] near Appleton House. South of Flitcham Hall [715 264], the formation disappears beneath the Quaternary deposits of the Babingley River and does not reappear until the Roydon area.

Babingley River to the River Little Ouse

Between Roydon and East Winch, the Carstone crops out as a continuous strip of low ground in which it is deeply weathered and largely obscured by drift deposits and sandy wash derived mainly from the formation itself. Outliers occur on higher ground at Brow of the Hill and Middleton, but these too are mostly covered by drift deposits.

Fragments of phosphatised ammonites, including *Cheloniceras* and *Tropaeum* derived from the basal bed of the Carstone, were collected loose as field brash [6922 2350] at Hall Farm, Roydon, close to the presumed westerly limit of the Carstone outcrop in this area. The presence of till and thick sandy wash makes the evidence uncertain, but the Carstone appears to be faulted against Leziate Beds and Dersingham Beds close to this locality. A small sand pit [6945 2349] at the farm is in soft brown and white sand with limonite-cemented veins and is probably deeply weathered and leached Carstone. Similar lithologies are exposed in small excavations [7002 2464 to 7018 2466] dug in the Carstone feature near Gorse Moor, Roydon; they were probably made for building sand.

A ditch [7028 2443] near Gorse Moor exposed the following deeply weathered section at the top of the formation:

	Thickness m
GAULT	
Very sandy and silty clay (Bed G 1) passing up into mottled buff and grey clays with a few belemnites (Beds G 2 and G 3), and then into red-mottled clay (Bed G 4)	c.2
CARSTONE	
Sandstone, brown, ferruginous, medium-grained, sparsely pebbly sand with limonite-cemented ribs; clayey in top 15 cm with nobbly, cream-coloured and pale brown, sandy, pebbly, phosphatic concretions up to 5 cm across	c.1

An almost identical section was exposed in a drainage ditch [7179 2098] close to the Gaywood River at Grimston.

Between the Gaywood River and the valley of the Middleton Stop Drain, where the solid deposits disappear beneath a thick Quaternary sequence that underlies the valleys of the River Nar and its tributaries, the Carstone outcrop again occupies low ground, much of it covered by till or head. There is no exposure in this area and the formation is known only from augering and as brash in the deeper drains and on the steeper valley slopes.

The following section was proved between 22.00 and 30.48 m in the Gayton Borehole [7280 1974]:

GAULT (see p.120 for details)

CARSTONE

	Thickness m	Depth m
Clayey pebbly sand and soft sandstone, dark greenish grey; line of pale brown, sandy and pebbly, phosphatised burrow infillings at top; bioturbated throughout; pebbles mostly less than 6 mm and made of clear or milky quartz; larger pebbles of quartz, quartzite, rotted ironstone and rotted chalky limestone common at several levels; strongly ferruginously cemented in part, mostly in association with 'chamosite' and limonitised 'chamosite' ooliths; traces of cross-bedding and vertical *Skolithos*-type burrows at several levels; coarsely glauconitic in lower part	8.43	30.43
Pebble bed; pebbles of vein quartz, quartzite, dark brown phosphate including an ammonite fragment, pyritised sandstone, ironstone, soft grey siltstone and angular mudstone fragments, up to 50 mm across and set in a glauconitic clayey sand matrix; very irregular base with pebbly sand infilling possible desiccation cracks in underlying Snettisham Clay, and vertical burrows of sand extending down for more than 50 cm into the clay	0.05	30.48

SNETTISHAM CLAY (see p.91 for details)

Carstone has been worked in a small quarry [6834 1914] in an outlier that caps the eastern part of the hill on which Brow of the Hill is built. The face, now deeply weathered, exposes 2.5 m of gently curved, trough cross-bedded, fine-grained, ferruginous sandstone at the base of the formation. The junction with the underlying Snettisham Clay is marked by a spring line, but is obscured by sandy wash.

This junction and up to 1 m of the basal part of the Carstone has been exposed from time to time on the northern side of Brow of the Hill at the top of the southern face [6844 1934] of Leziate Sand Pits. The basal bed is a soft, fine- to medium-grained, brownish green, glauconitic sand that weathers to rusty brown, and rests on a slightly irregular, burrowed surface (cf. *Arenicolites* and cf. *Skolithos*) of Snettisham Clay. The lowest 5 to 15 mm contains patchy concentrations of pebbles including ironstone (from the Snettisham Clay), ferruginous sandstone (Dersingham Beds), deeply weathered glauconitic phosphatic ironstone (?Mintlyn Beds), soft white sandstone (Leziate Beds), quartz, chert, rotted limestone and other ironstones.

The full thickness of the Carstone is exposed in a large erratic block within the till in the same face, about 700 m to the west-south-west [6777 1918], where the section below was measured. The thickness has probably been reduced by shearing and contortion within the erratic mass. The lithologies of the Gault, Carstone and the underlying beds suggest derivation from a nearby outcrop, possibly from the Brow of the Hill outlier.

GAULT, grey clay

CARSTONE

	Thickness m
Sandy pebbly clay with scattered small phosphatic nodules (0.1 m), passing down into deeply boxstone weathered, fine- and medium-	

grained, ferruginous, rusty brown sandstone with a few pebbles; pebble bed at base with quartz, quartzite, phosphate, chert and ironstone, resting with iron-stained contact on Dersingham Beds sands ... 5.9

The Carstone forms a large outlier that caps the higher ground between Middleton, East Winch and Blackborough End. Much of the western part of this outcrop is overlain by till. Mr B Young noted three old quarries [6820 1634; 6846 1606; 6830 1532] in the East Winch area that show up to 3 m of deeply weathered, cross-bedded, ferruginous, fine-grained sandstone. The following section was recorded in the Carstone beneath an irregular covering of till, from 1.5 to 5 m thick, in an excavation on the crest of the ridge [6705 1616] at Middleton:

	Thickness m
Sandstone, very pebbly, coarse-grained, ferruginous, soft gingery brown; planar base except for numerous vertical, ferruginously cemented, 6 mm-diameter tube-shaped burrows (cf. *Skolithos*), that penetrate the full thickness of the underlying bed, and a few larger burrows that carry down pebbly sand	0.60
Sandstone, fine-grained, gingery brown, ferruginous, soft, flat-bedded; upper layers deeply iron-stained	0.90

The section was deeply weathered but was observed over a distance of about 20 m, in which its character remained constant. The similarity of the pebbly upper bed, and its associated burrows, to the basal bed of Carstone at nearby Blackborough End (see below) suggests the possibility that the lower bed at Middleton may be part of the Dersingham Beds that have become deeply weathered and iron-stained.

The most extensive exposures of Carstone, apart from those at Hunstanton beach, occur in a series of pits worked for hard core (Carstone), sand (Leziate Beds) and aggregate (Glacial Sand and Gravel) at Blackborough End. The largest workings occur between Blackborough [673 143] and Foster's End [683 146]. The best sections occur along the northern face of the pits [6724 1469 to 6774 1485] where the following composite succession was recorded:

	Thickness m
Sandstone, frost-shattered, deeply weathered, pale brown to rusty brown, medium- and coarse-grained, pebbly, ferruginously cemented, with dense dark brown limonitic staining coating many surfaces	up to 1.7 seen
Sandstone, massive, fine-grained, yellow-brown to rusty brown, ferruginously cemented, with dense dark brown limonitic cement coating joint and bedding surfaces and tending to form boxstone structure; traces of planar cross-bedding at several levels in beds mostly less than 0.5 m thick;	up to 1.5 seen
Obscured by tip	c.1
Sandstone, greyish and brownish green, soft, ferruginous, glauconitic, medium- and coarse-grained sand; pebble bed in lowest 10 to 20 cm resting on relatively flat but intensely burrowed surface of soft white sands (Leziate Beds); pebbles predominantly rounded and subrounded, soft, clean, white quartz siltstone,	

mostly 10 to 40 mm across, some of which are bored by cf. *Trypanites*; quartz, ironstone and limestone pebbles also present; basal bed let down into large, irregular, low-angle ?crustacean burrows and into closely spaced vertical, double-barrelled, ferruginously cemented, tube-shaped burrows (cf. *Arenicolites*), some of which extend down for more than 1 m (Plate 10).

	Thickness m
	0.4 to 0.5

The depth of mechanical weathering in the upper part of the Carstone in the sections varies considerably over short distances and does not everywhere coincide with the lithological change noted here. A description of the frost-involutions, ice-wedge casts and other cryoturbation features that affect the Carstone at this locality were given by Evans (1973).

The basal 2 to 3 m of the Carstone, again with cf. *Skolithos* and other burrows carrying them down into the underlying Leziate Beds sands, were recorded in a large sand and hardcore pit [668 149] at Blackborough End. Up to 6 m of Carstone was worked there in early Victorian times for building stone (Rose, 1836).

The Carstone forms an inlier, surrounded by low drift-covered ground, at West Bilney Hall and Warren, Pentney. A smaller inlier forms the low rise on which Pentney Priory is sited. Mr Young noted that, at the time of the survey, Carstone was worked for hardcore in a large shallow pit [699 144] near West Bilney Hall, that exposed up to 3 m of deeply weathered ferruginous sandstone, pebbly in part and with traces of cross-bedding. A small degraded quarry 200 m to the south-west [6965 1418] presumably worked the Carstone as a building stone.

The Carstone emerges from beneath the wide drift-filled valley of the River Nar as a small outlier capping a low rise at Shouldham Warren and as a continuous outcrop, mostly on drift-covered low ground, that runs from Shouldham to Shouldham Thorpe. The formation then disappears beneath a broad tract of till and reappears about 7 km to the south in the Denver-West Dereham area. There, the outcrop is extensive and large parts of it are drift-free, but the formation is more argillaceous and less massive than in more northerly areas and has not been worked for building purposes. There are, in consequence, very few exposures in the area.

Up to 2 m of deeply weathered, pale brown, ferruginous Carstone sands rest with an almost planar contact on Leziate Beds in an old sand pit [6836 1133] at Shouldham Warren. There, the basal bed of the Carstone contains only rare, widely spaced pebbles, but elsewhere on the outlier pebbly sandstone debris is common at this level. The main outcrop of the formation has been traced from Bottom Fen, near Shouldham, to Foddestone Gap, Shouldham, by means of pebbly sandstone fragments in the soil, a feature and a spring line (the lower boundary) and by fragments of clayey green sand with sandy phosphatic nodules (the upper boundary). Debris from close to the upper boundary is relatively common because it crops out on such poorly drained low ground that there are numerous deep ditch sections. The junction with the Gault was exposed in ditches at Bottom Fen [6951 0977] and Abbey Farm [6910 0952; 6885 0928], Shouldham. This junction is obscured by thick drift deposits beneath Shouldham village and all the ground to the west of it.

The following section in the Carstone, which is probably typical in lithology and thickness of the Shouldham area, was proved between 45.03 and about 50.6 m in the Marham Borehole [7051 0803]:

	Thickness m	Depth m
Sand, brownish grey, glauconitic, fine- and medium-grained; yellowish green- and brown-mottled, bioturbated, weakly cemented; small pebbles, mostly up to 6 mm, at several levels; wisps of grey or green clay as burrow infillings or linings at many levels; traces of cross-bedding at one level; burrowing ubiquitous	2.65	47.68
Core loss, mostly in soft sands; base of formation placed on geophysical logs	c.2.9	50.6

The formation can be continuously traced by means of fragments in the soil and in ditches, and features from Ryston Park, Denver, via Bexwell and West Dereham to the edge of Methwold Fens. From there southwards, it lies entirely beneath drift deposits and is known only from boreholes.

The full thickness of the formation was exposed, albeit deeply weathered and much disturbed by cryoturbation, in a ditch section [6370 0243 to 6407 0237] at Stonehills Farm, Bexwell, where the following composite section was measured:

	Thickness m
GAULT, grey clay	
CARSTONE	
Sandy clay with sandy and finely pebbly, phosphatised burrow infillings	0.2
Sand, glauconitic, fine- and medium-grained, soft; brown, greenish grey and greenish brown; clayey in part; mottled and presumably bioturbated	c.0.4
Sand, clayey, soft, limonitic; orange- and yellow-brown	0.5
Pebbly clay, sandy in part; mottled yellow-brown, grey and dark green; with well-rounded pebbles of quartz, chert, ironstone phosphate	0.15
MINTLYN BEDS (greenish grey, fine-grained sands)	seen 0.3

The section is of particular interest in that the basal bed of the Carstone is clayey and almost identical in lithology to that of the formation at Hunstanton, and unlike that of most of the intervening outcrop.

The most southerly recorded outcrop of the Carstone is that described by Casey and Gallois (1973, pp.10–11) in the excavations dug for the Fenland Flood Relief Channel in 1961–62 between Roxham Farm and Wissington railway bridge, West Dereham [TL 639 995 to 662 996]. The composite section is as follows:

	Thickness m
CARSTONE	
Sand, very clayey, brownish grey	0.30
Phosphatic nodules (up to 150 mm diameter) in pebbly sand; gritty, grey-skinned, black-centred; some nodules pebble-studded, others veined with calcite; *Douvilleiceras mammillatum* (Schlotheim) and *Beudanticeras newtoni* Casey	0.15 to 0.20
Sandstone, argillaceous, dark grey, pebbly, with horizontal burrows (15 mm diameter) infilled with dark grey, partly phosphatised material; brownish-black phosphatic nodules stud the surface of the sandstone which weathers to cement colour	0.20
Sand, coarse-grained, brown with wisps of grey clay	0.25 to 0.40

	Thickness m
Sandstone, dark grey (weathering light brown), with wisps of clay and small pebbles, including polished ironstone; phosphatised in places; rubbly bedding; a few black phosphatic nodules with impressions of fossils at base, including *Leymeriella* sp.	1.20
Pebbly, coarse-grained sand and sand-rock in clayey matrix; dark grey and brown, with green streak	0.20
Sandstone, micaceous, medium-grained, grey, flaggy	0.10
Sand, clayey, dark grey	0.10 to 0.15
Sandstone, micaceous, medium-grained, grey	0.05
Sand, coarse-grained, dark grey, and clay with abundant pebbles (up to 15mm diameter); becoming indurated into cross-bedded sandstone in lower part	0.60
Sandstone with few pebbles, micaceous, grey; passing down into	0.60
Sandstone, conglomeratic with pebbles up to 20 mm diameter; reddish brown	1.40
Basement bed; band of pebbles and nodules with residual boulders of underlying Mintlyn Beds up to 0.6 m in length, pebbles up to 75 mm in diameter; nodules green-coated, spherical or oval, commonly enclosing wood, many derived from Mintlyn Beds (with *Hectoroceras*); rolled pieces of Lower Aptian ammonites; Lower Albian brachiopods and rare Hauterivian *Craspedodiscus*; nodules die out locally	0.03 to 0.20

MINTLYN BEDS (see p.73 for details)

The Carstone thins steadily southwards from West Dereham; it consists of 1.7 to 2.7 m of clayey ferruginous sandstone beneath Methwold Fens, and is only a thin pebbly sand at the base of the Gault south of the River Little Ouse (Gallois, 1988).

RED CHALK

The type section of the Red Chalk is the continuous cliff (Front cover and Plate 11) that runs from Hunstanton Promenade [6725 4130] almost to Old Hunstanton [6786 4238], a distance of about 1.2 km. The formation is readily accessible at the northern end of the section where it falls to beach level, by climbing cliff falls between there and the lighthouse, and in large fallen blocks.

Between Hunstanton and Snettisham, the Red Chalk forms a thin but persistent marker band that is one of the most easily mapped in Britain. At times when the fields are ploughed, its soil brash, together with that from the adjacent Carstone and Chalk, forms a colour and lithological contrast that is only slightly less muted than that in Hunstanton Cliffs. Between Snettisham and Dersingham, the colour of the Red Chalk becomes paler in a southerly direction until it is predominantly pink and cream coloured. At Dersingham, within a distance of less than one kilometre, the Red Chalk limestones pass laterally into red, pink and cream-coloured, highly calcareous clays with nodules and thin bands of pink and cream limestone in their upper part. The calcareous clays with thin limestones pass rapidly southwards at San-

dringham into the brown sandy clay and grey, cream and brown silty clays of the Gault.

The Red Chalk thickens north-westwards from Hunstanton (1.1 m of sandy limestone with marl wisps) into The Wash (3.2 m of limestone with marl wisps in Borehole 72/78), and Lincolnshire (5.5 m of interbedded limestone and red clay in the Skegness Borehole).

The Red Chalk has an extensive subcrop in Norfolk. It was proved in the deep onshore hydrocarbon-exploration boreholes at North Creake, South Creake, Saxthorpe, East Ruston and Somerton, in numerous offshore hydrocarbon boreholes (e.g. 78/22-2; Rhys, 1974) and in the BGS Trunch Borehole. It is presumed to underlie the whole of the county north of a line from Sandringham to Great Yarmouth.

Wiltshire (1859) and Seeley (1864a; 1864b) divided the Red Chalk at Hunstanton into three beds of limestone separated by two prominent seams of red sandy marl. There are numerous minor local variations in the sequence, but this general five-fold division can be recognised throughout the cliff exposure. The sequence, based on sections in the vicinity of the lighthouse [675 420], and its correlation with the sequence proved in the Hunstanton Borehole, is shown graphically in Figure 34.

The limestones contain a variety of primary sedimentary structures, including burrows, borings and probable stromatolitic laminations, but many of these have been considerably modified by secondary lithification, cementation and mineral migration. Some idea of the complexity of these processes can be gained from the descriptions given by Jeans (1980).

The Red Chalk at Hunstanton has yielded a remarkably rich and diverse fauna (Plate 12), much of it from fallen blocks. However, the great variety of echinoderms, polyzoans, brachiopods, bivalves and ammonites, many beautifully preserved specimens of which can be seen in museum collections, reflect more than 150 years of patient collecting, rather than an obvious abundance in the rock. Le Strange (1974, pp.47–48), basing his observation on more than 50 years collecting at Hunstanton, notes that the commonest fossils in the Red Chalk are *Inoceramus*, *Neohibolites* and the brachiopod *Moutonithyris dutempleana* (d'Orbigny). Cidarids and stems of *Torynocrinus* are relatively common, but other echinoderms are rare; well-preserved ammonites and nautiloids are rare; corals, except for the solitary *Podoseris*, and gastropods are very rare.

Most of the fauna preserved in museum collections probably came from loose blocks and is not stratigraphically identified, other than as Red Chalk. Some material is identified in terms of the three limestone divisions of Wiltshire (1859), but with varying degrees of reliability. Owen (1979, p.580) has summarised the ammonite evidence provided by these collections and has shown that Wiltshire's three limestones (lettered from the top downwards) can be correlated with the standard zonal/subzonal scheme for the Albian as follows:

Division A: *varicosum* and *auritus* subzones of the *inflatum* Zone; ?*rostratum* Subzone of the *dispar* Zone

Division B: *cristatum* and *orbignyi* subzones of the *inflatum* Zone

Plate 11   Carstone, Red Chalk and Lower Chalk, Hunstanton Cliffs.

At their highest point, near the lighthouse, the cliffs are up to 18 m high and expose cross-bedded ferruginous sandstones and pebbly sandstone at the top of the Carstone, the full thickness of the Red Chalk, and most of the Lower Chalk. The distinctive grey gritty chalk of the Totternhoe Stone (arrowed) forms a marker band in the upper part of the cliff. (A 10847).

?Division B or C: *nitidus* Subzone of the *loricatus* Zone; *intermedius* and *meandrinus* subzones of the *lautus* Zone
Division C: *?lyelli* and *?spathi* subzones of the *dentatus* Zone

Detailed collecting of the in-situ, non-ammonite assemblages by Mr Morter has enabled him to make detailed correlations with assemblages from the Gault of Norfolk, Suffolk and Cambridgeshire (see p.114). His conclusions concerning the zones and subzones present at Hunstanton are closely similar to those of Owen; so too were Andrews' (1983) conclusions based on a study of the foraminifera and ostracods. All three authors agreed with earlier studies that have suggested that the Red Chalk at Hunstanton is a very condensed, but almost complete representative of the Middle and Upper Albian.

Details

Hunstanton

The following composite section, shown graphically in Figure 34, is based on exposures in the cliffs in the vicinity of the lighthouse [675 420]:

Thickness
m

LOWER CHALK (see p.130 for details)

RED CHALK
Bed 5: limestone, hard, fine-grained and chalk-like; mottled pink to deep brick-red, intensely bioturbated, with burrows mostly picked out by paler infillings and darker coloured linings; nodular-texture when weathered; persistent purplish red-stained irregular surface within

| | *Thickness* |
| | m |

upper part of bed separates Bed 5a from Bed 5b
(see below); scattered sand grains and tiny
angular pebbles common below this surface
and increasing with depth; a few belemnites;
irregular base   0.38 to 0.43

Bed 4: mudstone, dark red and brownish red,
highly calcareous, earthy-textured and with
numerous sand grains and tiny angular pebbles,
mostly of brown-stained quartz and ironstone;
abundant belemnites; irregular base with local
shearing and loss of bed   up to 30mm

Bed 3: limestone, firm, fine-grained, chalk-like
but with common sand grains, small angular
pebbles and belemnites; pink and red burrow-
mottled; burrow-linings, infillings and pressure-
solution wisps of deep red sandy calcareous clay
occur throughout; weathering to nodular and
rubbly texture; irregular base   0.41 to 0.43

Bed 2: Clay, dark red, sandy, highly calcareous;
localled sheared out and replaced by irregular
purplish red to brown ferruginously stained
surface   up to 20mm

Bed 1: limestone, soft, fine-grained, very sandy;
passing down irregularly, with much
bioturbation, into pink and brown highly
calcareous sand; sharp lithological break at base
at almost planar, but burrowed surface   0.23 to 0.28

CARSTONE (see p.108 for details)

Mr Morter has correlated the in-situ non-ammonite fauna at
Hunstanton with assemblages from the East Anglian Gault
which occur in association with ammonites and belemnites.
The in-situ fauna, although only a small part of the total fauna
known from the Red Chalk at Hunstanton, can therefore be
used indirectly to allocate the divisions of the formation to
zones and subzones. The assemblages collected in situ to date
can be summarised as follows:

Bed 5b: probably *Mortoniceras rostratum* Subzone.
Stromatolites indet. (reported by Jeans, 1973), abundant
sponges indet, *Concinnithyris* cf. *subundata* (J Sowerby), *Ornatothyris* cf. *obtusa* (J de C Sowerby), *O.* cf. *pentagonalis* (Sahni), *O.*
spp., *Rectithyris* aff. *bouei* (d'Archaic), *Aucellina gryphaeoides* (J de
C Sowerby *non* Sedgwick., ?=*A. coquandiana* d'Orbigny &
Auctt.), *Aucellina krasnopolskii* (Pavlov), *A.* spp., *Ceratostreon
rauliniana* (d'Orbigny), *Neohiholites praeultimus* Spaeth, ?*N. men-
jailenkoi* Gustomesov, *N.* spp. indet. and *Cyclocrinus variolaris*
Seeley.

Bed 5a, upper part: *Callihoplites auritus* Subzone.
Stromatolites indet. (cf. Jeans 1973, 1980), abundant poorly
preserved sponges, 'Rotularia' cf. *umbonata* (J Sowerby), *Concinnithyris* cf. *subundata*, *Moutonithyris dutempleana* (d'Orbigny), *M.
dutempleana* variant (elongate/labiate foramen), common *M.*
aff. *oroseina* (Dieni & Middlemiss), *Birostrina* cf. *concentrica*
(Parkinson) probably derived in pebbles, *Ceratostreon rauliniana*
(d'Orbigny), *Inoceramus lissa* (Seeley) (=*I. tenuis* auctt.), common *Pycnodonte (Phygraea)* aff. *vesicularis* (Lamarck), *Neohibolites
ernsti* Spaeth, *N. oxycaudatus* Spaeth and *N.* spp.

Bed 5a, lower part: *Hysteroceras varicosum* Subzone.
Sponges indet., adherent bryozoans on brachiopods, *Kingena
spinulosa* (Davidson & Morris), *Moutonithyris dutempleana*, *M.* cf.
ichnusae (Dieni & Middlemiss), *M.* aff. *oroseina*, *Birostrina* cf. *concentrica*, *B. transversa* (Seeley), inoceramid chips indet., *Mortoniceras (Deiradoceras)* cf. *cunningtoni* Spath common *Neohibolites ern-*

Plate 12   Fossils from the Red Chalk at Hunstanton (all × 1).

1a–b *Anahoplites intermedius* Spath (BMNH C76151)
2a–b *Tetragramma brongniarti* Agassiz (BMNH E18152)
3a–b *Moutonithyris dutempleana* (d'Orbigny) (BGS GSM116480)
4 *Torynocrinus canon* Seeley (BMNH E51722)
5 *Neohibolites minimus* (Miller) (BGS GSM92759)
6a–b *Euhoplites microceras* Spath (BMNH C74989)
7a–b *Podoseris mammiliformis* Duncan (Fig. 7a (BGS GSM115037); Fig. 7b (BGS GSM115034))
8a–b *Euhoplites loricatus* Spath (BGS GSM87950)
9 *Bathrotomaria* sp. (BGS GSM115002)
10 *Inoceramus anglicus* Woods (BGS GSM58422)

(1a–b, 2a–b, 4 and 6a–b: photographs courtesy of Dr H G
Owen, The Natural History Museum, London)

sti, common *N.* cf. *ernsti*, *N. minimus* (Miller), common *N. oxycaudatus*, *N.* spp., cidarid radioles and *Holaster (Labrotaxis)* cf. *latis-
simus* Agassiz.

Bed 4: *Hysteroceras varicosum* Subzone.
Flucticularia cf. *sharpei* Ware, 'Rotularia' cf. *umbonata*, *Kingena
spinulosa*, *Moutonithyris dutempleana*, *Platythyris diversa rubicunda*
Cox & Middlemiss, *Birostrina* cf. *concentrica*, *Neohibolites ernsti*,
N. cf. *ernsti*, *N. ernsti-oxycaudatus* gr. and *N. minimus* (many
worn).

Bed 3, upper part: probably *Hysteroceras orbignyi* Subzone.
Moutonithyris dutempleana, *Terebratulina* cf. *martiniana* (d'Orbigny), *Birostrina* cf. *concentrica*, *B. sulcata* (Parkinson) (possible
fragments), *Turnus* sp., thick inoceramid chips of I. ?*anglicus*
gr., *Neohibolites minimus*, common *N. oxycaudatus*, *N.* spp., *Holaster (?Labrotaxis)* cf. *perezii* and *Holaster* sp.

Bed 3, middle part: *Hysteroceras orbignyi* and probably *Dipoloceras
cristatum* subzones.
Kingena spinulosa, *Moutonithyris dutempleana*, *M.* aff. *oroseina*,
common *Neohibolites minimus* and chips of *Inoceramus anglicus*.

Bed 3, bottom part: possible *Euhoplites nitidus*, *E. meandrinus*
and *Anahoplites intermedius* subzones.
Moutonithyris cf. *dutempleana* and *M. dutempleana* variant. Probable source of some ammonites preserved in ironstone in museum collections.

Bed 2: No fauna collected

Bed 1: *Hoplites dentatus* Zone.
Moutonithyris dutempleana, *M.* cf. *dutempleana*, *M.* aff. *dutempleana*, *Birostrina concentrica*, 'Ostrea' *papyracea* and *Neohibolites
minimus*.

Hunstanton to Sandringham

Between Hunstanton and Dersingham, the Red Chalk caps a
prominent feature and yields much distinctive soil brash. Southwards from Snettisham, the brash becomes predominantly pink
and cream coloured, but an abundance of nodular limestone fragments enclosing numerous belemnites, sand grains and small pebbles makes the boundaries of the formation easy to map. At Dersingham, within a distance of less than a kilometre between Hill
House Farm [693 309] and Doddshill [695 300], the Red Chalk
limestones pass laterally into red, pink and cream-coloured, highly
calcareous clays with nodules and thin bands of pink and cream
limestone in their upper part. These beds provide little soil brash
and their outcrop is largely covered by downwash.

The basal beds of the Red Chalk and the junction with the Carstone are exposed in the highest part of Snettisham Carstone quarry [6851 3496]. The section is deeply weathered and affected by cryoturbation: it appears to have representatives of Beds, 1, 2 and 3 of the Hunstanton section, with the same lithologies as, and similar thicknesses to those at Hunstanton. The full thickness of the Red Chalk was formerly exposed in a nearby Carstone quarry [6879 3474] at Norton Hill, Snettisham, but the section there is now completely degraded.

GAULT

The outcrop of the Gault in Norfolk can be traced from Sandringham southwards as far as West Dereham, at which locality the formation passes beneath the Recent deposits of Methwold Fens. It occupies low ground that is commonly covered by a veneer of sandy wash derived from the adjacent Chalk escarpment, with the result that the lower part of the formation, being farthest from the escarpment, is exposed only in the deeper drains and the higher part is very rarely exposed. In Victorian times the lower part of the formation was worked extensively at outcrop in the county, but on a small scale, for brickmaking and marl, and in the West Dereham area for agricultural phosphate. All the exposures are now degraded, and few specimens have been preserved. Continuous cores have been obtained from boreholes drilled close to the Gault outcrop at Gayton and Marham. Cores were obtained from the subcrop at Mundford and Great Ellingham.

At the most northerly point at which it can be recognised as a discrete argillaceous formation, in the Sandringham area, the Gault consists of about 2 m of pink and cream-coloured highly calcareous clay. This passes rapidly southwards into brown sandy clay and grey, cream-coloured and brown silty calcareous clays with thin beds of white and pink nodular chalky limestone, similar to parts of the Red Chalk, in their upper part. These limestones are present as far south as Gayton, but nowhere south of Sandringham do they form a major constituent of the Gault. Thin bands of white chalky limestone also occur at Bilney and Pentney, but these are at lower stratigraphical levels in the Gault.

The formation thickens steadily southwards from Sandringham and is 9.0 m thick at Gayton, 11.6 m at Marham and about 17.8 m at Hockwold. This apparently regular southward thickening is a small part of regional variation that is more complex, and as yet incompletely understood. The Gault thickens southward at outcrop in Suffolk and Cambridgeshire to about 25 m in the Soham Borehole [TL 5928 7448], about 30 m at Cambridge and about 48 m in the Duxford Borehole [TL 469 457]. It thins rapidly eastwards from its outcrop as it oversteps the Palaeozoic rocks of the London Platform. In the Norfolk subcrop, boreholes at Breckles, Mundford, Lexham, Norwich (Colman's Factory), Rocklands and Great Ellingham indicate the presence of a broad tract of ground lying between the London Platform (to the south) and the Red Chalk depositional area (to the north) in which the Gault was probably originally about

15 to 20 m thick (Figure 35), but where it has locally been reduced by erosion at the base of the Chalk.

These regional variations in thickness result from a number of interacting factors and their pattern will undoubtedly become more complex as more borehole data becomes available. The two most important factors affecting its preserved thickness in Norfolk are the attenuation of the whole of the Gault as it passes northwards in Norfolk into the Red Chalk, and variations in the amount of erosion that has occurred at the base of the Chalk.

At outcrop in Norfolk, the Gault can be divided into two parts on gross lithology, and these were termed the Lower and Upper Gault by De Rance (1868). The Lower Gault consists predominantly of medium and dark grey mudstones in which, according to analyses quoted by Perrin (1971, pp.140–149), illite and kaolinite are the dominant clay minerals. The Upper Gault consists mostly of pale grey mudstones with high calcium carbonate contents in which smectite is the dominant clay mineral (Perrin, 1971).

The Lower Gault and the lower part of the Upper Gault of Norfolk are made up of small-scale rhythms, mostly 1 to 2 m thick. Each rhythm begins with a medium or dark grey, shelly, pebbly, silty mudstone or muddy siltstone, rich in inoceramid prisms, oysters, belemnites, exhumed phosphatised burrowfills and water-worn phosphatic pebbles, that rests on a partially phosphatised and glauconitised, burrowed surface. This basal bed passes up into medium and pale grey calcareous mudstones by a decrease in the coarser clastic (including bioclastic) content and an increase in calcium carbonate. This lithological change is commonly accompanied by a decrease in faunal diversity and numbers. The upper part of the Upper Gault shows weakly developed rhythms, but is uniformly paler and more calcareous than the Lower Gault and contains fewer major erosion surfaces. Hill (1900, pp.330–335) recognised that much of the calcium carbonate in the Upper Gault was derived from coccoliths and this has been confirmed by later workers such as Black (1972–75).

De Rance (1868, 1875) subdivided the Folkestone type section of the Gault into 11 lithological units that could be distinguished by their faunas. Several of these units have erosion surfaces at their bases, and this largely explains why the boundaries of most of the units corresponded with ammonite zonal/subzonal boundaries for many years (e.g. Casey, 1966).

The main difficulty in attempting correlations within the Gault is that of recognising the bases of those individual rhythms that are erosive and which may locally cut out all or part of the underlying rhythms. A study of the Gault sequences proved in cored boreholes in East Anglia as part of the present work has shown that correlations can be made with confidence over relatively large distances by comparing the sequences of lithological and faunal events. These comparisons have enabled a standard sequence of 19 distinctive beds (G 1 to 19) to be devised that is applicable to the Gault of the central and northern parts of East Anglia (Figure 36).

The bases of 13 of these beds have been taken at erosion surfaces that mark the bases of complete or incomplete rhythms. In a nearshore area, such as that in which much of the Gault of East Anglia was probably deposited, these sedimentary breaks commonly coincide with faunal changes. The majority of the zonal/subzonal boundaries are thus coincident with bed boundaries in this region. The remaining 6 beds are so lithologically distinctive as to suggest that they are also related to widespread sedimentary events. All except one of the 19 beds have been shown to be persistent over large areas of East Anglia. Bed G 19, the youngest in the sequence, has been recorded only in the Gayton Borehole.

Local lithological variations are present in Norfolk, especially in the King's Lynn/Wash district. Thin beds of dark red to pink mudstone occur at several levels. The most intensely coloured of these, Bed G 4, can be traced southwards as far as Cambridgeshire, via the cored boreholes at Gayton, Marham, Mundford and Hockwold, its colour becoming fainter as the distance from the Red Chalk area of deposition increases. A less obvious lithological change takes place in the opposite direction. Close to the edge of the London Platform the Lower Gault is predominantly silty and grey; northwards from there the beds at any particular level become paler, more calcareous and less silty. In the Upper Gault, thin beds of very calcareous mudstone that are present in Beds G 14, G 16 and G 17 in Cambridgeshire pass northwards into laterally impersistent thin, muddy chalky limestones at Gayton (at the base of Bed G 14), Pentney (at the base of G 16) and Bilney (in G 17). The two younger beds were referred to as the 'Pentney Limestone' and the 'Bilney Limestone' respectively by Seeley (1861b).

The Gault contains a rich marine fauna dominated by bivalves: it includes relatively common solitary corals, serpulids, gastropods, scaphopods, ammonites, belemnites, crinoids and echinoids, together with an abundant microfauna of ostracods and foraminifera. Ichnofossils are represented by common *Chondrites*, possible thalassinoid burrows and a variety of trails. The modern zonal and

subzonal scheme for the Gault in England is based on ammonite assemblages from the work of Spath (1923a, 1923b, 1926), subsequently modified by Casey (1961b, 1966,) and Owen (1971, 1976). The ammonite assemblages can be supplemented by the use of other groups, notably belemnites (Spaeth, 1971, 1973), inoceramid bivalves, (Kauffman, 1976), ostracods (Wilkinson and Morter, 1982), foraminifera (Carter and Hart, 1977; Price, 1977) and coccoliths (Black, 1972–75). The zones and the subzones for that part of the Albian which is represented by the Gault in Norfolk have been determined by Mr Morter and are shown in Figure 36.

The recognition of zones and subzones is difficult in borehole cores in the Gault, despite the generally fossiliferous nature of the formation, because of the limited amount of material available from any particular horizon. Additionally in the present district, where the sequence is poorly exposed and attenuated in comparison with other parts of England, very few of the characteristic ammonite assemblages have been recognised at outcrop. The subzonal/zonal scheme has, however, been applied to the Gayton and Marham borehole sequences by comparing them with the standard faunal and lithological sequence worked out for East Anglia. A detailed description of the lithological and faunal sequence of the standard succession, bed by bed, is given elsewhere (Gallois and Morter, 1982).

Correlation of the most complete Gault sequences yet proved in Norfolk, those in the Gayton, Marham and Mundford C boreholes, with one another and with the Red Chalk sequence at Hunstanton is shown in Figure 37.

Details

Sandringham to River Nar

The lateral passage from the Red Chalk into the Gault is rapid and takes place in the vicinity of Home Farm [700 290],

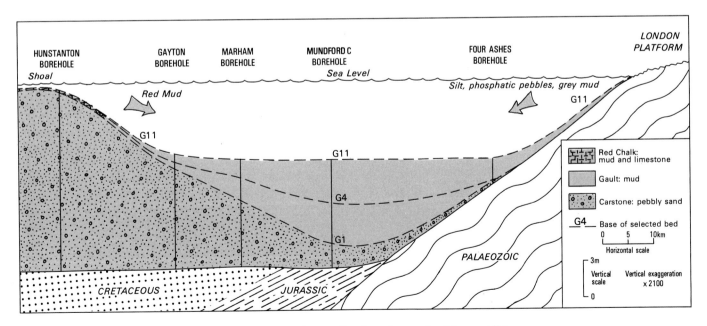

Figure 35 Presumed depositional environments immediately prior to the Upper Albian transgression.

Sandringham. There, the predominant Red Chalk lithologies of limestone and red clay pass into the Gault lithologies of brown sandy clay and grey, cream and brown silty clay. Between Sandringham and Flitcham, the Gault thickens rapidly southwards and the Carstone thins by a corresponding amount (Figure 37). Thin beds of pink and white, nodular, chalky limestone similar to those of the Red Chalk, continue to be present in the top of the Gault as far south as Gayton, but they form only a minor part of the Gault everywhere south of Sandringham.

Shallow pits [7001 2816 and 7008 2805] in cream and pink calcareous clays at York Cottage, Sandringham, are close to the northern limit of the Gault and are the most northerly recorded workings in the formation. These were probably dug for agricultural marl. Similar small marl pits are common on the Gault outcrop; other examples occur at Appleton House [7061 2745], Flitcham Hall [7168 2652] and in the intervening area [e.g. 7118 2698; 7128 2671]. The Gault succession at Flitcham Hall can be summarised, on the basis of augering and soil fragments, as follows:

	Thickness m
CHALK	
Hard nodular and shelly chalks	c. 1.0
GAULT	
Clay, white, cream and pink; highly calcareous, with bands of hard, nodular pink and cream chalky limestone	c. 1.5
Clay, brown, silty in part; belemnites common at several levels	c. 1.5
Clay, brown, sandy; passing down into Carstone	seen c. 0.5

Between Flitcham and Roydon, the Gault is overlain by drift deposits associated with the present and preglacial courses of the Babingley River. The basal beds of the formation form a small faulted outlier at Hall Farm, Roydon, but this is largely covered by drift deposits and there is no exposure. The main outcrop runs almost continuously southwards from Gorse Moor, Roydon via Roydon village, Grimston and Gayton to Gaytonthorpe Common, mostly as low ground with a veneer of pocketed sandy or gravelly wash. Throughout this area the formation can be roughly divided into two parts; a lower part which weathers to medium grey and brown clays and an upper, more calcareous part which weathers to pale and very pale grey clay with thin bands of soft argillaceous limestone. The outcrop is studded with numerous small, mostly water-filled pits that were probably dug for agricultural marl and subsequently served as a water supply. Over 70 such pits were recorded between Roydon and Gaytonthorpe Common.

The full thickness of the Gault crops out on the sides of a small hill capped by an outlier of Chalk at Gorse Moor, Roydon. Jukes-Browne and Hill (1887, p. 550) recorded the following sequence in a borehole drilled in the floor of an old chalk pit [7008 2426] near the top of the hill:

	Thickness m
CHALK	
Creamy white limestone; yellower towards base; exposed in pit	1.7
GAULT	
Tough grey argillaceous marl with belemnites	3.0
Yellowish and red-stained marly clay	0.5
Bluish grey clay becoming darker and sandy at base	2.1
CARSTONE	
Brown sand	seen 1.1

Comparison of this sequence with the local field evidence and the sequence proved in the Gayton Borehole (see below) suggests that part of the limestone allocated to the Chalk should be placed in the Gault (probably Bed G 19). The basal beds of the Gault and junction with the Carstone are exposed in a ditch section [7028 2442] on the north side of the hill. This shows brown sandy clays (Bed G 1) overlain by buff and grey mottled clays (Beds G 2 and G 3) and red mottled clays (Bed G 4). Old marl pits on the sides of the hill [7031 2436; 7010 2404; 7023 2406] are in brown and grey clays in the middle part of the formation.

A small but persistent feature, subsidiary to that formed by the basal beds of the Chalk, can be traced in the upper part of the Gault from Gorse Moor to Grimston Church. This is formed by pale grey, highly calcareous clays containing several thin (10 cm thick) impersistent bands of soft, argillaceous, chalky limestone. Some of these contain numerous *Neohibolites* and inoceramid bivalves, and rare ammonites indicative of the *cristatum* to *auritus* subzones (Beds G 14 to G 17).

Red and green mottled clays with abundant *Neohibolites minimus* (Bed G 4) overlain by a thin bed of fossiliferous soft muddy limestone with *N. minimus* and *Inoceramus concentricus* (?Bed G 5) crop out in the railway cutting [7024 2369 to 7029 2382] north-east of Roydon station. The following section in the basal beds of the Gault was recorded in a trench [7077 2278] close to the Three Horse Shoes Inn, Roydon:

		Thickness m
HEAD		
Sandy and clayey wash with scattered angular flints; some Carstone and Gault fragments; irregular base		1.2 to 1.5
GAULT		
Clay, brown and grey mottled, deeply weathered		0.6 to 0.9
Bed No.		
?G 5	Silt, pale grey, calcareous with numerous burrows infilled with dark grey, more clayey silt	0.15
G 4	Clay, irregularly interbedded dull reddish brown and grey; numerous *Neohibolites minimus*; scattered small white-coated spherical and tube-shaped phosphatic nodules	0.5
G 3	Clay, stiff, grey, with band of white-coated, brown-centred, phosphatic nodules at base, up to 5 cm across, many enclosing *Hoplites* spp. including *H. (H.) dentatus* (Dentatus Nodules)	0.9
G 1	Clay, dark grey, with lenses, wisps and burrow infillings of green Carstone-lithology sand	0.3
CARSTONE		
Sand, dark green, medium-grained, with small limonite-coated pebbles, and wisps and burrow infillings of dark grey clay		0.3

Grey and brown clays and sandy clays at the base of the formation were formerly worked for brickmaking at Pottrow [710 224] and adjacent to the site of a Roman villa [7185 2165] at Grimston. Spoil from these basal beds, notably sandy clays (Bed G 1), Dentatus Nodules (Bed G 3) and red-mottled clays (Bed G 4), is commonly thrown out from the deeper drainage ditches in the area.

A 12 cm-thick highly fossiliferous, soft argillaceous limestone crops out within a sequence of pale grey, highly calcareous clays in the bank of the Gaywood River [7212 2037] at Well

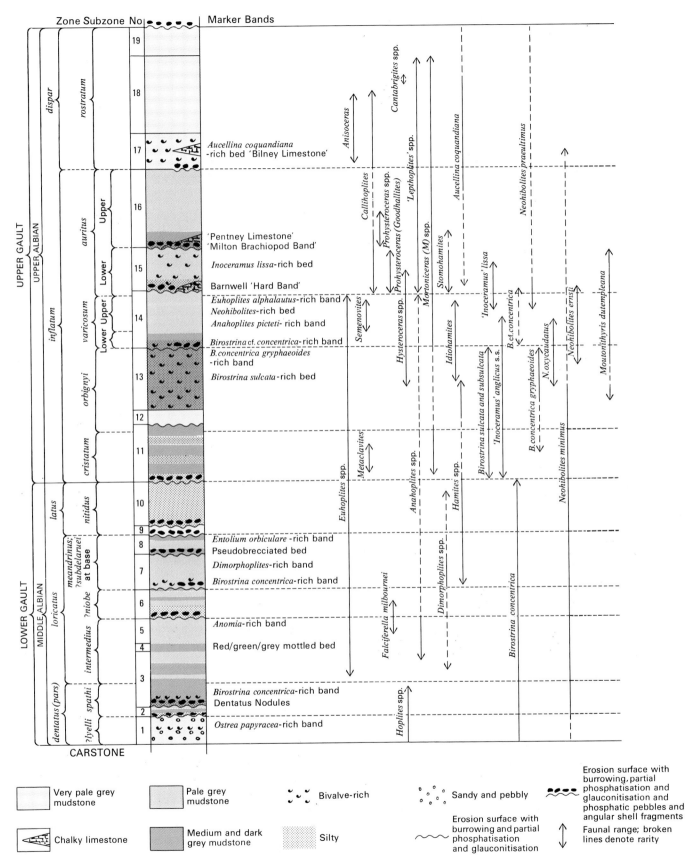

Figure 36 Generalised vertical section for the Gault of the district.

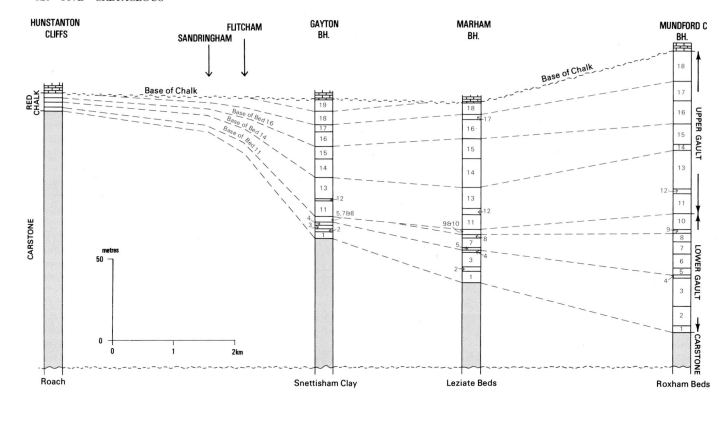

Figure 37   Correlation of Red Chalk and Gault sequences in west Norfolk.

Hall. The fauna of the limestone includes abundant *Euhoplites* and '*Inoceramus*', and common *Neohibolites* that correlate with Bed G 14 of the standard sequence.

The full thickness of the Gault was cored in the Gayton Borehole, where the following sequence was proved between 13.03 and 22.00 m:

	Thickness m	Depth m
CHALK (see p.135 for details) Hard bioturbated limestone with green-coated phosphatic pebbles at base		
GAULT		
Bed No.		
G 19 Limestone, soft, bioturbated, chalky, almost white but with disseminated pale yellow and orange staining; shelly, with common *Neohibolites* spp. including *N. praeultimus*, and *Aucellina coquandiana*; marked lithological change at base	0.51	13.54
Mudstone, soft, pale grey, highly calcareous; bioturbated throughout; sparsely shelly with *Neohibolites praeultimus* the only common fossil	0.25	13.79
G 17 and 18 Mudstone, soft, bioturbated; pale, medium and brownish grey, highly calcareous; band of soft, argillaceous, chalky limestone (0.11 m thick) in lowest part; shelly in lower part and		

at several levels in upper part, with abundant *Aucellina coquandiana* other fauna includes *Nielsenicrinus* aff. *cretaceus, Kingena spinulosa, Plicatula radiola gurgitis, Terebratulina martiniana, Entolium orbiculare* and *Neohibolites praeultimus*; passing down into

	1.30	15.09
G 16 Mudstone, soft, bioturbated, pale and dark grey; band of argillaceous chalky limestone (0.08 m thick) in lower part of bed; sparsely shelly, with *Aucellina coquandiana* the only common fossil; *Inoceramus* chips and fish debris also present; phosphatic pebble bed at base, with green-coated phosphatised mudstone pebbles up to 37 mm diameter, and with common brachiopods including *Moutonithyris dutempleana*	0.94	16.03
G 15 Mudstone, soft, bioturbated, pale and medium grey in upper part, reddish brown and pale green in middle part and silty medium grey in lower part; fossiliferous at several levels with *Isocrinus legeri, Nielsenicrinus cretaceus, Pycnodonte, Moutonithyris dutempleana* and other terebratulids, and *Inoceramus* chips; passing down into	0.81	16.84
G 14 Mudstone, soft, silty, bioturbated, medium grey; passing down into		

		Thickness m	Depth m

pale and medium grey bioturbated mudstones with band of pale grey, shelly, chalky limestone (0.10 m thick) in lower part of bed; belemnites, including *Neohibolites ernsti* and *N. minimus minimus* abundant in upper part of bed, and *Birostrina concentrica* abundant in chalky limestone and lower part of bed; other fauna includes *Monticlarella, Moutonithyris, Pycnodonte* and *Neohibolites oxycaudatus*; glauconitised surface at base — 1.17 — 18.01

G 13 Mudstone, soft, bioturbated, medium grey, silty; passing down into siltstone with much shell debris (0.3 m) and then into pale and medium grey bioturbated mudstones; shelly at several levels especially in top part of bed; *Birostrina sulcata* common throughout; other fauna includes *Nielsenicrinus cretaceus, ?Metaclavites, Plicatula minuta, Terebratulina martiniana, Birostrina concentrica gryphaeoides, Neohibolites minimus minimus* and in upper part of bed only, *Euhoplites* cf. *trapezoidalis, Hysteroceras* cf. *orbigny* and *Mortoniceras (Deiradoceras).* cf. *albense*; passing down into — 1.34 — 19.35

G 12 Limestone, very soft, and muddy, bioturbated, pale grey, with darker grey infillings of *Chondrites* and other burrows; sparsely fossiliferous — 0.16 — 19.51

G 11 Mudstone, soft, bioturbated, interbedded pale and medium grey; silty in lower part with band of medium grey siltstone at base; shelly in lower part, less so in upper; fauna includes *Isocrinus legeri, Nielsenicrinus cretaceus, Kingena spinulosa, Moutonithyris* sp. *Plicatula minuta, Terebratulina* cf. *martiniana,* `Anomia' sp., *Birostrina subsulcatus, B. concentrica gryphaeoides,* `Inoceramus' anglicus, Pseudolimea gaultina, Neohibolites minimus minimus* and cirripede valves; phosphatic pebble at base with rolled and worn belemnites resting on intensely burrowed surface — 1.04 — 20.55

G 5, 7 and 8 (condensed) Mudstone, soft, intensely bioturbated, pale green, grey and yellowish grey, with some streaks of dull reddish brown; burrow infillings of overlying silt extend down almost to base of bed; becoming very pale grey, highly calcareous in lower part; phosphatic pebble beds at two levels; rare limonite-coated sand grains (cf. Carstone) scattered throughout; sparse fauna with *Birostrina concentrica gryphaeoides* and

Neohibolites minimus minimus the only common fossils; passing down into — 0.33 — 20.88

G 4 Mudstone, soft, intensely bioturbated, pale green, dull reddish brown and grey, with rare sand grains; shelly in upper part with common *Birostrina*; other fauna includes *Kingena spinulosa, Pycnodonte (Phygraea)* aff. *vesicularis, Pseudolimea gaultina* and common *Neohibolites minimus minimus*; bed of green and red phosphatised mudstone pebbles at base resting on burrowed surface — 0.23 — 21.11

G 3 Mudstone, soft, bioturbated, medium grey; burrow infillings, wisps of shell debris and limonite-coated sand grains present throughout and becoming more abundant with depth; sparsely shelly with *Pycnodonte (Phygraea)* sp. and *Neohibolites* including *N. minimus minimus* and *N. minimus pinguis* the only common fossils; pebble bed at base with hard, black, phosphatic nodules, some enclosing *Hoplites* spp. including *H. (H.) dentatus* (Dentatus Nodules) — 0.25 — 21.36

G2 Mudstone, soft, medium grey, silty, with scattered grains and burrow concentrations of sand; shelly, with much shell dust and many larger fragments and whole shells; fauna includes *Moutonithyris dutempleana,* common *Birostrina concentrica, Neithea* aff. *quinquecostatus, Pseudolimea gaultina, Rastellum* sp. and *Hoplites (H.) dentatus*; plaster of *B. concentrica* at base resting on burrowed surface — 0.28 — 21.54

G 1 Mudstone, soft, intensely bioturbated, pale grey, with much limonite-coated sand and scattered small pebbles of quartz and rotted ironstone and, at one level, clasts of pale grey, soft mudstone; sparsely shelly except for concentration of *Birostrina concentrica* at one level; *Pycnodonte* sp., *Neohibolites minimus* and *Hoplites (H.)* sp. also present; passing down into — 0.46 — 22.00

CARSTONE (see p.110 for details)
Intensely bioturbated, clayey, pebbly, soft sandstone with large phosphatised burrow infillings

Between Gayton and Gaytonthorpe Common much of the Gault outcrop is covered by 1 to 2 m of gravelly sand wash. Parts of the formation are exposed from time to time in the deeper drains. One such section [7123 1766 to 7155 1770] near Wizard Lodge showed the following ascending sequence: pebbly sandy brown clay (Bed G 1), brown and grey sandy clays (Beds G 2 and G 3), purple and red mottled clay (Beds G 4 to G 8), pale grey calcareous clays (Beds G 11 to G 13), and with a thin band of *Birostrina*-rich muddy limestone (Bed G 14).

River Nar to River Little Ouse

The Gault is largely covered by drift deposits associated with the valleys of the River Nar and its tributary, the Gaytonthorpe stream, between Gaytonthorpe Common and Marham. Relatively drift-free areas of outcrop occur at East Walton, West Bilney, Pentney and near Narborough, but these are all on low, poorly drained ground and there is no exposure.

Whitaker and Jukes-Browne (1899, p.24) noted the basal beds of the Gault, probably beds G 1 to G 5 from their description, beneath pocketed sandy wash in a ditch [approx. 711 123] at Pentney Hall (now Ashwood Lodge). Rose (1835, p.180) had earlier recorded thin (0.15 m thick) shelly limestones in the Gault in boreholes at Pentney and West Bilney (subsequently the 'Pentney limestone' and the 'Bilney limestone' of Seeley, 1861b, p.236). Mr Morter has correlated the faunas of the Pentney limestone with that in Bed G 16 in the Gayton Borehole, and that of the Bilney limestone with that in Bed G 17.

The Gault crops out on low, but relatively drift-free ground between Marham and Shouldham. As elsewhere, the outcrop is characterised by a large number of small pits dug for marl. The lower part of the formation was exposed in a drain [6951 0978 to 6985 0978], about 2.7 m deep, adjacent to Button Fen, Marham. The following composite section was measured there:

	Thickness m
PLEISTOCENE	
Cryoturbated mixture of flint-rich sand, Gault-rich till and soliflucted Gault clay; very irregular cryoturbated contact with	0.5 to 1.5
GAULT	
Bed No.	
Clay, structureless, deeply weathered, mottled pale yellowish brown and pale and medium grey; a few phosphatic pebbles, *Neohibolites* and inoceramid fragments	c. 2.0
?G 11 Clay, deeply weathered, yellowish brown and medium grey, silty; *Neohibolites* and inoceramid fragments at several levels; band of widely spaced concentrations of small phosphatic pebbles and concretions with thick chalky cortices and hard coffee-coloured centres; many bored and striated by ?grazing animals	0.50
G 8 and 9 (condensed) Clay, pale grey with some dull red mottling, weathering to pale yellow	0.15
?G 5 and 7 (condensed) Clay, medium grey	0.25
G 4 Clay, mottled brick red and medium grey weathering to mottled yellowish brown with red streaks; widely spaced small- and medium-sized phosphatic concretions, mostly burrow infillings, with red-stained cortices and greenish grey centres	0.25
G 3 Clay, medium and dark grey, with a few wisps and burrow concentrations of Carstone-lithology sand; bed of closely spaced phosphatic nodules and pebbles at base with white cortices and coffee coloured centres; many enclose almost whole *Hoplites* spp., including *H. (H.) dentatus*, that fall apart at the sutures on extraction; some enclose *Birostrina* and pectinids; pebble bed rests on very shelly surface with rotted ghosts of numerous *Birostrina* and oysters, *Entolium* and concentrations of sand	0.15
G 2 Clay, medium grey, sandy, becoming progressively more so with depth; band of	

	Thickness m
closely spaced, small rounded, coffee-coloured with white cortices, sandy dense phosphatic nodules and pebbles at base	0.35
G 1 Clay, brownish grey, very sandy becoming more so with depth; small quartz and ironstone pebbles common in lower part; intensely bioturbated; band of closely spaced, small, rounded, sandy, phosphatic nodules (as Bed G 2) in middle part of bed; minor lithological change at base accompanied by burrowing	0.75
CARSTONE	
Sand, very clayey, finely pebbly; greyish and brownish green and dark greenish grey when unweathered, rusty brown when weathered; fine- and medium-grained, with dark grey burrow infillings; band of medium (up to 0.05 m long), pale brown with white cortices, sandy and pebbly phosphatic nodules, mostly in-situ phosphatisation of tube-shaped vertical burrows, at top of bed; band of larger (up to 0.1 m) nodules of same lithology occurs 0.45 m below top of bed	seen 1.0

The lower part of this sequence, the Carstone and Beds G 1 to G 4, was also exposed in a nearby ditch [6886 0935 to 6885 0921] at Abbey Farm, Shouldham. There, the Gault is considerably disturbed by cryoturbation and although the main marker bands are present, the bed thicknesses are much more variable.

The full thickness of the Gault was cored between 33.43 and 45.03 m in the Marham Borehole. This sequence is thicker, 11.6 m compared to 9.0 m, than that proved in the Gayton Borehole, but is lithologically and faunally similar, with the exception that the highest part of the Gault at Marham is less calcareous and contains fewer muddy limestone bands. Correlation of the two sequences is shown in Figure 37.

Between Shouldham and Crimplesham, the Gault outcrop is overlain by thick Pleistocene deposits. The formation has a broad outcrop in the Crimplesham, Ryston and West Dereham areas before passing southwards beneath the recent deposits of Methwold Fens. With the exception of very small inliers at Stubb's Hill and Shrubhill, the latter close to the county boundary, the Gault has no outcrop in Norfolk south of West Dereham.

The lower part of the Gault, overlain by glacial deposits, forms a small outlier at Ryston and a narrow outcrop between Ryston and Crimplesham. Up to 3 m of these beds were formerly worked for brickmaking in pits [6335 0139 (Brick Kiln Wood); 6350 0127] at Home Farm, Ryston. The basal beds and the junction with the Carstone are commonly exposed in ditches [e.g. 6371 0245; 6392 0269] between Ryston and Crimplesham, but are much disturbed by glacial action and cryoturbation and no accurate thickness measurement is possible. The sandy phosphatic nodules at the top of the Carstone and the Dentatus Nodules form useful marker bands. The red mottled clays (Bed G 4) that are so obvious in more northerly areas have not been recorded at outcrop in this area.

The basal beds of the Gault were well exposed in the West Dereham area in Victorian times in several large shallow pits dug for agricultural phosphate. No exposure has survived, but it is clear from the descriptions of Teall (1875), Keeping (1883) and Jukes-Browne (MS quoted in Whitaker et al., 1893, p.13) that the main source of phosphate was the large sandy nodules at the top of the Carstone. The Dentatus Nodules and other phosphatic nodules and pebbles in the basal 3 m of Gault, usually described as 'pale marly clay', were also worked.

CHALK

The Lower Chalk and the lower part of the Middle Chalk form a low but impressive escarpment which marks the eastern edge of the King's Lynn district from Hunstanton to the River Nar. The face and crest of this escarpment are largely free from wash and other drift deposits, in contrast to much of the low ground at its foot, and it is capped by the hard nodular chalks of the Melbourn Rock. The latter, the base of which marks the junction of the Lower and Middle Chalk, provides a good mapping horizon, and it was taken to define the eastern limit of the area surveyed. The Lower Chalk and the Melbourn Rock were formerly well exposed in Chalk pits in the district, but most of these have become degraded or have now been backfilled. The stratigraphy of the Chalk of the region and the key sections were described by Peake and Hancock (1961). Parts of the sequence are well exposed in Hunstanton Cliffs and in working pits at Hillington, Gayton and Marham (Wisbech district), and cored boreholes were drilled at these three localities as part of the present survey to enable the complete sequences to be determined there.

The Chalk of the King's Lynn district straddles the junction of the 'northern' and 'southern' faunal and lithological provinces of the English Chalk, and includes features common to both. The northern succession has much in common with correlatives in Germany, and the southern succession with the sequences in the Anglo-Paris Basin. Between the River Wissey and the River Nar the Chalk thins steadily northwards and consists largely of southern province rhythms of grey marly chalks and harder white chalks (the Chalk Marl of southern England), overlain by more uniform grey and white chalks (the Grey Chalk of southern England). The sequence continues to thin northwards from the River Nar until, in the vicinity of Heacham, it is thinner than at any locality elsewhere in eastern England. Between the River Nar and Hunstanton the Lower Chalk sequence can be closely matched with the northern province sequence described by Jeans (1980): many of the marker bands shown in the generalised vertical section (Figure 38) can be matched with the northern sequence and some take their names from localities in the Lincolnshire Wolds. It should be noted, however, that the distinction between the southern and northern provinces cannot be clearly defined and that the main marker bands, including the Inoceramus Beds, Totternhoe Stone, Nettleton Stone, Plenus Marls and Melbourn Rock can be traced from Lincolnshire throughout the present district and the remainder of Norfolk into Cambridgeshire without any break or marked change in lithology.

The major lithological differences occur in the overall nature of the chalks between these marker bands. Thus, although the Totternhoe Stone divides the Lower Chalk in the present district into two parts which stratigraphically roughly correspond with the Chalk Marl and Grey Chalk of southern England, the argillaceous rhythms and sponge-bearing (commonly ammonite-rich) hard chalk bands so characteristic of the southern chalk are absent. The equivalent beds in the district are hard porcellanous and gritty chalks, the Paradoxica Bed and the Inoceramus Beds, and nodular chalks with a few wisps of marl. Above the Totternhoe Stone the faintly rhythmic chalks of the Grey Chalk are represented in the King's Lynn district by a much thinner sequence of mostly smooth-textured chalks with several prominent thin marl seams and common marl wisps. At the top of the Lower Chalk, the Plenus Marls are much attenuated in the present district in comparison with their southern equivalents.

Much of the Chalk is composed of low-magnesium (less than 0.5 per cent Mg) calcium carbonate derived mostly from coccolithic algae (as disaggregated calcite plates, coccoliths and coccospheres), with foraminifera, bivalve shells especially Inoceramus, calcispheres, echinoderm plates and bryozoan fragments forming important constituents at some levels. The clay-mineral content of the Chalk varies from less than 1 per cent in the purest chalks up to 30 per cent in the more marl-rich layers. Jeans (1968) has described the clay minerals in the Lower Chalk of Cambridgeshire and Norfolk as composed largely of smectite and illite, and Harrison et al. (1979) have suggested that the smectite may be partly of wind-borne volcanic origin. This suggestion is supported by the fact that the thicker marl seams are laterally persistent and can be used as stratigraphical marker bands over very wide areas. However, Wray and Gale (1992) consider the bulk of the clay minerals in the marls in the Middle Chalk (Turonian) in southern Britain to be fluvial in origin, having been deposited at times of low sea level. Jefferies (1963) had previously suggested the same origin for the Plenus Marls. Small amounts, almost always less than 1 per cent, of clay- and fine-silt-grade quartz are present at some horizons and this too is probably wind-borne in origin. Phosphatic minerals, in the form of secondary phosphatisation and pebbles of phosphatised chalk, are conspicuous at some levels, notably in association with the erosion surfaces which mark the bases of the Paradoxica Bed, Inoceramus Beds and Totternhoe Stone. Glauconite occurs as coatings of pebbles and hardened chalk surfaces in the same situation, but more sparingly. No flint was recorded in the Lower Chalk or Melbourn Rock in the district.

The fauna of the Chalk shows that it was deposited in a sea of normal salinity. In addition to the comminuted shell debris which occurs in the matrix, ammonites, bivalves, brachiopods and echinoids are common at some horizons. Hancock (1976) suggested on the basis of the faunal and petrological data, that the purer chalks were deposited in between 100 and 600 m of water, but Burnaby (1962) using benthonic foraminifera and Kennedy (1970) using gastropods have suggested that parts of the Lower Chalk were deposited at depths of less than 50 m.

In the King's Lynn district the presence of angular shell debris, pebbles and intense bioturbation in the Paradoxica Bed, Inoceramus Beds and Totternhoe Stone suggest that they were deposited at shallower depths and in higher-energy environments than those of the surrounding chalks. Each of these beds has an erosive base, and there is evidence to suggest that the currents which deposited the Totternhoe Stone locally scoured large channels in the underlying chalk. Rhythmic sedimentation is so widespread in the Chalk, especially in the Lower Chalk where it is characterised by variations in the clay mineral/carbonate ratios, that Hancock and Kauffman (1979) have suggested that this resulted from worldwide eustatic changes in sea level. The erosive beds within the Lower Chalk in the district may therefore represent larger sea-level changes superimposed on the background rhythms. The steady overall upward diminution of coarser material in the Lower Chalk and the upward disappearance of siliciclastic material probably reflect gradual submersion and erosion of the land areas that supplied these materials. Gale (1989) has suggested that small-scale rhythmic sedimentation in the Lower Chalk is related to climatic cycles caused by variations (Milankovitch cycles) in the orbit of the sun with periodicities of about 20 000 and 100 000 years.

Lower Chalk

The Lower Chalk of the district is an estimated 14 m thick at Heacham. It thickens slowly northwards from there to about 16 m thick at Hunstanton, and then thickens more quickly beneath The Wash to more than 50 m where it re-emerges in the Lincolnshire Wolds. Southwards from Heacham it thickens steadily to 24 m in the Gayton/Hillington area and, in the Wisbech district, to about 30 m at Marham, much of the thickening occurring in the beds below the Totternhoe Stone (Figure 38).

Several prominent marker bands persist in the Lower Chalk throughout the district, and these can be readily used to correlate the short sections exposed in the chalk pits in the area (Figure 38). Detailed correlations, for example between individual thin marl seams, marl wisps or fossil bands, are often hard to make because of the poor state of weathering of many of these old sections, combined with minor glaciotectonic activity (usually seen as small displacements on many joints and bedding planes). The Lower Chalk sequence can be summarised as follows.

PARADOXICA BED

The Paradoxica Bed, so named because of its ramifying network of *Thalassinoides paradoxicus* (Woodward) burrows (Plate 13) which were formerly referred to the sponge '*Spongia paradoxica*' but which are now known to be crustacean (callianassid) burrows, consists of intensely bioturbated, hard, porcellanous, chalky limestone with phosphatised Red Chalk and chalk pebbles and possible stromatolite columns at its base, resting on an irregular surface cut in the Red Chalk. Jeans (1980) has described the contact with the Red Chalk, the basal bed of the Paradoxica Bed, and the complex sequence of early lithification processes that affected both horizons. The Paradoxica Bed represents a highly condensed sequence not unlike the Red Chalk, and contains within it several erosion surfaces (hardgrounds) which rest on lithified calcitised and in part glauconitised chalks (chalkstones). Because of its hardness it is persistent as field brash throughout the district and can be readily mapped out. It is 0.43 m thick in Hunstanton cliffs, and probably 0.2 to 0.5 m everywhere else in the district.

At Hunstanton, the Paradoxica Bed can be divided into a lower and an upper part, separated by a glauconitised erosion surface, throughout most of the southern and middle parts of the cliff section. At the northern end, at St Edmund's Point [677 424], the upper bed is either absent or is represented only by remnants welded to the base of the overlying Inoceramus Bed. The lower part of the Paradoxica Bed has yielded unidentified sponges, *Capillithyris woodi* Middlemiss, *Concinnithyris* aff. *subundata* (J Sowerby), *Kingena* cf. *arenosa* (d'Archiac), *Monticlarella carteri* (Davidson), *Moutonithyris* cf. *obtusa* (J de C Sowerby), *Nerthebrochus* cf. *robertoni* (d'Archiac), *Ornatothyris* cf. *pentagonalis* Sahni, *Terebratulina etheridgei* Owen, *Aucellina gryphaeoides* (J de C Sowerby), *A. krasnopolskii* (Pavlov), *A. uerpmanni* Polutoff and *Neohibolites ultimus* (d'Orbigny).

There is disagreement among specialists regarding the determination of the (predominantly small) terebratulids from the Paradoxica Bed. They are difficult to develop from the hard matrix, and it is unclear whether the small specimens represent juveniles, small ecomorphs or small species. They may include species described from the Cenomanian Tourtia of Belgium by d'Archiac (1847). The comparatively well-preserved brachiopod fauna that occurs in a ferruginous marl at the extreme base of the Paradoxica Bed is probably significantly different from that characterising the main mass of the unit, and probably does include some Tourtia species.

There are numerous additional records of fossils from the 'Chalk Basement Bed' in Whitaker and Jukes-Browne (1899, pp.57–61) based on collections made by H G Fordham; it is not known whether this material is still extant. Of particular note are the species recorded as '*Holaster laevis, H. subglobosus, Serpula annulata, Serpula umbonata, Ostrea vesicularis, Pecten orbicularis, Plicatula inflata, Spondylus striatus* and *Ammonites* spp'.

The upper part of the Paradoxica Bed contains large, biplicate terebratulids (*Tropeothyris?*) and '*Inoceramus*' ex gr. *crippsi?* Mantell; Dr A B Smith has recorded (unpublished MS) *Calliderma smithiae* (Forbes) and *Hirudocidaris uniformis essenensis* (Schlüter) from the middle and upper part of the Paradoxica Bed.

The co-occurrence of *Neohibolites ultimus* and *Aucellina* spp. in the lower part of the Paradoxica Bed represents the eastern England expression of the *ultimus/Aucellina* event recognised by Ernst et al. (1983) in northern Germany. In East Anglia, the Paradoxica Bed expands southwards into the Porcellaneous Beds of south west Norfolk and Cambridgeshire (Morter and Wood, 1983; Gallois, 1988, p.60) which yield *Aucellina* spp., '*Inoceramus*' ex gr. *anglicus* Woods (probably *I. comancheanus* Cragin) and poorly preserved ammonites including heteromorphs. The Paradoxica Bed and the Porcellaneous Beds can be assigned to the 'lower part' (*sensu* Gale and Friedrich, 1989) of the *Neostlingoceras carcitanense* Subzone, although the eponymous subzonal ammonite has not so far been recorded from eastern England.

BEDS BETWEEN THE LOWER INOCERAMUS BED AND
TOTTERNHOE STONE

Both the Lower Inoceramus Bed and Upper Inoceramus Bed consist of grey, gritty, bioturbated, shelly chalks with

Figure 38 Generalised vertical section of the Lower Chalk.

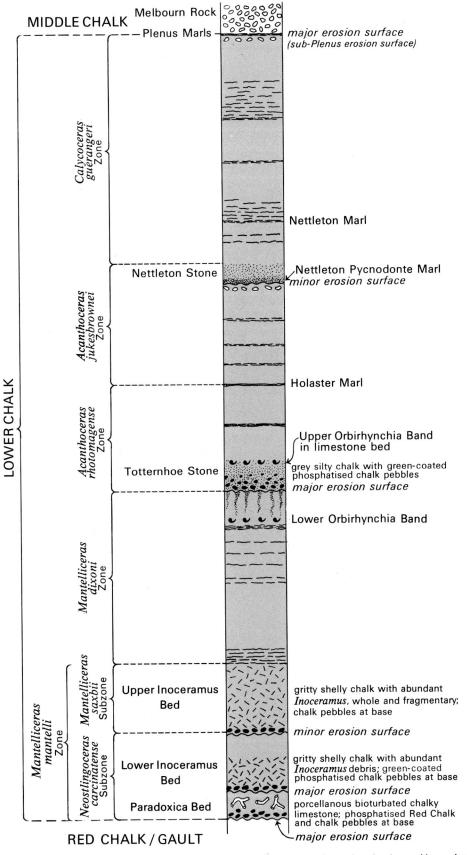

Not to scale: see text for details of thicknesses and discussion of zonal and subzonal boundaries

Key as for Figure 39

Plate 13 Burrows in the Paradoxica Bed (basal Chalk), Hunstanton Cliffs .

The basal bed of the Chalk is a dense, splintery pale pink and cream limestone with a ramifying network of branching and intertwining burrows. These latter were formerly thought to be casts of a sponge, '*Spongia paradoxica*' after which the bed is named, but are now known to be the feeding and dwelling burrows of small callianassid crustaceans. (A 10848).

many whole and fragmentary '*Inoceramus*', the gritty texture being due to abundant '*Inoceramus*' prisms. Both beds rest on erosion surfaces and both contain green-coated, glauconitised and phosphatised chalk pebbles in their lowest part. In the present district these pebbles are larger and more abundant in the lower bed, which here overlies a major break in sedimentation. The pebbles are common as field brash throughout the district, and are especially helpful in the Gayton area where they are readily distinguishable from soft chalky limestones in the highest part of the Gault (Bed G 19). The Lower Inoceramus Bed is 0.46 m thick in Hunstanton Cliffs.

The Lower Inoceramus Bed contains a rich and diverse fauna which includes *Parsimonia antiqua* J de C Sowerby, *Serpula* aff. *umbonata* (J Sowerby), *Capillithyris woodi*, *Grasirhynchia grasiana* (d'Orbigny), *Monticlarella? brevirostris* (Roemer), *Nerthebrochus?* sp., *Ornatothyris pentago-*

nalis, O. cf. *rara* Sahni, *Terebratulina?* sp., *Eopecten* sp., '*Inoceramus*' *crippsi crippsi* Mantell [abundant and rock forming], *Limaria* cf. *elongata echinata* (Etheridge), *Plagiostoma globosum* J de C Sowerby, *Plicatula inflata* J de C Sowerby, *Rastellum colubrinum* (Lamarck), *Spondylus* sp., *Neohibolites ultimus* and *Sciponoceras* sp. Dr A B Smith (unpublished MS) has recorded *Glyptocyphus difficilis* (Agassiz), *Hirudocidaris uniformis essenensis, Holaster* sp. nov. [common at base], *H. nodulosus* (Goldfuss), *Hemiaster griepenkerli, Tetragramma variolare* (Brongniart), and *Callidermasmithiae*.

The entry of abundant '*Inoceramus*' *crippsi crippsi* at the base of the Lower Inoceramus Bed is an inter-regional bioevent that permits correlation of this unit with the beds assigned to the informally named 'higher part' of the *Neostlingoceras carcitanense* Subzone in the Folkestone standard succession (Gale and Friedrich, 1989). At Folkestone, these '*Inoceramus*' *crippsi*-rich beds are characterised by *Tropeothyris? carteri* (Davidson), *Rastellum colubrinum*, *Hypoturrilites*, inflated *Mantelliceras*, (common), *Schloenbachia varians* (J Sowerby) and *Sharpeiceras* spp. The correlation is made between the condensed eastern England succession and the basinal Anglo-Paris Basin succession, not with the succession in Cambridgeshire.

The Upper Inoceramus Bed, 1.07 m thick at Hunstanton, contains abundant 'Inoceramus'; there are few records of other fossils. The specimen [British Museum C29622] of *Mantelliceras cantianum* Spath (cited by Kennedy, 1971, p.57) may be from this bed; *Calliderma smithiae*, *Hirudocidaris uniformis essenensis* and *Tetragramma variolare* have also been recorded. The inoceramids are predominantly thin-shelled forms, and may include '*Inoceramus*' *reachensis* Etheridge and '*I.*' *crippsi hoppenstedtensis* Tröger. Jeans (1968) recorded a stratigraphically important horizon, which can be traced into Lincolnshire, Humberside and Yorkshire, rich in large, poorly preserved turrilitoid ammonites (*Hypoturrilites?*) that rest on a glauconitised surface in the lower part of the bed. Tentative correlation with successions in southern England suggests that the Upper Inoceramus Bed may belong with the *Mantelliceras saxbii* Subzone of the *mantelli* Zone.

The beds between the Upper Inoceramus Bed and the Totternhoe Stone consist of off-white and cream-coloured, mostly slightly marly chalks of variable hardness, with marl wisps and stylolitic surfaces at many levels. Fossils are generally scarce, although some horizons are gritty due to *Inoceramus* debris. Scattered *Inoceramus* ex gr. *virgatus* Schlüter occur in marly and nodular chalks that overlie the Upper Inoceramus Bed at Hunstanton; they are indicative of the lower part of the *Mantelliceras dixoni* Zone. An horizon of nodular chalk with common '*I.*' ex gr. *virgatus* is overlain by fossiliferous marly chalks rich in *Orbirhynchia mantelliana* (J de C Sowerby), close below the Totternhoe Stone. These marly chalks, which also contain *Limaria* sp., *Temnocidaris dissimilis* (Woodward), *Tetragramma variolare*, *Tiaromma michelini* (Agassiz) and *Hyposalenia clathrata* (Woodward) form an important marker band which has been traced throughout much of eastern England (the Lower Orbirhynchia Band of Jeans (1980)). Similar bands with *Orbirhynchia* occur high in the Chalk Marl in southern England. The Lower Orbirhynchia Band of Jeans probably correlates with the lowest of three *O. mantelliana* bands recognised by Gale (1989) in the Lower Chalk at Folkestone, within the higher part of the *Mantelliceras dixoni* Zone of the Lower Cenomanian.

TOTTERNHOE STONE TO NETTLETON STONE

The Totternhoe Stone is the most conspicuous marker band in the Lower Chalk in the district: it is well exposed in Hunstanton Cliffs and forms a prominent bench in many of the chalk pits described below. It consists of silty, intensely bioturbated grey chalk (about 0.7 m thick at Hunstanton) with pebbles of hard phosphatised, calcitised and glauconitised chalk in its lower part, that rests erosively on the underlying beds. Many of the pebbles in the Totternhoe Stone have bored surfaces. There is evidence at Gayton and elsewhere in East Anglia to suggest that it locally occupies channels which have cut out much of the underlying chalk sequence (Figure 39). The Totternhoe Stone typically rests on chalks that have not been hardened, and burrows infilled with grey gritty Totternhoe Stone matrix extend down 0.5 to 1.0 m into the underlying chalk.

Well-preserved fossils are difficult to collect in situ in the Totternhoe Stone at Hunstanton, but its lithology is so distinctive that material in fallen blocks and in collections can readily be identified as being from that horizon. It has yielded *Acanthoceras* and *Schloenbachia* indicative of the *Turrilites costatus* Subzone of the *Acanthoceras rhotomagense* Zone, together with the belemnites *Actinocamax primus* Arkhangelsky and *Belemnocamax boweri* (Crick) and a brachiopod assemblage which includes *Modestella geinitzi* (Schloenbach), indicative of the Middle Cenomanian. In addition, the following fauna has been recorded (by Messrs Morter and Wood during the present survey and by Drs A S Gale and A B Smith and Mr D C Ward (unpublished MSS)) from the Totternhoe Stone of the district, mostly from Hunstanton: *Micrabacia coronula?* Goldfuss, *Onchotrochus serpentinus* Duncan, *Glandifera rustica* J Sowerby, *Hamulus*, *Serpula umbonata*, *Capillithyris squamosa* (Mantell), *Concinnithyris*, *Grasirhynchia martini* (Mantell), *Kingena concinna* Owen, *Modestella geinitzi*, *Terebratulina* aff. *imbricata* Owen, *T. protostriatula* Owen, unidentified gastropods, *Entolium* sp. [common], *Euthymipecten beaveri* (J Sowerby), indet. inoceramid fragments, *Neithea sexcostata* (S Woodward), *Oxytoma seminudum* Dames [common], *Limaria* sp., *Cretiscalpellum glabrum* (Roemer), *Calliderma smithiae*, *Crateraster* sp., *Metopaster* sp., '*Nymphaster*' *radiatus*, *Holaster subglobosus* (Leske), *Hyposalenia clathrata* (Woodward), *Salenia petalifera* (Desmarest), *Temnocidaris dissimilis* (Woodward), *Tetragramma variolare*, *Tiaromma michelini*, *Hexanchus*, *Scapanorhynchus*, *Squalicorax falcatus* (Agassiz), *Squalus* sp. and *Synechodus* sp.,

The fauna is similar to that of the 'Cast Bed' of the Folkestone succession, close to the base of the Middle Cenomanian. The incoming of the belemnites permits correlation with the widespread 'primus event' of Germany (Ernst et al., 1983).

In the present district, a band of hard chalky limestone immediately above the Totternhoe Stone is rich in *Austiniceras*, together with *Orbirhynchia mantelliana* (the Upper Orbirhynchia Band of Jeans (1980)). This last is the correlative of the highest of the three *O. mantelliana* bands in southern England, where it is overlain by a widespread mid-Cenomanian non-sequence.

The chalk above the Upper Orbirhynchia Band is hard, off-white and slightly marly, with wisps and thin seams of greenish grey and olive-green marl at several levels. Some of these can be correlated over short distances within the district (Figure 38), but there are too few good exposures and too many correlation uncertainties due to joint movements to demonstrate their lateral persistence. At Heacham and Snettisham the most promising of these potential marker bands, referred to as the Holaster Marl in Figure 38, is rich in *Holaster* and '*Inoceramus*' ex gr. *atlanticus* (Heinz) indicative of the *Acanthoceras jukesbrownei* Zone.

NETTLETON STONE TO PLENUS MARLS

The Nettleton Stone consists of hard and very hard, cream-coloured chalk with patches of pinkish brown phosphatisation in its upper part; grey and brownish grey, slightly gritty, marly chalk in its middle part; and brownish grey marl with common *Pycnodonte* in its lowest part (the Nettleton Pycnodonte Marl). This last rests on an irregular surface of hardened nodular chalk. The similarity of the lithology and its relationship to the underlying beds has resulted in the Nettleton Stone being commonly mistaken for the Totternhoe Stone in the past. However, the presence of a hardened surface and the absence of burrows that extend down into the underlying chalk, are characteristic of the Nettleton Stone. The Nettleton Stone has been shown to be laterally persistent at outcrop throughout Lincolnshire (Jeans, 1980) and Norfolk. It has yielded rare *Acanthoceras*, which probably indicate that it lies within the *A. jukesbrownei* Zone. The Nettleton Pycnodonte Marl yields '*Inoceramus*' ex gr. *atlanticus* and can be correlated with an influx of *Pycnodonte* in the Lower Chalk of northern Germany (the Pycnodonte Event of Ernst et al., 1983) that has been dated as *jukesbrownei* Zone age.

The beds above the Nettleton Stone consist of alternations of hard off-white and very hard, cream-coloured, slightly marly chalks with marl wisps and thin seams at several levels. The Nettleton Marl forms a conspicuous marker band throughout the district, a little above the Nettleton Stone (Figure 38). In weathered sections, the marl at the base of the Nettleton Stone could easily be mistaken for the Nettleton Marl were it not for the presence of common *Pynodonte* in the former.

A persistent horizon of nodular chalk a few centimetres below the Nettleton Marl, and the horizons adjacent to it, yield *Orbirhynchia* sp. nov. Another brachiopod-rich band occurs in pockets in the nodular chalks with green-coated surfaces immediately beneath the erosion surface at the base of Plenus Marls: this band contains *Ornatothyris rara* Sahni and *O. latissima* Sahni. Ammonites are very rare in the beds between the Nettleton Stone and the Plenus Marls in the district. However, the non-ammonite fauna, typically oysters and thin-tested echinoids such as *Holaster trecensis* Leymerie, *Camerogalerus cylindricus* (Lamarck) and *Tylocidaris asperula* (Roemer), can be matched with that of the *Calyoceras guerangeri* Zone of southern England.

The Plenus Marls are represented in the district by only a few centimetres of marl and marly 'smooth griotte' chalk (*sensu* Mortimore, 1979) which rests on hardened, nodular chalk with green-stained surfaces. These beds have been well exposed in chalk pits at Barrett Ringstead, Heacham, West Newton and Hillington where they are largely protected from weathering by the overlying Melbourn Rock. At many localities, however, they have been subjected to tectonic and glaciotectonic shearing and are now reduced to wisps or thin seams of rusty brown clay.

Jefferies (1963) in a detailed study of the Plenus Marls throughout the Anglo-Paris Basin, divided the marls into eight beds which can be distinguished by a combination of lithological and faunal features throughout much of the depositional basin. In the present district, where the sequence is greatly condensed, few of the eight beds can be recognised with confidence. Tentative correlations be-

tween the three best sections in the district with well-exposed sections at Welton, Lincolnshire [255 868] and Marham, south-west Norfolk [7051 0803], are shown in Figure 40. The sub-Plenus erosion surface is well developed throughout the district. Jefferies (1963) recorded *Entolium membranaceum* (Nilsson) and *Plicatula barroisi* Peron at Hillington, and large *Ostrea vesicularis* Lamarck at Hillington and Heacham. The distinctive *Actinocamax plenus* (Blainville) was not recorded, which suggested to Jefferies that the lower part of the Plenus Marls of the standard sequence might be absent. He recorded a much richer fauna from the more complete sequence at Marham, including '*Discoidea cylindrica* (Lamarck), *Offaster sphaericus* Schlüter, *Concinnithyris burhamensis* Sahni, *C. subundata* (Sowerby), *Orbirhynchia wiesti* (Quenstedt), *Ornatothyris* sp., '*Rhychonella*' *lineolata* Phillips *carteri* Davidson and large *Ostrea vesicularis* Lamarck', indicative of beds 1, 3 and 8 of the standard sequence with possible representatives of Beds 4, 5, 6 and 7. Where better developed elsewhere in southern England, the Plenus Marls yield ammonites indicative of the *Metoicoceras geslinianum* Zone.

Middle Chalk

The basal bed of the Middle Chalk, the Melbourn Rock, caps a prominent feature and was taken to mark the eastern limit of the area surveyed. The crest of this feature is mostly drift-free and the Melbourn Rock can readily be traced in the soil brash: large quarries have been dug into the feature at Barrett Ringstead, Heacham and Hillington. The Melbourn Rock consists of up to 4 m of hard and very hard nodular chalks ('nodular griotte' of Mortimore, 1979) with a few marl wisps. The degree of nodularity and the individual marl wisps can be used to effect detailed correlations between the few well-exposed sections in the district (Figure 40). Shell debris is common at many levels in the Melbourn Rock, including *Inoceramus* and other bivalves, brachiopods and echinoids. The stratigraphically important ammonite *Mammites nodosoides* (Schlüter), indicative of the *M. nodosoides* Zone,

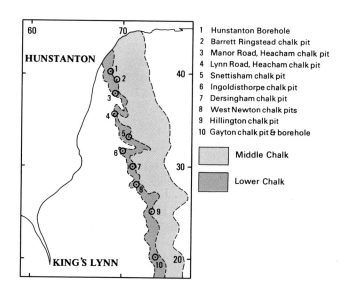

1 Hunstanton Borehole
2 Barrett Ringstead chalk pit
3 Manor Road, Heacham chalk pit
4 Lynn Road, Heacham chalk pit
5 Snettisham chalk pit
6 Ingoldisthorpe chalk pit
7 Dersingham chalk pit
8 West Newton chalk pits
9 Hillington chalk pit
10 Gayton chalk pit & borehole

Middle Chalk

Lower Chalk

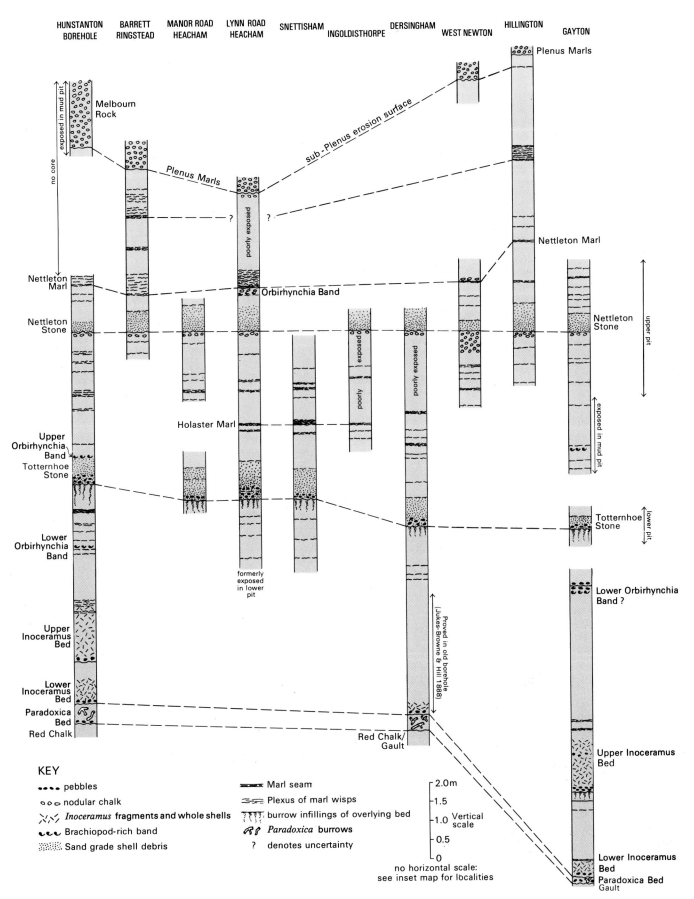

Figure 39 Correlation of Lower Chalk sequences.

first appears a little above the base of the Melbourn Rock in the district, and is common at several horizons.

DETAILS

Hunstanton to Flitcham

The lower half of the Lower Chalk, including the junction with the Red Chalk, is continuously exposed in Hunstanton Cliffs [6725 4131 to 6794 4242] for a distance of about 1.3 km (Front Cover). The southern part of the section is mostly inaccessible, but the beds below the Totternhoe Stone can be reached by scrambling up debris cones in much of the northern part. The low dip brings the basal beds down to beach level at the northern end of the section. The Totternhoe Stone and the higher beds cannot be examined in situ except by scrambling up the more degraded parts of the cliff, but much of this part of the sequence can be examined in large fallen blocks. The Paradoxica Bed and the Lower Inoceramus Bed form prominent marker bands at the base of the Chalk throughout the section. The Totternhoe Stone can be traced as a band of markedly greyer chalk in the middle part of the cliffs in the central part of the section.

The following composite sequence was measured in the vicinity of the light-house [6752 4212]; the faunas are described on pp.124–127 and the main lithological features are shown graphically in Figure 38:

	Thickness m
Soil and subsoil; brown calcareous silt with small angular chalk pebbles	0.3
Chalk, off-white, soft, with traces of greenish grey marl; thinly bedded but much disturbed by cryoturbation and weathering; thin marl at base	1.5
Chalk, off-white, soft, slightly argillaceous, with sub-parallel, planar, thin bedding picked out by very thin wisps of greenish grey marl; concentration of marl wisps, 20 mm thick, in middle part of bed	1.02
Chalk, off-white, soft, argillaceous, with closely spaced laminae of greenish grey marl	0.18
Chalk, off-white, soft, with planar, thin bedding picked out by very thin marl wisps	0.41
Chalk, complexly interbedded and interlaminated with dark greenish grey marl and marly chalk	0.10
Chalk, off-white, hard, with a few marl streaks; marl wisp at base	0.46
Chalk, greyish white, very hard; passing down, with irregular bedding surface at base, into	0.30
Totternhoe Stone: chalk, pale and medium grey burrow-mottled, weathering in part to brownish grey; becoming darker grey with depth; slightly argillaceous; silty and sandy textured due to much included comminuted shell debris; becoming both coarser and more argillaceous with depth; pebbly in lowest 10 to 15 mm with many pebbles of green-coated phosphatised chalk ('Brassil'); irregular base with many burrowfills of grey sandy chalk extending down for up to 0.6 m	0.48
Chalk, off-white, soft with marl streaks throughout; burrowfills of grey gritty chalk extend down from overlying bed	0.59
Chalk, complexly interbedded and interlaminated with greenish grey marl and grey marly chalk	0.13
Chalk, off-white with rare marl wisps; prominent bedding plane in middle part of bed	0.61

	Thickness m
Chalk, off-white with numerous wisps of greenish grey marl	0.05
Chalk, off-white with scattered marl wisps	1.07
Chalk, off-white with numerous marl wisps	0.15 to 0.23
Upper Inoceramus Bed; chalk, hard, pale grey, gritty textured and shelly with many *Inoceramus* fragments; shelly at many levels with rich diverse fauna including common terebratulids and rhynchonellids; prominent bedding plane 0.48 to 0.51 m above base; pebbly in lowest part; resting on irregular surface of	1.12 to 1.14
Chalk, off-white, hard with marl wisps forming weak bedding planes; passing down into	0.37 to 0.46
Lower Inoceramus Bed; chalk, grey, gritty with much *Inoceramus* debris; intense bioturbation picked out by pale grey and off-white, smooth-textured chalks; prominent band of hard, dense glauconitised and phosphatised chalk pebbles in basal bed; rests on irregular highly indurated surface	0.92 to 0.95
Paradoxica Bed; limestone, hard, splintery; intensely bioturbated white, creamy white, yellow, pink and red with ramifying network of 'Paradoxica' burrows; stylolitic surfaces at several levels; irregular base with blebs of deep red hematitic staining on lower surface; rests on very irregular indurated surface of Red Chalk	0.43

RED CHALK (see p.113 for details)

The following sequence was proved in the Hunstanton Borehole drilled at Lodge Farm [6857 4078]; the highest 1.8 m was proved in the mud-pit excavation:

	Thickness m	Depth m
Soil and subsoil; clayey calcareous sand with angular chalk fragments; cryoturbation structures	0.5	0.50
MIDDLE CHALK		
Melbourn Rock; hard, nodular chalks; frost damaged and with much iron-staining	1.3	1.80
LOWER CHALK		
Plenus Marls; soft iron-stained clay	10 mm	1.81
No sample; rock-bit drilling	3.16	4.97
Chalk, very hard, cream-coloured, with several stylolitic surfaces coated with greyish green marl; passing down into	0.20	5.17
Chalk, hard, off-white, slightly marly and marly, with marl wisps concentrated in basal part to form a thin seam containing echinoid debris and common oysters; irregular base	0.25	5.42
Chalk, alternating hard and very hard, cream- coloured and off-white with some soft, more marl-rich bands; thin marl seams and wisps at several levels	0.88	6.30
Nettleton Stone; chalk, hard and very hard, cream-coloured with patches of pinkish brown phosphatisation, passing down into grey and brownish grey; becoming more marly and slightly gritty with depth; bioturbated throughout; gritty brownish grey marl at base with common *Pycnodonte* resting on an irregular, indurated surface	0.37	6.67

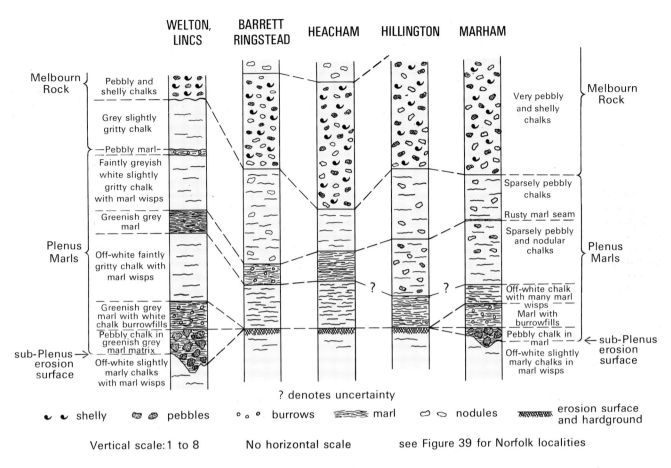

Figure 40   Correlation of Plenus Marls and Melbourn Rock sequences.

	Thickness	*Depth*
	m	m

Chalk, hard, off-white, slightly marly, with grey, greenish grey and olive-green marl wisps and thin seams at several levels, some coating stylolitic surfaces; band of very hard, white, nodular chalk with marl-coated surfaces at top of bed; patchy glauconitic and phosphatic cementation present at several levels — 3.16 — 9.83

Totternhoe Stone; chalk, hard and very hard, grey and brownish grey; complexly bioturbated; becoming gritty and sandy textured in lower part; glauconite-stained pebbles of splintery, pinkish brown, phosphatised chalk up to 15 mm across common in lower half of bed; very irregular, indurated surface at base — 0.66 to 0.74 — 10.53

Chalk, hard and very hard, cream-coloured, with rare wisps of grey marl; burrow infillings of brownish grey sandy chalk from the overlying bed extend throughout the bed; gritty, with much *Inoceramus* debris in basal part — 0.64 — 11.17

Chalk, moderately soft becoming harder with depth; off-white becoming cream-coloured in lower part with some pinkish brown phosphatised patches; slightly marly throughout, with marl wisps and thin seams at many levels; stylolitic

surfaces and hard, locally indurated bands become more common with depth; concentration of thin marl seams at base — 2.24 — 13.41

Upper Inoceramus Bed; chalk, hard, pale grey, gritty-textured and shelly with many *Inoceramus* fragments; other shells include terebratulids and rhynchonellids, common at some levels; at base, very shelly bed with probable pebble of white chalk rests on irregular surface — 1.45 — 14.86

Chalk, hard and very hard, off-white, slightly marly with common marl wisps; intensely bioturbated brownish grey in lower part; sharp colour change at base — 0.71 — 15.57

Lower Inoceramus Bed; chalk, grey, gritty, with *Inoceramus* debris; intensely bioturbated, with a variety of burrows and trails picked out by paler grey and off-white smooth textured chalks; patchy phosphatisation, especially in lower part of bed; pebbles of very hard, green-stained phosphatised chalk up to 25 mm across common in two bands in lower part of bed; irregular indurated surface at base — 0.68 to 0.72 — 16.27

Paradoxica Bed; limestone, hard, splintery; white, creamy white, pink and yellow mottled due to intense bioturbation including 'Paradoxica' burrows, some of which are infilled with the overlying grey shelly chalk and penetrate top 0.1 m of

	Thickness m	Depth m

bed; numerous stylolitic surfaces and core
much broken; core break at base with
blebs of deep red hematitic staining on
basal surface of bed   0.55 to 0.59   16.84

RED CHALK

Lower Chalk has been worked in small pits near St Edmund's Point [6799 4228], adjacent to Park Road [6743 4059] and adjacent to Lynn Road [6787 3979], Hunstanton, and Middle Chalk has been worked in small pits at Lodge Farm [6844 4082] and in Hunstanton Park [6934 4148; 6928 4205]; these are now completely degraded. The Totternhoe Stone [6788 3966] and Melbourn Rock [6784 3985] were exposed in shallow excavations when the Lynn Road was widened near The Firs.

The junction of the Lower and Middle Chalk is exposed in an old chalk pit [6892 4003] at Barrett Ringstead Farm. The following composite section is present; it is shown graphically and its correlation with other sections in the district is indicated in Figure 39:

	Thickness m

Soil and subsoil; brown calcareous silt with
angular chalk pebbles   0.3

MIDDLE CHALK

Melbourn Rock: chalk, white and greyish white
shelly chalks with some yellowish brown
staining; frost-shattered and disturbed by
cryoturbation, but with traces of rubbly and
nodular bedding; becoming harder and less
disturbed with depth; very thin marl wisp
forming bedding plane at base   1.2
Chalk, hard, white and greyish white, nodular;
moderately shelly with common Inoceramus
fragments   0.85
Chalk, greyish white with yellowish brown
staining; nodular, with complexly interbedded
wisps and laminae of brown-stained marl   80 to 100mm
Chalk, hard, white, nodular, with concentration
of small rounded chalk pebbles in top part;
shelly, with abundant Inoceramus fragments;
three prominent, closely spaced bedding
planes along marl wisps in middle part of bed;
becoming harder and more nodular with
depth; marl wisp forming bedding plane at
base   2.05
Chalk, hard, off-white, nodular, with scattered
wisps of dark grey marly chalk; shelly, with
abundant Inoceramus debris; bedding plane at
base   0.30
Chalk, moderately hard, cream-coloured;
sparsely nodular, with marl wisps flowing
around nodules; abundant Inoceramus debris;
bedding plane and marl wisp at base   0.25
Chalk, hard, off-white, intensely nodular; shelly
and with sandy texture due to abundant
coarse- and fine-grained Inoceramus debris;
bedding plane and marl wisps at base   0.30
Chalk, moderately hard, greyish white,
moderately nodular, with wisps of grey marl
and marly chalk throughout; Inoceramus debris
common; bedding plane at base   0.30

LOWER CHALK
Plenus Marls: soft, off-white, marly chalk with
closely spaced marl wisps (50 to 80 mm thick)
resting on chalk with widely spaced marl wisps;
irregular green ?glauconite-stained surface at
base   0.20
Chalk, soft, off-white, with marl wisps
throughout; highest part indurated, with
yellowish brown and some green staining;
irregular base resting on indurated surface   0.56
Chalk, soft, off-white, with marl wisps
throughout and with concentrations up to
80 mm thick of thicker marl wisps at two levels;
top 0.15 m of bed indurated and nodular with
yellowish brown staining; passing down into   2.35
Chalk, soft, off-white and greyish white, slightly
marly and with marl wisps and thin seams
scattered throughout; fauna includes
Camerogalerus cylindricus (Lamarck);
concentration of marl seams (Nettleton Marl)
in lowest 80 mm   0.59
Chalk, soft, off-white, irregularly bedded, with
marl wisps throughout; bedding plane at base   0.38
Nettleton Stone; chalk, soft, medium and pale
grey becoming darker with depth; silty
textured, becoming sandy and argillaceous
with depth; 50 to 100 mm-thick bed of grey
marly chalk at base with numerous small
white chalk pebbles and burrow infillings, and
abundant Pycnodonte; Terebratulina
protostriatula juv.? Owen, Terebratulina sp. and
Inoceramus shell chips also present; resting on
irregular indurated iron-stained surface   0.68
Chalk, soft, off-white, with streaks and wisps of
grey marl and marly chalk throughout; thin
marl seam with small chalk pebbles in middle
part of bed   seen 0.55

Lower Chalk is still worked at Manor Road, Heacham [682 387]. At the time of the survey, the workings consisted of a shallow upper pit that exposed the Nettleton Stone and about 2 m of the underlying chalk, and a lower disused pit that exposed a deeply weathered section in the Totternhoe Stone. The intervening beds were not exposed, but their thickness was estimated to be similar to that at Lynn Road, Heacham (Figure 39). Lower Chalk has been worked in small pits at Church Farm, Heacham [6852 3835; 6913 3778], but these are now degraded.

The full thickness of the Lower Chalk, with the exception of about 2 m at the base, and the lowest beds of the Middle Chalk were formerly exposed in the large pit at Lynn Road, Heacham [688 368] (Whitaker and Jukes-Browne, 1899, p.53). This pit, now backfilled, was formerly the best section in the district where the beds between the Totternhoe Stone and the Melbourn Rock could be examined, and the only one in which these sequences could be seen at a single locality. In addition, the Melbourn Rock was especially well exposed in an air-weathered face in the southeastern part of the workings. The composite sequence and its correlation with other sections in the district are shown graphically in Figure 39, and is described below.

The Totternhoe Stone, the Nettleton Stone and the Melbourn Rock formed prominent marker bands that enabled the various scattered exposures in the pit to be correlated. The pit was worked on two levels: the beds below the Nettleton Stone were best exposed in the southern face of the lower level [6875 3669], those above in the south-eastern face of the upper level [6879 3661]. Some of the sections, especially those at the level of the Plenus Marls, were disturbed, probably by glacial tectonics, and not all showed the full sequence. The full thickness of the Lower Chalk at Heacham, assuming the beds below the Totternhoe Stone to be of similar thickness to those proved in the Hunstanton Borehole, is estimated from the present survey to be about 17.1 m, somewhat thinner than the 'nearly 60 ft' (18.3 m) estimated by Whitaker and Jukes-Browne (1899, p.53). The following composite section was measured

	Thickness m
Soil and subsoil; chalky sandy silt with angular and chalk pebbles	0.5

MIDDLE CHALK

Melbourn Rock: chalk, greyish white and off-white with some yellow and brown staining; traces of grey marl wisps; shelly in part, with *Mytiloides* fragments; frost shattered and with cryoturbation structures in highest part; iron-stained marl wisp at base — 1.7

Chalk, moderately hard, greyish white, with wisps of pale grey marl; flaggy and thickly bedded — 0.63

Marl, brownish grey, iron-stained in part; tectonically disturbed — 20 to 70mm

Chalk, hard, greyish white, gritty, shelly, with common *Mytiloides*; flaggy and rubbly bedded; lenticular pavement of recrystallised *Mytiloides* up to 20 mm thick at base — 0.43

Chalk, hard, greyish white, flaggy bedded but with bedding broken up by scattered nodular patches and chalk pebbles; *Mytiloides* fragments common throughout — 0.26

Chalk, hard, greyish white, very nodular and pebbly; scattered *Mytiloides* fragments; *Mammites nodosoides* in top part of bed; irregular bedding plane at base — 0.35

Chalk, hard, greyish white, gritty, with scattered *Inoceramus* fragments; nodular in part and with a few scattered chalk pebbles; passing down into — 0.41

LOWER CHALK

Plenus Marls: chalk, moderately hard, greyish white, with wavy bedding picked out by wisps of grey marl — 0.13

Marl, softened, rusty brown, with many small chalk pebbles and white elliptical burrow infillings of chalk; gritty texture due to comminuted shell debris (up to 25 mm); passing down into complexly interbedded marl seams and wisps and greyish white slightly marly chalk; planar base — 75 to 100mm

Chalk, moderately hard, greyish white, gritty, with wavy bedding picked out by wisps of brownish grey marl; irregular surface at base — 0.14 to 0.17

Chalk, tough, off-white, splintery and nodular, hard ground with numerous stylolitic surfaces; orange-brown and greenish grey marl coated surfaces occur throughout with concentrations on upper surface and in upper part; patchy cementation extends down 0.46 m from overlying surface; passing down into — 0.13 to 0.18

Chalk, moderately hard, off-white, slightly marly, with marl wisps throughout; flaggy bedded; wisp of rusty brown marl 0.66 m above base of bed; passing down into — 1.65

Chalk, moderately hard, greyish white and grey, with wavy bedding picked out by many prominent wisps of grey marl; passing down into up to — 0.33

Nettleton Marl, grey and greenish grey, complexly interbedded with marly chalk; fauna includes *Orbirhynchia* sp. nov; irregular, locally hardened surface at base with rare small oysters — 50 to 76mm

Chalk, moderately hard, off-white, with wavy bedding picked out by marl wisps; highest part locally iron-stained, nodular and patchily hardened; *Orbirhynchia* sp. nov. occurs above and below the horizon of nodular chalk; marl wisp and bedding plane at base — 0.38

Nettleton Stone: chalk, tough, grey, massive, becoming sandy and argillaceous with depth; very argillaceous at base with *Pycnodonte*-rich marl resting on irregular iron-stained indurated surface — 0.46

Chalk, hard, nodular, patchily cemented, with iron-stained and marl-coated surfaces; becoming off-white, slightly marly and less indurated and nodular with depth; passing down into — 0.23

Chalk, moderately hard, off-white, slightly marly, with marl wisps at several levels and very thin seams forming prominent bedding planes at two levels — 1.00

Chalk, off-white, and pale grey marl complexly interbedded in two bands 0.05 and 0.13 m thick, separated by 0.15 m of off-white slightly marly chalk with common ferruginous stained sponges — 0.30

Chalk, moderately hard, off-white, slightly marly and very marly, with marly wisps at several levels including a 75 mm-thick concentration of wisps in middle part of bed that forms a prominent bedding plane — 0.84

Marl and marly chalk complexly interlaminated, soft, pale grey; laterally variable between finely laminated very marly chalk and thin marl seam; small oysters common and crushed *Holaster* locally abundant; *Kingena concinna* Owen, *Terebratulina* sp., '*Inoceramus*' ex gr. *atlanticus* (Heinz), cidarid radioles and *Holaster subglobosus* also present; small pebbles or burrow infillings of white chalk common — 20 to 70mm

Chalk, moderately hard, off-white, with *Holaster subglobosus*; as above — 0.20

Marl, soft, yellowish brown; laterally variable from thin seam to plexus of wisps; fauna includes *Kingena concinna?*, *Concinnithyris* sp.? and small pycnodonteine oyster — 10 to 75mm

Chalk, moderately hard, pale grey; tougher, sparsely nodular and cream-coloured in highest part; marl wisps common in lower part; irregular slightly hardened surface at base — 0.76

Totternhoe Stone; hard, pale and medium grey, becoming darker with depth; massive, with single bedding plane in lower third of bed; sandy and clayey, becoming more so with depth; fauna includes terebratulids of uncertain affinity; pebbly in lowest 0.18 m with many small pebbles of green-coated, hard, yellow and yellowish brown, phosphatised

	Thickness m
chalk; *Inoceramus*, terebratulids and echinoid spines common in lower part of bed; rare large ammonites including ?*Austiniceras* in upper part; *Acanthoceras* and *Schloenbachia* occur as glauconitised moulds in the basal part, with *Entolium* sp. and *Oxytoma seminudum* Dames; irregular base rests on indurated surface; *Thalassinoides* burrow systems infilled with grey gritty Totternhoe Stone sediment extend down into underlying beds	0.74
Chalk, moderately hard and hard, off-white and cream-coloured; marl concentrations occur at several levels including four prominent very thin seams that give rise to bedding planes	seen 1.8

Lower Chalk has been worked in three large pits at Eaton [6962 3583] and Snettisham [6983 3321; 6925 3490], all of which are now called 'Limekiln Plantation'. Only the last-named has any exposure: it shows about 2 m of deeply weathered and frost-shattered chalk. This is presumably the locality mistakenly referred to by Whitaker and Jukes-Browne (1899, p.57) as 'E.N.E. of the church' (NNE correct). The best exposure of Lower Chalk in the Snettisham area is the old pit at Redbarn Road [6963 3392], but this too is being rapidly infilled with rubbish. About 9 m of strata, in which the Totternhoe Stone formed a prominent marker at about the middle of the section, were formerly exposed (Whitaker and Jukes-Browne 1899, p.53). The lower part of the pit has been infilled in recent years and only the Totternhoe Stone and the beds above were visible at the time of the present survey (Figure 39).

The Nettleton Stone and about 3 m of beds above and below it are exposed in a large pit at Ingoldisthorpe [6925 3210], but the sections are so deeply weathered and disturbed by movements along joints, that no accurate measurement is possible. The depth of the pit suggests that the Totternhoe Stone was formerly exposed in its floor.

The Totternhoe Stone, Nettleton Stone and the intervening beds are still exposed in Dersingham Parish Pit [7013 3092], although the higher beds are deeply weathered. Jukes-Browne and Hill (1887, p.560) recorded the beds below the Totternhoe Stone, partly in a section and partly in a borehole, as 5.3 m (17.5 ft) thick. The Paradoxica Bed and the Lower Inoceramus Bed were both present. Correlation with other sections in the district is shown in Figure 39.

A large old pit in Dersingham Wood, Sandringham [6987 2936], was worked on two levels, with the Totternhoe Stone forming the floor of the upper level. This bed was exposed at the time of the first geological survey (Whitaker and Jukes-Browne 1899, p.51), but the section is now completely degraded.

The upper part of the Lower Chalk and the basal beds of the Middle Chalk were worked in pits near the water tower [7043 2775; 7062 2762] and Appleton House [7141 2722; 7149 2706], West Newton. The first of these is the largest and covers the greatest stratigraphical range. The more resistant parts of the sequence are still exposed and these include the Nettleton Stone and the Melbourn Rock (Figure 39). The nodular indurated chalk underlying the Nettleton Stone is thicker and better developed there than at any other locality recorded in Norfolk. The second water-tower pit exposed only the Nettleton Stone; it must have been excavated largely in the beds between this bed and the Totternhoe Stone. The pits near Appleton House are small, but both expose the highest part of the Lower Chalk, including the Plenus Marls, and the lower part of the Melbourn Rock.

Lower Chalk was worked in pits on either side of the Babingley River at Flitcham [7245 2663; 7294 2604]. The northern pit is completely degraded, but appears from its position and shape to have worked the Totternhoe Stone and the beds immediately above and below it. The southern pit lies below the level of the Totternhoe Stone and is also completely degraded, except for a small exposure of grey gritty chalk, possibly the base of the Upper Inoceramus Bed, resting on an indurated surface of white chalk. This latter pit contains the remains of an old kiln.

Flitcham to Gaytonthorpe Common

The full sequence of the Chalk between the Nettleton Stone and the Melbourn Rock is well exposed in a working chalk pit at Hillington [723 244], although parts of the sequence, notably the Plenus Marls, are affected by glacial tectonics and are reduced in thickness. The present workings are an extension of much older workings referred to by Rose (1835, pp.277–728) as 'Congham', and in which Whitaker and Jukes-Browne (1899, p.51) recorded about 11 m of the upper part of the Lower Chalk. The older workings were probably 2 to 3 m deeper than those now exposed, but the higher levels, which now expose the Plenus Marls and most of the Melbourn Rock, have been considerably extended by the modern working. Correlation of the section with others in the district is shown in Figure 39. The Nettleton Stone forms a good marker in the lower face, and the Melbourn Rock occupies much of the middle and upper faces. The Plenus Marls seem to be present as a persistent thin bed of pebbly iron-stained marl, as noted by Peake and Hancock (1961, p.33). However, this band locally contains *Neohibolites minimus* (Miller) (derived from the Gault) and pockets of glacial sand, and has been emplaced by fluid injection during the Pleistocene along a horizontal joint formed at the base of the Melbourn Rock, and fed by low-angle joints which may penetrate the full thickness of the Lower Chalk.

The following composite section was recorded:

	Thickness m
Soil and subsoil; sand with angular chalk fragments	0.6
Cryoturbation bed; angular chalk fragments set in a chalk paste with pockets and lenses of sand and broken flints up to	0.9
MIDDLE CHALK	
Chalk, hard, thickly and rubbly bedded, becoming nodular in lower part; greyish white and white but with much yellow and yellowish brown staining adjacent to joints; shelly, with abundant *Inoceramus* fragments	2.4
Not exposed	c. 0.1
Melbourn Rock; chalk, hard, nodular grey and greyish white; gritty textured; shelly, with abundant *Mytiloides* fragments; rarer brachiopods and echinoids; *Mammites* and *Lewesiceras* common at one level; *Mytiloides mytiloides* (Mantell) and *Inoceramus* aff. *apicalis* Woods? also present; *Austiniceras*?, *Orbirhynchia* sp. and *Monticlarella jefferiesi* Owen occur just above the Plenus Marls	1.1
LOWER CHALK	
Plenus Marls: clay, soft, rusty brown and chocolate-coloured; intensely sheared; locally sandy and passing into white or greyish brown sand; both clay and sand presumed largely	

	Thickness m
glacial in origin and wholly or partially replacing a thin marl seam	10 to 20mm
Chalk, hard, greyish white, sparsely nodular; slightly gritty, with *Inoceramus* fragments; passing down into	0.10
Chalk, moderately hard, off-white, slightly marly, with marl wisps throughout picking out laminar and wavy bedding; thin marl seam at base resting on green-stained irregular surface	0.12
Chalk, moderately hard, off-white, slightly marly, with marl wisps picking out bedding and bioturbation structures; harder, nodular and indurated in top 0.1 m with nests of *Ornatothyris* including *O. latissima* Sahni, *O.* aff. *rara* Sahni, *Camerogalerus cylindricus* and a flat-based holasterid, possibly *Holaster trecensis* Leymerie	1.90
Chalk, soft to moderately hard, off-white, with wisps of olive-grey marl throughout; concentrations of thin marl seams at top and bottom of bed giving rise to prominent bedding planes	0.60
Chalk, moderately hard, off-white, slightly marly, with marl wisps picking out bedding and bioturbation structures; two minor and two prominent concentrations of marl wisps (including Nettleton Marl) in lowest part of bed; passing down into	4.65
Nettleton Stone; chalk, hard, grey weathering to brownish grey; *Concinnithyris* aff. *bulla* (J de C Sowerby)in lower part of bed; becoming more argillaceous and darker grey with depth; circular burrow infillings of white chalk prominent in upper part of bed; crushed '*Inoceramus*' ex gr. *atlanticus* also present in basal part; very marly chalk rich in pycnodonteine oysters rests on an irregular, indurated surface at base	0.76
Chalk, hard, nodular, indurated, off-white and white, with secondary yellow staining	0.15
Chalk, moderately hard, off-white, slightly marly, with marl wisps throughout fauna includes '*Inoceramus*' cf. *atlanticus*	1.3

Lower Chalk has been worked in small pits at Gorse Moor, Roydon [7008 2427], Grimston [7252 2123; 7288 2206; 7185 2222] and Well Hall, Gayton [7220 2015], but these are now completely degraded. The first of these formerly exposed a 'very hard creamy white limestone' and the Lower Inoceramus Bed (Jukes-Browne and Hill, 1887, p.550). The limestone was described as 1.7 m (5.5 ft) thick; the field evidence, and that from the nearby Gayton Borehole (see p.118), suggests that the top part of this bed is the Paradoxica Bed and that the remainder is a muddy chalky limestone at the top of the Gault (Bed G 19). Soft chalky limestones occur at several levels in the upper part of the Gault in that area, but they break down rapidly in the soil and are readily distinguishable from the large fragments of hard, splintery Paradoxica Bed that are common in the local soil brash. Green-coated phophatised chalk pebbles and large fragments of Lower Inoceramus Bed are also common in the soils of the Roydon to Gayton area, and these provide additional marker bands close to the Gault–Chalk junction.

The lithological contrast between the massive, well-jointed limestones of the basal Chalk and the soft chalky limestones and marls of the Gault is reflected in a marked change in permeability. This causes strong springs to be emitted from the base of the Chalk, notably near Grimston Church [7210 2194; 7214 2167; 7221 2150], Sowshead Spring [7257 2088], Well Hall, Gayton [7259 2030], and Gayton Hall [732 189].

Lower Chalk has been worked in a complex of pits on the north side of Gayton village: the most easterly of these [727 197] is still in work. Whitaker et al. (1893, pp.34–35) recorded a total of about 15 m of Lower Chalk in the older pits with the Totternhoe Stone, the lowest bed seen, probably present in the floor of the most westerly pit [725 197]. Whitaker recorded a most unusual occurrence of pebbles and boulders of granitic, gneissic and basaltic rocks 'in the Chalk' at this locality, and likened them to the stones in the Cambridge Greensand of Cambridgeshire. Pockets of drift with far-travelled erratics occur at the top of some of the working faces, and similar material is common in the area. These occurrences, combined with the possible presence of glaciotectonic features such as those at Hillington chalk pit, mean that the field relationships of any stones found apparently in situ in the Chalk at Gayton would need to be carefully examined before their method of emplacement could be determined.

The following composite section was proved in the working pit and the Geological Survey Gayton Borehole [7280 1974] drilled in its floor. Correlation of the sequence with others in the district is shown in Figure 39. The present pit is shallow and the only prominent marker band is the Nettleton Stone.

	Thickness m
Chalk pit	
Soil and subsoil; chalky paste with angular chalk pebbles and local concentrations of glacial sand and bleached flints; cryoturbated in part	up to 1
Chalk, moderately hard but much affected by weathering; off-white and greyish white, with wisps and thin seams of marl at several levels, including possible Nettleton Marl; thin marl seam at base	1.50
Nettleton Stone; hard, massive, grey; gritty- textured in lower part; clayey and silty, becoming more so with depth; irregular base resting on indurated surface	0.38
Chalk, hard, white, with patchy yellow iron- staining in upper part; nodular in top part with yellow-stained indurated nodules coated with marl from overlying bed; becoming softer with depth; thin marl wisp at base	0.23
Chalk, moderately hard, off-white; as above with marl wisps and thin seams at several levels; shelly band rich in terebratulids near base of bed (proved in mud-pit excavation only)	3.50

	Thickness m	Depth m
Borehole		
No sample, rock-bit drilling (about 1.5 m over lap with bed above)	3.35	3.35
Chalk, moderately hard, but with some softer and, especially in lower part, more indurated bands; off-white becoming whiter where more indurated and greyer where more marly; slightly marly throughout and with lamination and bioturbation picked out by marl wisps and more marly chalks at many levels; minor erosion surface with green-coated chalk pebbles resting on hardened surface in top part of bed; shelly, with		

	Thickness m	Depth m

common *Inoceramus* fragments at some levels; very shelly band rich in *Inoceramus* and brachiopods in top part of bed; manganese staining common on joint surfaces; 75 mm-thick olive-green marl seam close to base of bed; passing down with bioturbation into — 4.30, 7.65

Upper Inoceramus Bed; chalk, hard, massive, grey, gritty, intensely bioturbated; shelly, with common *Inoceramus* fragments; band of small brown phosphatised chalk pebbles in upper part of bed; concentration of closely spaced hard pinkish brown phosphatised chalk pebbles with green ?glauconitic-stained skins at base of bed, resting on an irregular hardened surface — 1.43, 9.08

Chalk, very hard, partially phosphatised and recrystallised, yellowish and brownish white; becoming softer, off-white and cream-coloured, and more marly with depth; marl seam and *Inoceramus*-rich band in upper part of bed; lamination and bioturbation picked out by marl wisps and very thin seams at several levels; marl seam at base — 1.83, 10.91

Lower Inoceramus Bed; chalk, hard, massive, grey, gritty, intensely bioturbated, with much *Inoceramus* debris; patches of glauconitisation and phosphatisation at several levels; band of very hard, green-coated phosphatised chalk pebbles, up to 50 mm across, at base resting on an irregular, indurated surface — 0.33, 11.24

Paradoxica Bed; limestone, very hard, brittle, pinkish and yellowish brown, intensely bioturbated; band of small (up to 20 mm across), soft, brown, phosphatic pebbles at base; junction with bed below much disturbed by coring but thought to be very irregular and bioturbated — 0.16, 11.40

GAULT (see p.120 for details)

The Totternhoe Stone has been exposed in the most westerly pit from time to time since the present survey. Dr Gale (unpublished MS) has recorded a section that showed 0.65 m of Totternhoe Stone, the base of which was 5.05 m below the base of the Nettleton Stone (Figure 39). He recorded the following fauna from the Totternhoe Stone: *Hamulus, Serpula umbonata, Concinnithyris, Kingena concinna, Orbirhynchia* sp., *Terebratulina nodulosa, T. protostriatula, Aequipecten* sp., *Camptonectes* sp?, *Entolium*, indeterminate inoceramid fragments, *Oxytoma seminudum* (common), *Plagiostoma globosum, Plicatula inflata, Spondylus latus, 'Teredo', Actinocamax primus, Schloenbachia* in pebble preservation, *Cretiscalpellum glabrum, Calliderma smithiae, Stereocidaris* sp., *Phymosoma* sp., *Lamna* sp. and *Squalicorax falcatus*.

The presence of the Upper Orbirhynchia Band in the Totternhoe Stone, rather than above it as in other sections in Norfolk, combined with a well-developed glauconitised pebble conglomerate at the base of the Totternhoe Stone may indicate that the stone occupies a shallow channel at this locality. Dr Gale also recorded *Concinnithyris*, indeterminate inoceramids, *Pycnodonte* sp., *Calycoceras* sp., cidarid radioles, *Camerogalerus cylindricus, Holaster* sp. and a *Squatina*? tooth in the Nettleton Stone, and brachiopods, cidarid radioles and *Discoides subuculus* in the beds between the Totternhoe and Nettleton stones

Between Gayton and Gaytonthorpe Common, the Lower Chalk outcrop has been severely glaciated and occupies low ground, largely covered by a veneer of sandy and gravelly drift, adjacent to the broad, drift-filled Nar valley. The junction with the Gault can, however, be mapped with confidence because fragments of Paradoxica Bed, and green-coated phosphatised chalk pebbles from the base of this bed and from the base of the Lower Inoceramus Bed, are common in the spoil from the deeper ditches. These beds were observed in situ in temporary excavations in Gayton village [7278 1900], where the junction with the Gault is marked by a strong spring, and at Wizard Lodge [7215 1762; 7192 1757]. At this last locality loose, rounded blocks of green-stained Paradoxica Bed, up to 0.2 m across, were present: these appear to have been derived from the basal pebble bed of the Lower Inoceramus Bed.

SIX
Quaternary

Quaternary deposits crop out over the whole of the Fenland part of the district, about one third of the upland part and almost all of The Wash. They can be divided into those that were deposited in glacial, periglacial and temperate climates during the latter part of the Pleistocene period some 150 000 to 10 000 years ago, and those that have accumulated in a temperate climate during the last 10 000 years. The younger deposits, the Recent marine and freshwater sediments that form Fenland and The Wash, overlie large parts of the Pleistocene sequence and consequently have a much greater area of outcrop in the district.

PLEISTOCENE DEPOSITS

The Pleistocene history of East Anglia has been a subject of controversy for more than 100 years, largely because of the patchy distribution of the deposits and the inconclusive nature of much of the evidence. Only a fragmentary record is present in most areas and the poorly exposed nature of the deposits usually make it difficult for correlations to be made, even over relatively short distances. The absence of a reliable method for determining the ages of Pleistocene deposits more than about 46 000 years old (the practical limit for radiocarbon dating) and the scarcity of diagnostic faunas or floras compound the problem.

Few Pleistocene deposits contain indigenous shelly faunas, and those that do have assemblages made up of long-ranging forms that are of little stratigraphical value. Amino-acid geochronology has been used to estimate the relative ages of shells, but with varying success (see below). Vertebrate remains are relatively common in some Pleistocene deposits, although often indicative of specific climatic conditions, but they are rarely stratigraphically useful.

The most determined attempts to provide a biostratigraphical framework for the Pleistocene have been based on pollen assemblages. Although many individual forms are long-ranging it has been claimed that the pollen spectra of deposits formed during temperate episodes can be distinguished from one another. Attempts have also been made to use mammals, bivalves, gastropods, foraminifers, ostracods and beetles to construct climatic 'curves' in the hope that they will prove to be distinctive.

In recent years measurements of the ratio of the protein amino acid L-isoleucine to that of the non-protein amino acid D-alloisoleucine (the D/L ratio) in marine and nonmarine shells has been used to estimate the ages of Pleistocene deposits, the rate of epimerisation from the L to the D form being thought to have been constant throughout the past 800 000 years (Bowen, 1989). The method is not, however, without its problems because the rates of epimerisation differ between marine and nonmarine forms, and between different genera in the same environment. In addition, some deposits contain shells which have D/L ratios which differ by greater amounts than can be accounted for by differences in their epimerisation rates alone. This may indicate reworking of some of the shells, but it might also be due to contamination or other errors that are as yet poorly understood in the method.

The knowledge that Pleistocene history is one of repetition of cold and warm climates led to the proposal (Mitchell et al., 1973) that the Pleistocene in Britain should be divided into three glacial phases (Anglian, Wolstonian and Devensian) and two interglacial phases (Hoxnian and Ipswichian). These are collectively referred to as the British Standard Stages. In the absence of diagnostic palaeontological or chemical evidence, the individual stages were not uniquely defined and, in practice, the application of the scheme has depended largely on the recognition of field evidence to enable the stratigraphical relationships of the different deposits to be determined.

Work on terrestrial Pleistocene deposits elsewhere in Europe and in North America, on marine Pleistocene deposits from the Atlantic and Pacific oceans, and on the chemistry of the Arctic and Antarctic ice sheets, has shown that the Pleistocene period was characterised by numerous climatic fluctuations of varying lengths. In addition, some cold phases had short warmer periods (interstadials) within them, and some interglacial phases contained short cold periods. Many of the cold phases may not have given rise to glaciations in Britain, and early glacial deposits may have been removed by later glacial or temperate erosion. The recognition of the British Standard Stages is, therefore, by no means simple, nor are they universally accepted. In Fenland, for example, the most widespread glacial deposit, the Chalky-Jurassic till (Chalky Boulder Clay of some authors), has been placed in the Anglian Stage (e.g. Mitchell et al., 1973) and the Wolstonian Stage (e.g. Straw, 1979). In the Midlands, the type section of the 'Wolstonian Glaciation' has been correlated with the 'Anglian' of adjacent areas on the basis at the field evidence (Sumbler 1983), and this has been supported by the limited amino-acid evidence available (Rose, 1991).

Because of these difficulties, the present account is confined to the description of the field evidence that enables the stratigraphy and a tentative history to be deduced for the Pleistocene deposits of the district. The sequence and its probable correlation with adjacent areas is summarised in Table 12 and is shown graphically in Figure 41. At least two glacial and two temperate episodes are represented by Pleistocene deposits in the district, but the deposits are mostly geographically sepa-

Table 12   Correlation of the Pleistocene sequences of the King's Lynn district with those of adjacent areas.

Deposits			Approximate age in years before present	Presumed sea level relative to present in King's Lynn district	Erosional features in the Fenland region
King's Lynn district	Southern and central Fenland	Humberside and Lincolnshire			
Solifluction deposits	Solifluction deposits; formation of cryoturbation and ground-ice features	Solifluction deposits; Roos Bog peats	10 000 to c.15 000	Rising from −100 to −30 m	Meltwater erosion of solid and older Pleistocene deposits
Hunstanton till and associated sand and gravel in northern part of district		Skipsea and Withernsea tills and associated sand and gravel; Dimlington Silts	c.15 000 to c.25 000	−100 to −120 m	
Solifluction deposits and ground-ice features in the south	Lowest terraces of Gt Ouse and Cam and Lark; Wretton Terrace of River Wissey	'Fen edge gravels'	c.25 000 to >40 000	Slightly lower than present	?Formation of Fenland 'islands'
Hunstanton raised beach; Upper Tottenhill Gravels	?March Gravels; higher terraces of Cam and Lark	Sewerby raised beach	>40 000	+3 to +5 m	Formation of Hunstanton buried cliff; meltwater erosion of older deposits
Lower Tottenhill Gravels; possible offshore till	?Highest terraces of rivers Cam and Ouse; Head Gravel	?Basement till of Bridlington		Lower than present	Extensive erosion of earlier Pleistocene deposits
Nar Valley Clay	?Salt marsh clay at March	not known	>40 000	Rising from −8.5 to +30 m	
Nar Valley Freshwater Beds	not known			Rising from −20 to −8.5 m	
Varved clays passing down into Chalky-Jurassic till with associated sand and gravel deposits	Chalky-Jurassic till with associated sand and gravel deposits	Chalky-Jurassic till with associated sand and gravel deposits		−100 m	Modification of fluvial valleys; formation of scour hollows and tunnel valleys; massive erosion of Cretaceous escarpments
Preglacial weathering produces				Falling from 0 to −100 m	Preglacial valley system

rated. The sequence contains three unconformities of unknown duration (Figure 41), and the relative ages of some of the deposits cannot therefore be directly demonstrated. A description of the deposits and the evidence on which the sequence is based is given below.

Chalky-Jurassic till

The oldest Pleistocene deposits in the district are the Chalky-Jurassic till[1] and associated sand and gravel. They infill a topography cut into the solid deposits and are made up almost entirely of Upper Jurassic and Cretaceous rocks derived from Lincolnshire and Norfolk. The till now occurs as dissected remnants of what may originally have been an almost continuous deposit. Thick till sequences (up to 80 m) are confined to deep valleys in the rock-head surface in the King's Lynn area and beneath Fenland and The Wash; thin patches (mostly less than 10m) of till occur on the interfluves and on some valley sides in the Castle Rising, Ashwicken and Middleton areas (Figure 42). In places the valley till is continuous with that extending across the interfluves and no evidence has been recorded to suggest that more than one phase of deposition of Chalky-Jurassic till occurred.

Gravels and gravelly sands composed of flint and the other more resistant erratics in the till occur in association with the boulder clay in the Babingley valley, Grimston Warren, Ashwicken, Middleton and Blackborough End (Plate 14). They have been worked on a small scale for hardcore (see Chapter Seven for details), but are now poorly exposed. Their relationship to the boulder clay is complex. On the north side of the Nar valley at Blackborough End [675 144], gravels are banked against a cliff cut in Sandringham Sands (Plate 15) and appear to be part of a lateral moraine which underlies the main mass of the valley till. In the central part of the valley at Wormegay [681 128], similar gravels form a ridge surrounded by till. Elsewhere, glacial gravels of apparently similar age occur beneath (e.g. at Leziate [677 192]) and on top of (e.g. in the Babingley valley [703 254]) the till, although this latter feature may result from preferential erosion of boulder clay from above and around the gravel. Contacts between the boulder clay and gravel can be sharp or gradational.

The Chalky-Jurassic till everywhere in the district is characteristically composed of slightly sandy clay with fine- and medium-gravel-sized erratics of chalk (usually more than 90 per cent of the clasts), flint, Upper Jurassic cementstone and shale, Lower Cretaceous sandstone and ironstone, Red Chalk, Carstone and rare farther-travelled materials. Although laterally very variable, the till shows a general increase in degree of sorting, being mostly very heterogeneous on the higher ground, relatively well sorted and more uniform in the lower valley sides, and locally becoming well sorted and passing laterally and upwards into varved clays in the valley floors. The lowest part of the till is commonly rich in locally derived materials, and there is evidence in the district, notably at Rising Lodge and Brow of the Hill, that very large masses (greater than $20 \times 20 \times 10$ m) of solid deposits have been incorporated into the lower part of the till. Erratic blocks of Upper Jurassic clay, several metres thick, were proved in several boreholes in Fenland.

On the Upper Jurassic and Lower Cretaceous outcrops, dark grey clayey tills, derived from the Upper Jurassic and containing Jurassic cementstone and Lower Cretaceous erratics, are overlain by paler grey, more calcareous clays rich in chalk and flint erratics. Evans (1975) has suggested that this lithological difference indicates a readvance of an upper (chalky)-till ice across stagnating lower-till ice. In many sections, however, there is no sharp junction between the two lithologies and it seems likely that the lower till merely represents a basal layer of slower moving ice. Blocks of densely pyrite- and

1 The term till is used in this account to describe all the deposits formed as the ground moraine of a continental ice-sheet. The bulk of this material is shown on the map as Boulder Clay: it includes within it pockets of sand and gravel of varying sizes. Where these latter are sufficiently large to be mapped, they are shown on the map as Glacial Sand and Gravel.

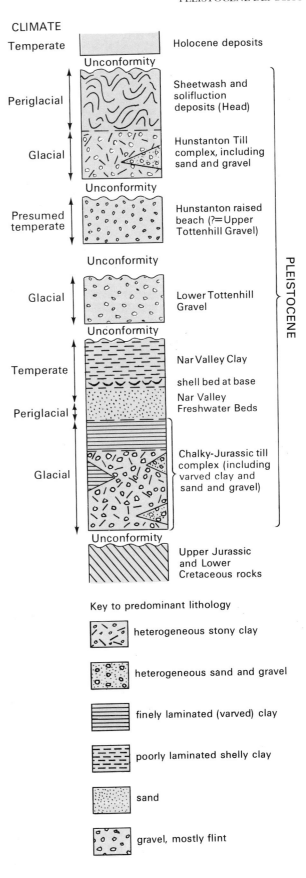

Figure 41 Summary of Pleistocene stratigraphy.

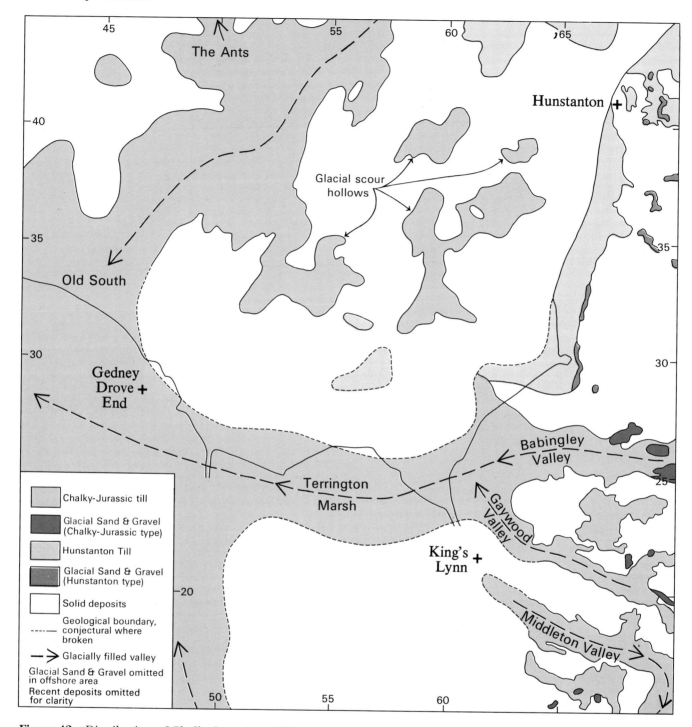

Figure 42 Distribution of Chalky-Jurassic and Hunstanton tills and associated sand and gravel.

calcite-cemented sandstone, up to 2 m across, derived from the basal bed of the Sandringham Sands but commonly referred to in the literature as 'Spilsby Sandstone', have been recorded in all the large exposures of Chalky-Jurassic till dug in recent years in the King's Lynn area.

The erratic content of the till suggests derivation from the north-west, from the broad Upper Jurassic clay vale lying between the escarpments of the Lincolnshire Limestone and the Chalk. A more westerly source would have included more Triassic and Lower and Middle Jurassic rocks. A more northerly source would have included more flint from the Middle and Upper Chalk, and could not have provided the Upper Jurassic clays because these rocks have only a very limited outcrop in the North Sea.

On the eastern edge of Fenland, between Wolferton and Tottenhill, the Lower Cretaceous escarpment is broken by four east-west valleys whose floors and sides are cut in Chalky-Jurassic till (Figure 42). At present these

Plate 14   Glacial sand and gravel, Blackborough End gravel pits [6760 1457].

Coarse, poorly sorted, flint-rich gravels are complexly interbedded with clean sands, dirty sands and, in
their lower part, heterogeneous clayey sandy gravels, in sections up to 8 m high. Imbrication, cross-
bedding and ripple-drift bedding are common at many levels, and cryoturbation structures occur within
(as seen here) and in the upper part of the sequence. The beds dip towards the modern valley and
probably pass beneath Chalky-Jurassic till. They appear to have been deposited as a lateral moraine
banked against a valley side cut in Sandringham Sands and Carstone. (A 11804).

valleys carry small streams, the Babingley River, the Gaywood River, the Middleton Stop Drain and the River Nar,
but the presence of up to 70 m of glacial deposits within
them (Figure 43) suggests that the valleys and streams
were formerly much larger.

The glacially filled Babingley valley rises near Great
Massingham and from there can be traced westwards, via
boreholes at Grimston, Congham and Hillington, to the
present-day Babingley valley at Castle Rising. West of Castle Rising, the valley passes beneath the Recent sediments of Fenland and its course is presumed from borehole evidence to continue almost due west beneath Terrington Marsh until it meets another glacially filled valley
in the vicinity of the River Nene.

The glacially filled Gaywood valley can be traced from
Roydon Common [690 218] southwards and then westwards beneath the valley of the present day Gaywood

River to Gaywood Bridge [638 214]. West of there, the
course of the glacially filled valley is hidden by Recent
deposits. There is insufficient borehole data to indicate
its position beneath the northern part of King's Lynn,
but it probably turns northwards to become a tributary of
the Babingley valley.

The glacially filled Middleton valley lies beneath the
present valley of the Middleton Stop Drain between
Fairstead and Middleton Towers Station. West of
Fairstead, its course is lost beneath Recent sediments
under King's Lynn, where there is insufficient borehole
data for it to be traced. East of Middleton Towers, the
valley rapidly broadens until, at East Winch, it merges
with a broad outcrop of glacial deposits, which is connected to the Nar Valley.

In the Cretaceous outcrop area, where it can best be
studied, the glacially filled Nar valley, which forms the

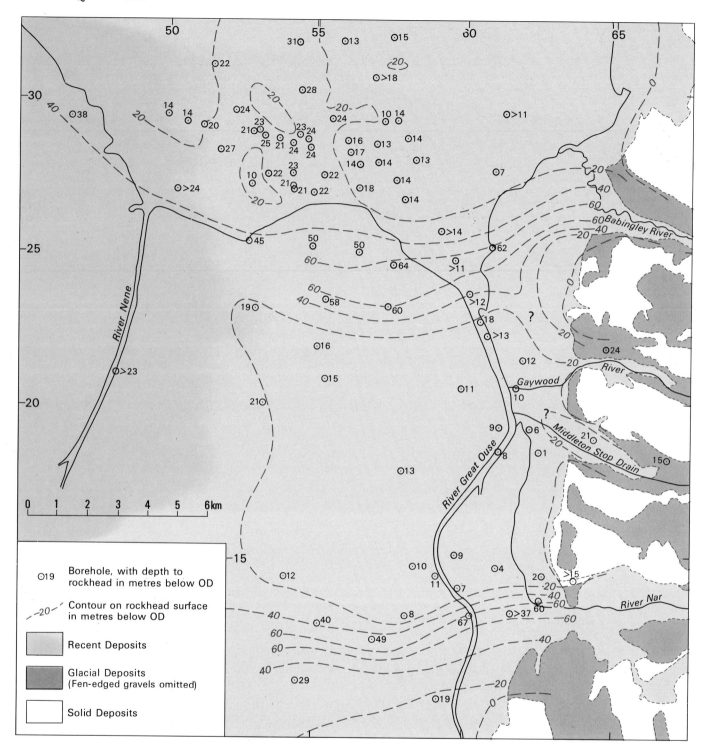

Figure 43   Rockhead contour map of King's Lynn area.

southern boundary of the present district, is the longest and broadest valley in west Norfolk. However, beneath Fenland, where it is cut in Upper Jurassic clays, it appears to be no wider or deeper than that of the Babingley River. The start of the glacially filled valley is hidden beneath the thick glacial deposits of the boulder clay plateau in the area of Litcham [886 177] and it is not

known whether or not the valley is a closed basin. West of Litcham, the course of the valley is well defined by boreholes at Lexham, Castle Acre and Narford [765 140]. Westwards from Narford its course beneath extensive spreads of Pleistocene and Recent deposits is uncertain, but it probably turns north-westwards at Narborough and then south-westwards at West Bilney to pass

Plate 15   Glacial sand and gravel banked against Leziate Beds, Blackborough End gravel pits
[6759 1460]
Flint-rich glacial gravels are banked against a series of small cliff-like features (arrowed) cut into soft white
and green, glauconitic Leziate Beds sands. The gravels dip towards the modern valley: their basal beds are
poorly sorted and contain large blocks of Carstone and other locally derived materials. (A11802).

beneath the Recent deposits of Fenland near Blackborough End. Boreholes at Wiggenhall St Peter, Lordsbridge, Spice Chase and West Walton show its course to continue westwards from Blackborough End to the Nene.

These four valleys are a small part of a complex concealed drainage system that underlies Fenland and The Wash (Figure 44). The positions of most of the glacially filled valleys are conjectural because of the scarcity of boreholes in the area. In southern Fenland, the River Wissey follows the course of a drift-filled valley that appears to be locally up to 70 m deep between West Tofts and Stoke Ferry. Boreholes at Boston, Crowland, Long Sutton, March, Salters Lode and Wisbech have proved thick drift sequences which may indicate the positions of valleys associated with a glacially filled Great Ouse drainage system. Horton (1970) has traced the drift-filled valley of the River Great Ouse from Stony Stratford as far north as Huntingdon. The possible course of the drift-filled valley of the Great Ouse beneath Fenland must be largely speculative. Drift sequences over 40 m thick have

been proved at Long Sutton, Tydd St Mary, Boston, Fosdyke and Wisbech.

Seismic profiling within The Wash proved the presence of several deep glacially filled valleys (Wingfield et al., 1979). These valleys, together with those in Fenland, have been interpreted (Gallois, 1979a) as part of a preglacial river system that was graded to a sea level about 100 m below that of the present day. This system, which consisted of the Great Ouse, Nene, Cam, Lark, Wissey, Little Ouse, Welland and part of the Trent, probably drained through a single valley, that of The Wash River or Llwnc Gwonow, which breached the Chalk escarpment in the central part of The Wash (Figure 44) beneath the present-day Well deep.

The reasons for interpreting this drainage system as one of fluvial origin, albeit glacially modified, rather than one of wholly subglacial origin can be summarised as follows:

1. The glacially filled valleys show no preferred orientation but appear to form a dendritic pattern draining towards a single outlet in The Wash.

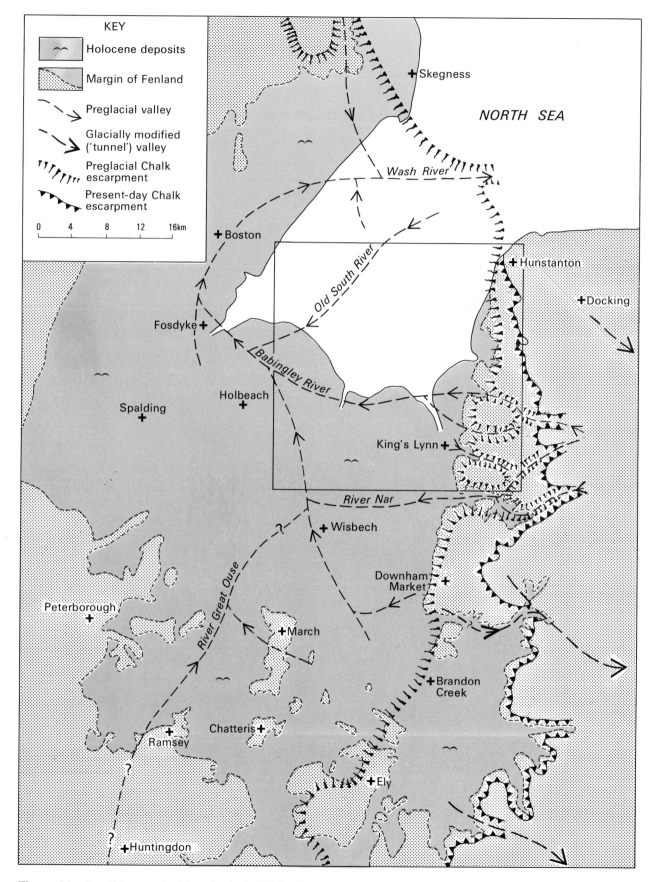

Figure 44   Possible preglacial and subglacial drainage system of Fenland and The Wash.

2. The glacially filled valleys of the Babingley, Gaywood, Nene and Ouse rivers are closed basins with outlets only into Fenland.

3. The cross profiles and long profiles of the glacially filled valleys beneath The Wash can be seen, in the seismic sections (Figure 45), to have gentle slopes and to show many of the subtleties of form that are associated with fluvial erosion and subaerial weathering.

4. The lithology of the infilling material in the valleys is remarkably constant, being Chalky-Jurassic till in the lower parts and water-sorted till and varved clay in the upper parts.

5. The contact of the till and the underlying Upper Jurassic clays is almost everywhere gradational and indicative of erosion by ice rather than by subglacial waters. Chalk-rich till commonly passes down into Jurassic clay-rich till that rests on in-situ Jurassic clay cut by polished shear surfaces.

Woodland (1970) has interpreted the borehole data from the boulder clay plateau areas of central and east Norfolk, adjacent to the district, as showing tunnel valleys cut into the Chalk surface. Cox and Nickless (1972) have used a similar explanation to account for the extensive spreads of glacial sand and gravel in the Norwich area, and Sparks and West (1965) considered the buried valley beneath the upper reaches of the River Cam to have resulted from subglacial erosion. Wingfield (1990, fig.3) has suggested that the glacially filled valleys beneath the Nar valley and The Well are incised hollows formed as plunge pools at the edge of sea-ice. There are well-documented examples of a large number of such hollows, up to 360 m deep and 30 km long, in the North Sea but their form and infilling materials are unlike that of the glacially filled Nar valley, and are probably unlike those of the more poorly known, Well-deep valley.

The apparent presence of both a preglacial valley system and a tunnel-valley system beneath the Chalky-Jurassic till of adjacent parts of Norfolk is not incompatible. Boreholes have proved drift-filled valleys more than 100 m deep in the area just east of the present day Nar–Wensum watershed (Woodland, 1970). Such valleys do not appear to extend westwards to breach the Chalk escarpment between Hunstanton and Stoke Ferry, but rise irregularly eastwards towards the position of a former ice front (Cox and Nickless, 1972). The present-day watershed, between streams flowing westwards to the Ouse drainage and eastwards to the Yare drainage, approximately marks the dividing line between the areas of presumed preglacial and subglacial drainage (Figure 44). This line lies a little to the east of what in preglacial times must have been an impressive Chalk escarpment comparable to those of the present-day Chilterns and the North and South Downs. The roots of this escarpment now occur as a belt of almost drift-free Chalk separating the patchily drift-covered areas of west Norfolk from the completely drift-covered till plateau of central Norfolk. Debris from this escarpment is scattered throughout the Chalky-Jurassic till of East Anglia.

The position of the former escarpment is indicated by the presence, at a number of localities, of exceptionally large erratics within the Chalky-Jurassic till. These erratics preserve parts of what was the local stratigraphy, at localities several kilometres west of the present day outcrops of the rocks they contain. At Rising Lodge, King's Lynn [668 230], there are transported masses of Gault and Chalk which are probably hundreds of metres in length, and at Leziate [677 192] the till encloses a raft of Carstone and Gault about 50 m long. Farther south, at Downham Market and Ely, similar large erratics of Cretaceous rocks occur in the Chalky-Jurassic till.

To the east of the presumed position of the former Chalk escarpment, the rivers Wissey, Lark and Cam have buried valleys which have been interpreted as subglacial valleys draining south-eastwards or southwards (Woodland 1970, pl. 1). To the west of the former escarpment, the rivers Babingley, Nar and Great Ouse appear to have preglacial valleys infilled by glacial deposits (Figure 44). The Chalk escarpment seems, therefore, to have been a temporary barrier to the ice movement and to have separated an area of predominant erosion in the west from one of deposition in the east. The preglacial valleys are interpreted as scarp-slope valleys that have suffered little glacial modification The 'tunnel valleys' are probably former dipslope valleys which have undergone extensive modification.

Contours on the rockhead surface in the southern part of the district, assuming a preglacial valley system to be present, are given in Figure 43. It should be recognised that the true rockhead surface is likely to have a much more complex shape than that indicated by the limited amount of available borehole data. Comparison of the subglacial topography seen in the offshore seismic sections with that of the land area on the eastern side of Fenland, suggests that the general shape of the surface shown in Figure 45 is probably correct but, in addition to the two main valleys, a number of short tributary valleys and scour hollows are probably also present. In the offshore area, lying within a concealed plateau of solid deposits which stretches from the Old South valley to the Norfolk coast, there are a number of pockets of till which rest on a rockhead surface at 20 to 40 m below OD. These features appear to be glacial scour hollows infilled with Chalky-Jurassic till (Figure 42).

The till in the valley sequences generally becomes less stony and better sorted in its upper part and, in the valleys of the Babingley, Middleton Stop Drain and Nar, becomes interbedded with varved clay before finally passing up into varved clay. A trench in the Middleton Stop Drain valley at Tower End [650 174 to 668 178] showed that varved clay in the central part of the valley passed laterally, via water-sorted almost stoneless till, into heterogeneous Chalky-Jurassic till in the valley side. When traced away from the valley floor towards the interfluve the stones in this till become larger and the till generally more heterogeneous. Elsewhere in the district, lenses of varved clay occur within the till, almost always in valleys.

Details

Babingley Valley

The glacially filled Babingley valley rises near Great Massingham [780 210] and from there runs westwards to Grimston as a

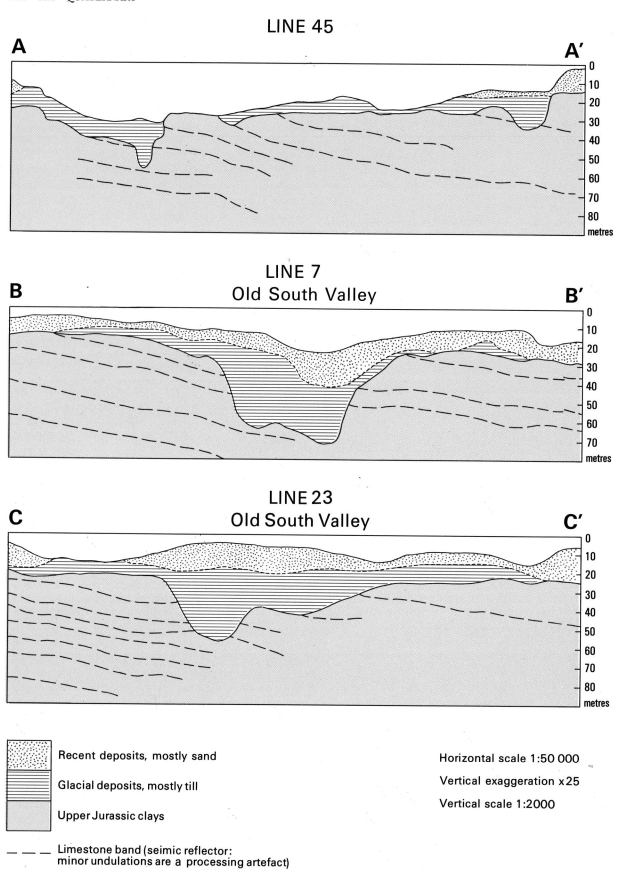

Figure 45 Profiles across buried, drift-filled valleys in The Wash.

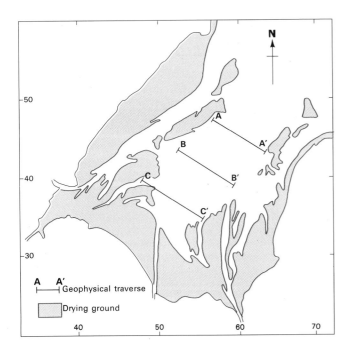

A   A'
├───┤ Geophysical traverse

▦ Drying ground

40   50   60   70

chalk-valley floored by flint-rich gravels, with patches of similar drift on the sides. Boreholes at Grimston [7352 2284], and Congham [7229 2326] proved up to 55 m of drift. They appear to be sited within a well-defined, steep-sided valley, because Lower Chalk crops out close to both boreholes. At Congham, the valley must turn northwards to follow a broad outcrop of till that links Congham with the present-day valley of the Babingley River: all possible southerly and westerly courses are blocked by an unbroken outcrop of Gault.

The glacially filled valley joins that of the present Babingley River near the old flax mill where a borehole [693 256] proved the drift to be more than 30 m thick. The easterly continuation of the present Babingley valley through Hillington to Harpley Dams appears to be a more recent feature, floored only by thin drift deposits.

Between the old flax mill and Castle Rising, the glacially filled valley is again well defined and 1.25 to 1.5 km wide. There, it is cut in Lower Cretaceous rocks and infilled by Chalky-Jurassic till and glacial sand and gravel. West of Castle Rising, the valley passes beneath the Recent sediments of the marshlands and its course is presumed to continue almost due west via boreholes at Vinegar Middle [6092 2505] and Ongar Hill [5767 2444]. Seismic profiles along the River Great Ouse (Floyd, 1973), between its outfall at Admiralty Point and Lynn Road bridge, were difficult to interpret in the area where the glacially filled Babingley valley is thought to pass beneath the river. A glacially filled channel, about 70 m deep, can be made out in the section between West Bank Beacon [598 241] and Admiralty Point [590 255], although not clearly enough for the shape of the valley to be determined. The glacially filled valley probably continues westwards from the Great Ouse to the River Nene in the vicinity of the Nene outfall; there are boreholes on its northern flank at Walker's Marsh [5268 2530] and Gedney Drove End [4665 2939]. The valley continues westwards towards Boston and may unite with the Old South glacially filled valley near Holbeach St Marks.

The limited amount of evidence available suggests that the cross profiles of the glacially filled Babingley valley are similar to those of the glacially filled valleys proved by seismic work within The Wash (Wingfield et al., 1979), being narrow and steep-sided in the Chalk, less narrow and steep in the Lower

Cretaceous and broad, with ill-defined boundaries, in the Upper Jurassic clays.

The Chalky-Jurassic till and associated sand and gravel have extensive outcrops on the lower slopes of the Babingley valley between Wolferton [650 270] and Hillington Park [720 260], and between North Wootton Marsh [640 250] and Hillington Village [710 255]. The till is everywhere poorly exposed and is mostly covered by a veneer of cryoturbated sandy and gravelly sand wash 1 to 3 m thick. Grey clay with many chalk and flint erratics is exposed in the deeper ditches and can be augered in the numerous marl pits that have been dug into the outcrop.

Large discrete masses of sand and gravel occur at St Felix's Church [665 262], near Cat's Bottom [675 267], at Vincent Hills [693 263] and upstream [704 253] from the Depot. The gravels at St Felix's Church form a raised, flat-topped area and may be a remnant of a former sheet of gravelly head, similar to that which has extensive outcrops upstream in the valley. The gravels near Cat's Bottom have been extensively worked [674 268; 683 270; 679 270] in pits up to 6 m deep. All are in flint- and chalk-rich gravels. The last locality showed small exposures of flint cobbles and chalk pebbles with rare lenses of sand, a few erratics of ferruginous Dersingham Beds, Carstone, quartzite and quartz, and rare metamorphic and igneous rocks. This erratic suite is very similar to that of the sand and gravel associated with the Hunstanton Till (see p.153), and it is possible that this outcrop is the most southerly deposit of that glaciation yet recorded. The junction of the gravels and the Chalky-Jurassic till is obscured by wash.

Sand and gravel has been worked in small pits [e.g. 694 264] at Vincent Hills, and continues to be worked sporadically in a shallow, partly flooded pit [704 258] upstream from the Depot. At this last locality, up to 3 m of gravel has been exposed, consisting of crudely bedded flint- and chalk-rich angular pebbles in a chalky and sandy matrix. An upper layer of dark brown sand with bleached flints, a decalcified derivative of the lower layer, rests with a very irregular cryoturbated contact on the chalk-rich gravels in most sections.

A striking feature of the glacial deposits in the lower part of the Babingley Valley, are clusters of ridges and hollows (2 to 3 m in differential height) which form an undulating topography known as 'hummocks and hollows'. Elsewhere at the edge of Fenland, Sparks et al. (1972) have suggested that similar structures have resulted from the freezing and thawing of large masses of ground ice. In the Babingley Valley, especially good examples occur near Whalley Farm [674 255 to 690 254] and upstream from the Depot [697 257 to 710 255] in which the ridges are composed largely of gravelly sand and the hollows are underlain by a veneer of sandy wash and peat which rests on Chalky-Jurassic till.

Castle Rising to South Wootton

There is no permanent exposure in the large outcrop of Chalky-Jurassic till centered on Rising Lodge, but temporary sections have revealed it to be laterally variable and to have a complex junction with the underlying Sandringham Sands. The cuttings [666 227 to 664 224] for the King's Lynn Bypass, south of Rising Lodge, proved up to 10 m of chalk- and flint-rich grey clay till with common large 'Spilsby Sandstone' erratics (calcareous sandstones from the Spilsby Sandstone and Roxham Beds). A borehole [6680 2300] at Rising Lodge proved 7 m of loose sand in an area where large pockets of sand, flint gravel and possible large erratic masses of solid deposits occur within the till. An old marl pit [6653 2365] north east of Rising Lodge appears to be in very disturbed brown and grey Gault clay; a similar pit [6711 2285], south east of the Lodge, contains till composed almost entirely of Gault and Lower Chalk lithologies. Whitaker and Jukes-Browne (1899, p.34) noted layers of pinkish marl in the

Gault in the first of these pits and presumed the Gault to be in situ in this and several other pits in the Rising Lodge area. The disturbed nature of the beds and their complex relationship to the till suggest that a large outlier of Gault, possibly in part capped by Lower Chalk, was formerly present at Rising Lodge, but has now been incorporated into the till.

Between Warren Farm [668 219] and Rising Lodge several broad topographical features can be traced around the hillside. On each of these, the steeper slopes are made of Chalk-rich or Gault-rich till and the intervening gentle slopes have a thick covering of loose, flint-rich, sandy head. These topographical features may reflect features in solid formations (Carstone, Gault and Chalk) concealed by a veneer of till.

An example of the irregular nature of the base of the till was recorded in a road cutting [6913 2423] at White Hills Wood, where a steep-sided pocket of till occurs within the Dersingham Beds, 150 m north-east of the main till outcrop. On its south-western side, the base of the till has an irregular 40 to 60° dip; on its north-eastern side, the junction is vertical. The bulk of the till consists of very heterogeneous, dark grey clay, sandy clay and chalky clay with many flint and chalk pebbles and common 'Spilsby Sandstone' blocks. Over part of the section, the highest part of the till consists of reddish brown clayey sand with much Carstone debris, and without chalk. This sandy till rests with an irregular, cryoturbated contact on the chalky till; it is reminiscent of the Hunstanton Till of more northerly areas, but is probably a decalcified derivative of the chalky till.

Gaywood Valley

The glacially filled Gaywood valley starts at Roydon Common [690 218] and runs southwards and then westwards beneath the valley of the present-day Gaywood River to Gaywood Bridge [638 214]. For most of its length, the glacially filled valley is about 1 km wide and probably has steep sides. A temporary exposure in 1965 on the King's Lynn Bypass [660 217], on the northern side of the valley, showed a vertical contact between Sandringham Sands and chalk-rich till. Field evidence indicates that very steep contacts occur throughout the length of the northern side of the valley. A cored borehole [648 216] close to the northern side of the valley at Reffley proved 30 m of Chalky-Jurassic till resting on Kimmeridge Clay.

West of Gaywood Bridge, the course of the glacially filled valley is hidden by Recent deposits and there are insufficient borehole data to indicate its position beneath the northern part of King's Lynn. Seismic profiles along the river Great Ouse clearly indicate Kimmeridge Clay at shallow depths for most of the section between West Bank Beacon [598 241] and Wiggenhall St Germans Bridge [596 141], and it therefore seems unlikely that the glacially-filled Gaywood valley passes beneath the Ouse. Only at Lynn Docks, where interference from quayside structures complicates the seismic records, might a narrow, glacially filled valley have remained undetected. The glacially filled Gaywood valley probably either turns sharply southwards at Gaywood to join the glacially filled Middleton valley beneath King's Lynn, or turns more gently northwards to become a tributary of the preglacial Babingley to the north of King's Lynn.

The Chalky-Jurassic till crops out on the lower slopes of the valley on the south side of Roydon Common, between Chilver House [682 200] and the King's Lynn Bypass [656 210], and between Grimston Warren [682 215] and Reffley Farm [641 216]. It is everywhere poorly exposed and mostly overlain by 1 to 3 m of flinty sand head, but there are numerous old marl pits dug into the till and it is exposed in many of the deeper drains. It consists of relatively uniform chalk- and flint-rich clay with common 'Spilsby Sandstone' boulders.

Sand and gravel associated with the till has been worked at Grimston Warren [6792 2133] and Waveland Farm [6765 2078] in pits up to 3 m deep. At the latter locality, reputedly dug for ballast for the adjacent railway line, small exposures were seen in flint-rich gravels which dip 10° south-east, and which were much affected by cryoturbation in their upper part. The gravels are lithologically varied, and range from angular flint grit to well-rounded flint cobbles, with complex lenticular bedding.

Middleton Valley

The glacially filled Middleton valley is well defined between Fairstead [645 190] and Middleton Towers Station [670 180]. It is about 0.5 to 0.75 km wide, probably steep-sided and cut through the Lower Cretaceous rocks into Kimmeridge Clay. A borehole in the middle of the valley near Fairstead [6424 1888] proved 21.5 m of Chalky-Jurassic till on Kimmeridge Clay. West of Fairstead, its course is lost beneath Recent sediments under King's Lynn. East of Middleton Towers, it rapidly broadens until, at East Winch, it merges with a wide tract of glacial deposits which is connected to the Nar valley. Boreholes at East Winch have recorded up to 72 m of glacial deposits resting on Kimmeridge Clay.

The apparent absence of thick drift deposits beneath King's Lynn, the indication from seismic profiling data that no glacially filled valley passes beneath the Great Ouse in that area and the apparent easterly deepening of the valley, all suggest that the glacially filled Middleton valley runs eastwards from Fairstead to East Winch and from there southwards to Blackborough End to become a tributary of the preglacial Nar.

Chalky-Jurassic till occupies the lower slopes of the valley between Mintlyn Farm [645 193] and Leziate [688 176], where it is again poorly exposed. A shallow valley infilled with till at Holt House Farm [675 184] connects the lower outcrop with an outcrop of very heterogeneous till that caps the ridge at Brow of the Hill [678 192]. Large exposures [6772 1916 to 6788 1922] in the face of the working sandpit showed a complex sequence of up to 10 m of till which included pale-coloured, chalky till overlying dark grey, clayey till-rich Jurassic clays; a large erratic mass of Carstone overlain by Gault (see p.110 for details); an erratic mass of Leziate Beds sands thrust over laminated clays; and large pockets of poorly sorted, coarse, flint-rich gravel.

Flint-rich gravels have been worked in a large, shallow pit at Ashwicken [696 187] and in several pits [e.g. 6940 1990; 6925 2005] in the narrow strip of gravel east of Chilver House.

The deep trench dug for the No. 4 Gas Feeder Main provided a continuous section between Fair Green and Middleton Towers Station. At Fair Green it showed Kimmeridge Clay-rich till with many chalk pebbles resting on Sandringham Sands [at 6583 1740]. When traced north-eastwards, the erratic content of the till became rarer and the individual stones smaller; the sandy clay matrix became better sorted, and wisps and lenses of silt and waterlaid till became more common, until [at 6605 1750] the till passed laterally into brown and grey stoneless clays with rare lenses of chalky till. These clays passed north-eastwards into reddish brown, laminated (varved) clays. Varved clays were seen beneath a cover of gravelly sand head in the gas-pipeline trench [6684 1778] at Middleton Towers Station, and in the deeper ditches in the adjacent central parts of the valley.

North Runcton to Middleton

There are numerous old marl pits, but no permanent exposure, in the large outcrop of till that extends from north of North Runcton to east of Middleton. As at Rising Lodge, there is evidence to suggest that the base of the till is irregular, and that the till is locally thin. An excavation [6707 1616] for a reservoir east of Middleton, showed only 1m of chalky clay till resting on a level surface of Carstone. A section 40 m east, showed 3 to 5 m of similar till on Carstone.

Nar Valley

The start of the glacially filled Nar valley is hidden beneath the thick glacial deposits near Litcham [886 177]; west of there, its course is well defined, with steep sides cut in Chalk. Boreholes within the valley, but close to the Chalk outcrop, at Lexham [8660 1715] and Castle Acre [8156 1529] proved drift deposits up to 80 m thick.

From Castle Acre to Narford [765 140], the glacially filled valley is generally well defined and 1 to 1.5 km wide. Westwards from there, at Narborough, it probably turns north-westwards to meet its tributaries, the Gaytonthorpe valley and the Middleton valley, in the West Bilney area. At West Bilney, it turns south-westwards and passes beneath the Recent deposits of Fenland near Blackborough End.

Several boreholes in the Setchey area have proved thick glacial deposits, mostly till overlain by varved clay, beneath the valley of the present-day Nar. A seismic survey along the Great Ouse proved these glacial sequences to be present everywhere between St Germans Bridge [596 141] and the railway bridge [597 104] at St Mary Magdalen. A borehole at Wiggenhall St Peter [6030 1319] was sited on the basis of this work and proved 69 m of drift deposits. Westwards from there, boreholes at Lordsbridge [5704 1244], Spice Chase [5513 1292] and West Walton [4913 1316] proved glacial sequences from 43 to 50 m thick, and are thought to be on the flanks of this preglacial valley (Figure 43).

The field and borehole evidence suggests that the glacially filled Nar valley is asymmetrical in cross profile in the Blackborough to Wiggenhall area, with a very steep northern side and a relatively gentle southern side. Almost vertical contacts between glacial deposits and the Sandringham Sands occur in Blackborough End Gravel Pits [676 145]. Between there and Setchey, the field relationships of the glacial deposits and the solid rocks suggest similar steep contacts. Boreholes at Setchey [627 138 and 6375 1330] and Lordsbridge [5704 1244] also suggest a steep-sided glacially filled valley.

The large outcrop of sand and gravel that extends from Crancourt Manor [687 154] to Blackborough [671 141] has been worked over a long period in extensive pits up to 6 m deep. The larger of these [682 146; 675 145] were sporadically in work at the time of the survey and contained a large number of exposures up to 5 m high. These showed cross-bedded, fine to coarse flint gravels interbedded with gravelly sands and some thick beds of sand (Plate 14). Lenticular cross-bedding, dipping south to south-east towards the adjacent valley, was present in most sections, and the highest 1 to 2 m of beds were everywhere affected by cryoturbation. On its northern side [673 145 to 677 147] the second of these pits has been extended back to work the Leziate Beds and Carstone. Throughout most of its length, the junction of the gravels and the Leziate Beds is a cliff-like feature cut into the soft sands (Plate 15).

Offshore area

The offshore seismic and sampling surveys for the Wash Feasibility Study (Wingfield et al., 1979) proved extensive subcrops of Chalky-Jurassic till beneath Recent sediments in the Old South valley and in glacial scour hollows between the valley and the Norfolk coast (Figure 42). Small outcrops occur in the floors of Old Lynn Channel [525 350], Gat Channel [490 395] and Daseley's Sled [585 360], but these are largely covered by a veneer of gravelly sand.

Nar Valley Beds

The Nar Valley Beds comprise a freshwater unit, the Nar Valley Freshwater Beds, overlain by a marine unit, the Nar Valley Clay.

The older formation is everywhere overlain and overstepped by the younger and has no outcrop in the district. It is known only from boreholes in the valleys of the Nar and the Middleton Stop Drain. The Nar Valley Clay crops out in both these valleys; at the edges of its outcrop it rests with marked unconformity on a variety of older rocks. Much of the outcrop is now concealed by thick deposits of sandy head and Recent deposits.

In the central parts of the Nar and Middleton Stop Drain valleys, where the Pleistocene sequence is most complete, the Nar Valley Freshwater Beds appear to rest conformably on varved clays associated with the Chalky-

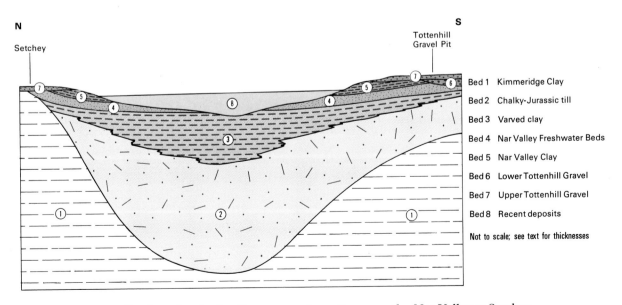

Figure 46 Generalised section in the Quaternary deposits across the Nar Valley at Setchey.

Jurassic till (Figure 46). Boreholes in the Nar valley near Setchey [6458 1394; 6377 1435] proved fine-grained, grey, silty sands and silts (basal part of the Nar Valley Freshwater Beds) resting on grey silts with widely spaced laminae of reddish brown clay (highest part of the varved clay sequence) which passed down into varved clay. The varved clay was probably deposited in a series of small, ice-dammed lakes during the retreat phase of the glaciation that produced the Chalky-Jurassic till. The Nar Valley Freshwater Beds, from which Stevens (1960) obtained a flora indicative of an ameliorating climate, may therefore represent the last stages of the glaciation which produced the Chalky-Jurassic till and the early stages of the subsequent interglacial stage. Lenses and discrete beds of freshwater peat are common in the highest part of the Nar Valley Freshwater Beds.

The Nar Valley Clay consists of finely laminated marine clays, silts and silty clays and has been proved at outcrop (Stevens, 1960) and in boreholes to overlie the Nar Valley Freshwater Beds. At East Winch, West Bilney and in the Nar valley, the base of the clay is marked by a shell bed composed almost entirely of the large oyster *Ostrea edulis* Linné; at many localities it rests on freshwater peat. The age of both the peat and the oyster bed have been shown by radiocarbon dating of samples (IGS C14/129 and IGS C14/222) from a borehole [6502 1419] at Setchey to be greater than 40 000 years.

At the mouth of the Nar valley, the junction of the marine and freshwater sequences in the central part of the valley is at about 8.5 m below OD. The junction rises towards the valley sides to about 4 m OD at Tottenhill Gravel Pit [633 115] in the south and 2.0 m below OD at Setchey village in the north. The junction also rises upstream. It is at more than 6 m OD in the valley side at Horse Fen, West Bilney [692 142] (Stevens 1960, p.295), and at about 13 m OD at East Winch [705 116].

The former brickpits in the Nar Valley Clay yielded a rich fauna of beautifully preserved bivalves and gastropods (see Table 13, based on Whitaker et al., 1893, pp.84–85), foraminiferida (Funnell in Lord and Robinson, 1978), ostracods (Lord and Robinson, 1978; Athersuch et al., 1985; Lord et al., 1988) and diatoms (Mitlehner, 1992) that indicate deposition in shallow, brackish-marine and marine environments. All but one of the bivalve and gastropod species still live in British waters, and almost all occur in the present-day Wash. Rose (quoted in Whitaker et al., 1883, p.84) noted that the clay also yielded bones and teeth of mammals, some now extinct, which presumably strayed onto the shallow mudflats of the Nar Valley estuary. These include teeth of '*Bos*' (?Aurochs), '*Elephas primigenius* Blumenbach' (probably *Mammuthus*, Mammoth), *Equus cabalus* Linné (probably *Equus ferus* Boddaert, Horse), '*Rhinoceras tichorhinus* Cuvier' (probably *Coelodonta antiquitatis* (Blumenbach), Woolly Rhinoceras) and antler fragments of '*Cervas elephas* Linné' (Red Deer).

The Nar Valley Clay was clearly deposited in a rising sea that transgressed eastwards across the Nar Valley Freshwater Beds and older Pleistocene deposits. Rose (1835–36) recorded exposures of shelly Nar Valley Clay at heights up to 24 m OD in the Narford area. This sug-

gests a minimum sea level at about 30 m OD for the maximum extent of the transgression if later isostatic or other differential land movement is assumed to be absent. Sparsely stony clays mapped as Nar Valley Clay by Whitaker et al. (1893, p.89) in the Ashwicken area [698 188] at heights greater than 30 m OD have been re-interpreted in the present survey as water-sorted till.

Stevens (1960, p.295) and Baden-Powell and West (1960, p.79) have described sections at Blackborough End Gravel Pits [684 145] in which gravels of presumed glacial origin overlie the Nar Valley beds. Whether these gravels are contiguous with the main mass of glacial sand and gravel (presumed to be part of the Chalky-Jurassic till), are part of the Tottenhill Gravels (see below), or are solifluicted material derived from the Blackborough End gravels is unclear because the sections are no longer visible. No correlative of the Nar Valley Beds has yet been described from other parts of Fenland or the adjacent region. This is surprising because the deposits have a large outcrop and subcrop in the Nar and Middleton Stop Drain valleys over a topographical range of more than 30 m and could be expected to occur extensively at similar heights around the whole Fenland margin from the Humber to Hunstanton. West (1991) has suggested that the absence of the Nar Valley Beds from the southern part of Fenland indicates that the region was still high ground at that time. The absence of the Nar Valley Beds in the valleys of the Babingley and Gaywood rivers remains inexplicable.

Details

The outcrops of the Nar Valley Beds in the district occupy low ground in the valleys of the Middleton Stop Drain and the River Nar and its tributaries, where they are almost entirely obscured by sandy head derived from the adjacent Cretaceous rocks and glacial deposits. Weathered brownish grey Nar Valley Clay with rotted shell fragments occurs in the deeper ditches and numerous old clay pits between Tower End [667 173] and Station Farm [699 169], at East Winch [695 162] and Horse Fen [692 142]. The extensive brickpits at East Winch, which yielded many of the beautifully preserved shells that can be seen in museum collections, are now completely degraded.

Tottenhill Gravels

Between King's Lynn and Stowbridge (in the Wisbech district), an almost continuous narrow strip of flint gravel forms an eastern limit to the Fenland Holocene deposits. North of Hardwick, beneath the urban areas of Gaywood and South Wootton, patches of similar gravel have been exposed from time to time. Whitaker et al. (1893, p.91) interpreted the bulk of this gravel as a terrace deposit marking the east bank of a forerunner of the River Great Ouse. However, there is no evidence that the Great Ouse, or indeed any other Fenland river, followed a course along the eastern edge of the Fens before its diversion there in Roman times. Nor is there any depositional evidence for a west bank of such a river. Whitaker (*in* Whitaker and Jukes-Browne, 1889, p.92) later suggested that similar patches of gravel near King's Lynn might have a marine origin comparable with that of the Hunstanton raised beach (see below). Few exposures have oc-

Table 13   Invertebrate fossils from the Nar Valley Clay

Foraminiferida
Ammonia beccarii (Linné)
Elphidium '*clavatum*' Cushman
Elphidium incertum (Williamson)
Elphidium macellum (Fichtel & Moll)
Elphidium williamsoni Haynes
Quinqueloculina seminulum (Linné)

Annelida
Pomatoceras triqueter Linné

Gastropoda
Aporrhais pespelecani (Linné)
Bittium reticulatum Linné
Buccinum undatum Linné
Hinia incrassata (Strom)
Hinia pygmaea (Lamarck)
Hydrobia (*Peringia*) *ulvae* (Pennant)
Hydropleura septangularis (Montagu)
Littorina littorea (Linné)
Littorina obtusata (Linné)
Lunatia alderi (Forbes)
Turritella communis Risso

Bivalvia
Abra alba (Wood)
Acanthocardia echinata (Linné)
Cerastoderma edule (Linné)
Chlamys varia (Linné)
Corbula gibba (Olivi)
Macoma balthica (Linné)
Mya arenaria Linné
Mysella bidentata (Montagu)
Mytilus edulis Linné
Ostrea edulis Linné
Scrobicularia plana (da Costa)
Spisula solida (Linné)
Spisula subtruncata (da Costa)
Tapes (*Ruditapes*) *decussatus* (Linné)

Ostracoda
Aurila arborescens (Brady)
Carinocythereis whitei (Baird)
Elofsonella concinna (Jones)
Hirschmannia '*tamarindus*' (Jones)
Leptocythere castanea (Sars)
Leptocythere pellucida (Baird)
Leptocythere spp.
Loxoconcha rhomboidea (Fischer)
Robertsonites tuberculatus (Sars)
Sarcicythereis bradyii (Norman)
Sarcicythereis punctillata (Brady)
Semicytherura cf. *S. sella* (Sars)

Cirripedia
Balanus sp.

Echinodermata
Psammechinus miliaris (Gmelin)

curred in these gravels in recent years, but those at West Winch, Stowbridge and Tottenhill have provided useful stratigraphical information.

The most complete section in this part of the Pleistocene sequence in west Norfolk at the time of the survey was that in Tottenhill Gravel Pit [632 116] at the mouth of the Nar valley. There, the Nar Valley Beds were unconformably overlain by a complex sequence of flint gravels (Figure 47) that have been termed the Tottenhill Gravels (Gallois, 1979b). The gravels can be divided into two parts. The Lower Tottenhill Gravel, up to 5 m thick, consists of medium and coarse, poorly sorted, mostly angular gravels and is characterised by cross-bedding which dips steeply (20° to 25°) to the south-east, and by frost wedge and cryoturbation structures. The gravels contain numerous pebbles of woody and reedy peat, soft brown clay, varved clay and Jurassic and Lower Cretaceous erratics, all presumed to have been derived from nearby outcrops of Nar Valley Beds and Chalky-Jurassic till. The Upper Tottenhill Gravel, about 3.5 m thick, consists of fine- and medium-grained gravels which are better sorted than the lower gravels and which have planar cross-bedding that dips westwards at 8° to 15°. The junction of the two units was marked by a well-defined, slightly uneven surface at 5.0 to 5.2 m above OD along which the upper gravels truncated cryoturbation and bedding structures in the lower gravels.

In the south-eastern part of the main pit [6330 1135], the Upper Tottenhill Gravel was overlain by up to 1 m of cryoturbated, dull brown and reddish brown, sparsely stony, sandy clay and clayey sand which passed southwards along the section into dirty clayey sand. This lithology is similar to the more sandy parts of the Hunstanton Till (see below), and although the stones are almost entirely angular flints, the clay and the underlying Upper Tottenhill Gravel also contain brown-stained quartzites of possibly Triassic origin and rare igneous rocks. These last include a distinctive quartz porphyry which also occurs in the Hunstanton till at Hunstanton (p.153). The relationships of the Pleistocene deposits at the mouth of the Nar valley, in the Setchey to Tottenhill area, are shown diagrammatically in Figure 47.

The lower gravels have no known outcrop, but the Upper Tottenhill Gravel appears to form part of the discontinuous narrow outcrop of gravel that runs northwards via Setchey and West Winch to Hardwick, and southwards to Stowbridge. The junction of the lower and upper gravels is at about 5 m above OD at Tottenhill, falling slowly westwards. Boreholes in the Nar Valley show that spreads of gravel that are probably contiguous with the Tottenhill Gravels occur beneath Recent deposits in the central part of the valley upstream from Setchey, but the relationship of these to the outcrops of glacial sand and gravel (both presumed to be associated with the Chalky-Jurassic till) is not known.

The Lower Tottenhill Gravel, by virtue of its included frost structures, indicates a cold phase younger than the interglacial Nar Valley Clay. The Upper Tottenhill Gravel appears from its field relationships to have formed either as a marine beach or as a lacustrine beach on the eastern shore of a fenland lake. In southern Fenland, shelly gravels at a similar height (the March Gravels) have been interpreted as river terraces debouching into a marine bay (Gallois, 1988). At North Wootton, South Wootton and Fairstead Estate, King's Lynn, patches of gravel lithologically similar gravel to the Upper Tottenhill Gravel occur

Figure 47 Relationship of the Nar Valley Beds to the Tottenhill Gravels at Tottenhill.

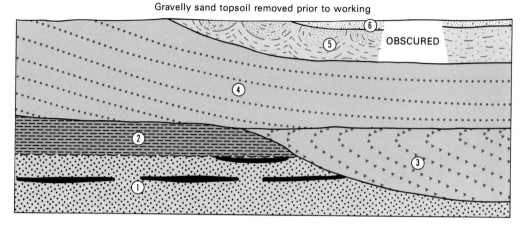

Bed 1 Nar Valley Freshwater Beds: sands with lenses of woody peat

Bed 2 Nar Valley Clay: dark grey marine clay with oyster-rich bed at base

Bed 3 Lower Tottenhill gravel: poorly sorted, steeply cross-bedded flint gravel with much debris derived from Nar Valley Beds and Chalky-Jurassic till

Bed 4 Upper Tottenhill Gravel: flint gravel with low angle cross-bedding; patchy calcrete at base: transgressive across cryoturbation structures in 3

Bed 5 ? Till: cryoturbated mixture of dull red-brown, sparsely stony, sandy clay and gravelly sand

Not to scale; see text for thickness   Bed 6 Gravelly sand: cryoturbated and irregularly piped into underlying bed

at heights of up to 14 m above OD; these are presumed to mark the positions of high strand lines of the same marine embayment (Plate 16). They are shown on the map as older storm-beach gravels. They are poorly exposed and neither they nor the Tottenhill Gravels have yielded faunal or floral evidence as to their age.

The workings at Tottenhill have been extended [636 115] to the east of the A10 road since the time of the survey, and now expose Tottenhill Gravels resting on Chalky-Jurassic till. The sedimentology of up to 14 m of gravel and sand has been described by Gibbard et al. (1991, 1992), who concluded that the deposits formed part of a fluvioglacial delta at the mouth of the River Nar. Lewis and Rose (1991, fig. 46) have interpreted the reddish brown silty clay in the cryoturbated layer that covers the whole site, as evidence of a period of temperate-climate soil formation, followed by cold-climate disturbance.

Details

North Wootton to Hardwick

Much of North Wootton, South Wootton and Gaywood are built on flat, terrace-like spreads of flint-rich gravel. These have been worked in shallow pits up to 3 m deep at North Wootton [640 241], South Wootton [641 226; 646 227] and Gaywood [6360 2055], but were not exposed at the time of the survey. A separate strip of gravel, on which the A10 road is built, runs from Hardwick southwards to West Winch.

The gravels appear to be the remnants of more than one deposit: patches of low-level gravels (at heights of 6 to 9 m above OD) at North Wootton [640 240], South Wootton [640 225],

Gaywood [636 205] and Hardwick [635 185] being separated from high-level gravels (at 12 to 15 m above OD) at South Wootton [646 224] and Gaywood [643 206]. Flint hand axes, ascribed by Wymer (1985) to the Palaeolithic, have been collected from one of the pits [641 226] in the low-level gravels at South Wootton.

West Winch to Stowbridge

A gas pipeline trench at West Winch [631 167] revealed 3.0m of cross-bedded sand and well-rounded flint gravel banked against a low cliff of Chalky-Jurassic till and soliflucted brown sandy till (Figure 48). A section on the opposite side of the A10 road, a few metres to the west, showed 3.5 m of sandy, fine-grained, well-rounded flint gravel with cross-bedding dipping gently eastwards. At Stowbridge [614 074], about 3 m of similar flint gravel rests on Kimmeridge Clay. The lower and upper surfaces of the gravels at West Winch (at about 4.0 and 7.6 m above OD) and Stowbridge (about 3.7 m and 6.7 m above OD) are at similar heights.

Hunstanton Till

Patches of reddish brown till, the Hunstanton Till, markedly different in lithology from the Chalky-Jurassic till, crop out on low-lying ground along the north Norfolk coast between Hunstanton and Blakeney, and along the eastern margin of The Wash between Hunstanton and Snettisham. The latter outcrop continues southwards beneath Holocene deposits into the south-eastern part of The Wash, and small patches occur beneath Holocene sediments in the offshore area (Figure 42). Outliers of the till occur at Hunstanton and Ringstead at heights of up to 33 m above OD.

Plate 16   Dersingham Fen and reclaimed marshlands: view north-west from Sandringham Warren [6802 2943].

Between Lodge Hill (far distance) and Dersingham Fen (foreground) the eastern limit of Fenland is a steep feature that was probably a Pleistocene cliff line cut in the Leziate Beds prior to the deposition of the Hunstanton Till. The low ground at its foot is now underlain by poorly drained sandy wash derived from the Leziate Beds (foreground) which is in part overstepped by Recent Terrington Beds clays (middle and far distance).   (A 10858).

The till is probably less than 10 m thick everywhere on the land area in the district. Its surface has a low relief and is poorly drained, and there are few exposures other than those in ditches. The till is usually fresh even in these shallow exposures. Glacial sand and gravel deposits associated with the till extend over a larger area than the till; they include the famous esker at Hunstanton Park, well-developed kames and a possible collapsed esker in the valley of the Heacham River, and a line of gravel mounds that extends as far south as Wolferton. The gravels are everywhere composed predominantly of flint and chalk, but they also contain common metamorphic, igneous and other far-travelled stones which enables them to be readily distinguished from the gravels associated with the Chalky-Jurassic till.

The Hunstanton Till is characteristically a dull reddish brown sandy clay (thought to be derived in part from Permo-Triassic marls that crop out on the sea bed in the nearshore zone between Aberdeen and Newcastle), which contains chalk and flint erratics together with 'Bunter' pebbles, Carboniferous sandstones and coals, Jurassic limestones and sandstones, schistose and gneissic metamorphic rocks, and a variety of igneous rocks. Among the more easily identifiable igneous rocks are Whin Sill dolerite from northern England, Cheviot porphyry from southern Scotland and rare rhomb porphyry from Norway.

Despite the emphasis in the literature on the more exotic erratics, chalk and flint are by far the most common stones in the till in the Hunstanton area. In a compre-

Figure 48 Junction of the Upper Tottenhill Gravel and Chalky-Jurassic till at West Winch.

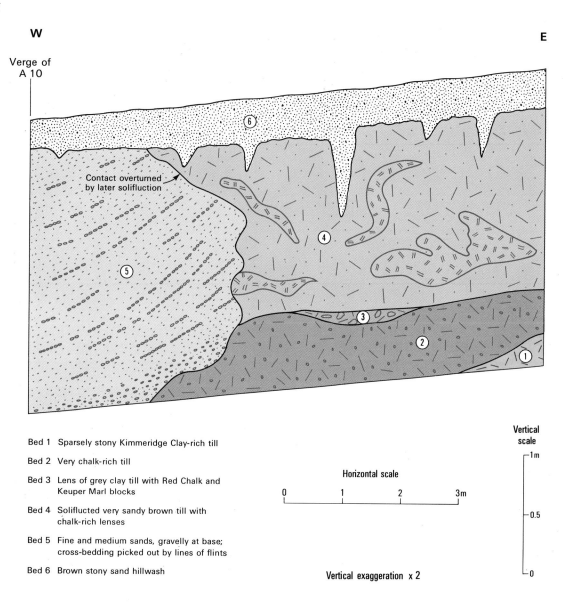

W

E

Verge of A 10

Contact overturned by later solifluction

Bed 1 Sparsely stony Kimmeridge Clay-rich till

Bed 2 Very chalk-rich till

Bed 3 Lens of grey clay till with Red Chalk and Keuper Marl blocks

Bed 4 Soliflucted very sandy brown till with chalk-rich lenses

Bed 5 Fine and medium sands, gravelly at base; cross-bedding picked out by lines of flints

Bed 6 Brown stony sand hillwash

Horizontal scale

0   1   2   3m

Vertical scale

1m

0.5

0

Vertical exaggeration x 2

hensive review of the distribution, lithology, origin and age of the till, England and Lee (1991) noted that an average of only 4.4 per cent of the erratics in the till at Hunstanton and Heacham are far travelled. Many of these can be matched with rocks which crop out in the eastern parts of northern England and southern and central Scotland; it is clear that the bulk of the Hunstanton till was deposited by an ice-sheet which flowed south-south-eastwards down the east coast, and which was fed by ice-streams that ran down from the Highlands, Southern Uplands and Pennines (Figure 49). Boulton et. al. (1977) have suggested that the till was deposited by a single, rapid ice surge.

The relationship of the Chalky-Jurassic till to the Hunstanton Till has yet to be directly demonstrated. There have been records of reddish brown Hunstanton-type till overlying grey Chalky-Jurassic-type till, for example in boreholes in The Wash (p.159). However, these need to be interpreted with caution. The Hunstanton Till is locally so rich in Cretaceous material as to be indistinguishable from the Cretaceous-rich parts of the Chalky-Jurassic

till and, where decalcified in its upper part, the Chalky-Jurassic till can itself give rise to a dull reddish brown sandy till that rests on its more normal lithology with a sharp, usually cryoturbated, contact.

At Hunstanton and Old Hunstanton, the Hunstanton Till overlies the Hunstanton raised beach, shelly chalk and flint gravels which appear to be banked against an old cliff cut in Chalk, Red Chalk and Carstone. At some localities, such as Old Hunstanton [681 425], the upper part of the beach deposit is interbedded with frost-shattered chalk debris (Figure 50). Former exposures at Hunstanton [e.g. 672 402] showed the beach deposits to contain marine gastropods and bivalves, and many far-travelled erratics that can be matched with those of the Hunstanton Till (Whitaker and Jukes-Browne, 1899, p.90). The abundance of these far-travelled stones in the raised beach suggests that there were, and possibly still are in the offshore area, extensive outcrops of an older till of similar lithology to the Hunstanton Till.

The molluscan fauna is similar to that of the present-day Wash or southern North Sea; when combined with the height of the gravels (maximum about 8 m above OD), the fauna suggests deposition in a climate similar to that of the present day at a slightly higher sea level than at present. It has been suggested (Whitaker and Jukes Browne, 1899) that the Hunstanton raised beach might be the correlative of the shelly March Gravels of southern Fenland and, on the basis of their field relationships, the Upper Tottenhill Gravels (Gallois, 1988).

However, such evidence needs to be treated with caution because of the repetitive nature of the climatic events that occurred in the Pleistocene. Many of the temperate phases, whether interglacial or interstadial, probably produced sea levels comparable to or higher than those of the present day, and each maximum level is likely to have left behind patches of its more resistant deposits, such as gravels. As yet, there is no sure way of distinguishing the ages of such deposits: correlations made on the basis of their heights above sea level are valid only on the assumption that they represent the deposits of a single temperate phase.

The age of the Hunstanton Till cannot be directly demonstrated, but in the Holderness peninsula in Humberside a series of tills (the 'Hessle', Skipsea and Withernsea), lithologically similar to the Hunstanton Till, overly a beach deposit and chalk cliff comparable to that at Hunstanton. At Sewerby, Humberside, Lamplugh (1888) described the beach as composed of shingle rich in far-travelled erratics (many the same as occur in the Hunstanton raised beach) together with a molluscan fauna that can also be matched with that from Hunstanton, and vertebrate bones.

The molluscan fauna is not diagnostic of age but it does suggest a similar climatic origin for both the Sewerby and Hunstanton raised beaches. The vertebrates from Sewerby include warm-climate forms of the elephant, hippopotamus and rhinoceros (Boylan, 1967), indicative of an interglacial rather than an interstadial phase (Catt and Penny, 1966, p.386). Catt and Penny (1966, p.387) have also recorded an older till, the Basement Till, beneath the raised beach at Sewerby and it is from this material that far-travelled erratics in the beach deposit are presumed to have been derived. If this interpretation is correct then, by analogy, one might expect to find an older till, lithologically similar to the Hunstanton Till, in the offshore area close to the north Norfolk coast. It has been suggested (Catt and Penny, 1966, pp.403–404) that the Basement Till of Holderness is equivalent to part of the Chalky-Jurassic till of Norfolk.

However, it seems more probable from the field evidence in Norfolk that the Hunstanton raised beach, and by analogy that at Sewerby, represents a temperate phase within a single glaciation and that this glaciation is separated from that which deposited the Chalky-Jurassic till of the feasibility study area by the Nar Valley Beds interglacial.

At Dimlington, Lincolnshire mosses in silts which lie between the Basement Till and the Skipsea/Withernsea Till have been radiocarbon dated at 18 240 ± 250 years and 18 500 ± 400 years before present (BP) (Penny et al., 1969). A date of 16 713 ± 340 BP from Kildale,

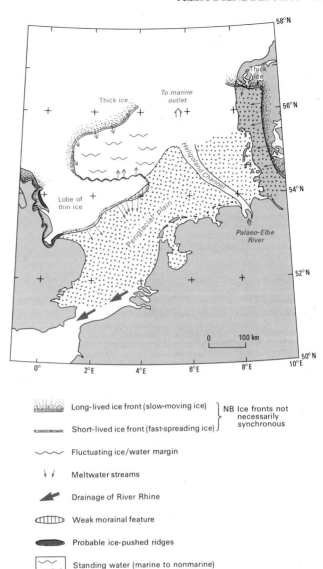

Figure 49 Generalised palaeogeography of the southern North Sea at the last ice maximum (Cameron et al., 1992, fig. 110).

north-east Yorkshire (Keen et al., 1984) from organic-rich sediments that postdate the equivalents of the Skipsea/Withernsea tills in that area, suggests that these tills, and by analogy the Hunstanton Till, were deposited during a cold phase that lasted less than 2000 years (England and Lee, 1991). This is in accord with the suggestion that the Hunstanton Till was deposited by a single ice surge.

Predominantly cold climates prevailed until the end of the Pleistocene, about 10 000 years ago, and gave rise to extensive erosional and depositional features in the district. Evidence of permafrost activity, probably in part contemporaneous with the deposition of the Hunstanton Till, is common on the chalk downlands between Hun-

Plate 17   Cambering in Chalk, Barrett Ringstead Chalk Pit [6891 4001].

The pit, which is cut into the northern side of a dry valley, exposed almost horizontal Lower and Middle Chalk in its back face. When traced towards the valley side, the chalk becomes progressively more broken. On the right, in-situ chalk dips at 20° to 30° towards the valley; on the left, detached imbricate blocks that have moved downslope have greater dips. Within the mass there is an upward grading from coarse to fine blocks due to frost shattering and transport sorting. Brown loams at the top of the section (top left) have travelled from the adjacent plateau and upper slopes as a solifuxion sludge. (A 10880).

stanton and Gayton. Frost stripes and polygons occur at many localities, and the chalk is extensively cambered and soliflucted in the valley sides (Plate 17).

Ringstead Downs, now a dry valley, was probably initiated by meltwater from the wasting Hunstanton ice sheet and was subsequently modified by snow meltwater each year until long after the ice had retreated from Norfolk. The valley of the Heacham River between Heacham and Eaton, although beyond the outcrop of the Hunstanton Till, contains pockets of glacial sand and gravel that show ice-contact features. There too, the valley has been modified by both ice and snow meltwaters. A patch of flint gravel up to 3 m thick at the mouth of the valley at Heacham village was probably deposited by these meltwaters as a fan-like terrace. Farther south, the dry valleys of

the chalk outcrop and their continuations across the Lower Cretaceous outcrop were all extensively enlarged by meltwaters during the latest Pleistocene.

The processes of cryoturbation, in combination with solifluction, meltwater and aeolian transport, remobilised many of the softer solid and drift deposits in the district to produce extensive spreads of predominantly sandy deposits that drape the lower slopes of the river valleys and the slopes adjacent to the fenland. These are shown on the map as Head. The most extensive outcrops occur between Snettisham and Wolferton, at and adjacent to Roydon Common, and in the Middleton valley where they are largely derived from the soft sands of the Leziate Beds. Most of the Head deposits in the district continued to be modified by hill-creep, solifluction and

Figure 50 Relationship of the Hunstanton Till to the Hunstanton Raised Beach at Hunstanton.

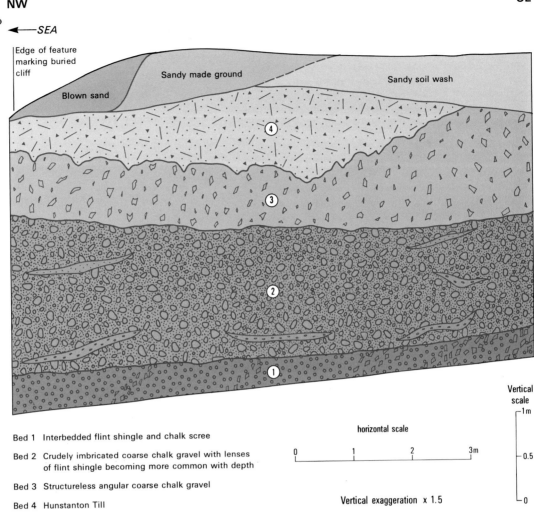

NW

←SEA

SE

Edge of feature marking buried cliff

Blown sand

Sandy made ground

Sandy soil wash

④

③

②

①

Bed 1 Interbedded flint shingle and chalk scree

Bed 2 Crudely imbricated coarse chalk gravel with lenses of flint shingle becoming more common with depth

Bed 3 Structureless angular coarse chalk gravel

Bed 4 Hunstanton Till

horizontal scale

0   1   2   3m

Vertical scale

┌1m

0.5

└0

Vertical exaggeration x 1.5

rain-wash until modern times, and they are therefore described with the Recent deposits (p.170).

Details

Hunstanton area

Patches of Hunstanton Till, rarely more than 3 m thick, crop out on the higher ground at New Hunstanton, in Hunstanton Park and on the seaward-facing slope east and north east of the park. Sand and gravel, mostly in the form of discrete mounds (kames) occurs in association with the till at Hunstanton and Ringstead Downs and in the Hunstanton Park esker.

A ditch section [6911 4252 to 6935 4310] at Old Hunstanton showed up to 2.5 m of typical Hunstanton Till, consisting of stiff, dull reddish brown, sandy clay with flint, chalk, Carboniferous limestone and sandstone, and igneous and metamorphic erratics resting on up to 1m of fine- and medium-grained sand with lenses of flint- and chalk-rich gravel. This latter may be a beach deposit, comparable to that recorded at Hunstanton. A nearby excavation [6875 4264] at Warren Farm showed a similar shingle deposit, overlain by up to 2m of Hunstanton Till, resting on a chalk surface which slopes gently seaward. Between this locality and the northern end of Hunstanton Cliffs, the till is banked up against a steep feature cut in chalk.

A trench [6811 4247] dug at the western end of this feature to try to expose the junction of the Hunstanton Till and the underlying beach deposits (Figure 50), exposed the following section:

	Thickness m
Blown sand, made ground and soil wash: blown sand banked up against seaward side of degraded artificial bank	up to 0.8
Hunstanton Till: stiff, dull reddish brown, sparsely stony clay; stones mostly fine- gravel-sized flints and chalks with rarer Carstone, Red Chalk, rotted sandstone (?Carboniferous), granite, rotted coal, rotted ironstone, vein quartz and unidentified igneous rocks; irregular, cryoturbated base with pipes of till extending down 0.3 m into underlying bed	up to 0.9
Cryoturbated chalk gravel: fine and medium gravel-sized, angular chalk fragments set in a matrix of chalk paste and dull reddish brown clay; larger chalk fragments joint bounded; jumbled mass devoid of bedding or sorting; irregular passage at base into	up to 0.9
Imbricated chalk gravel: sand grade to boulder-sized, angular, chalk fragments set in a matrix of poorly sorted sand and fine and medium gravel;	

larger blocks show imbricate structure dipping north-westwards (seawards); lenses of fine-grained sand in upper part of bed become replaced by thin interbeds of fine gravel in lower part; gravel composed of well-rounded chalk and flint with rarer Red Chalk, Carstone, vein quartz, ironstone, jasper, Jurassic mudstone, coarse feldspathic sandstone (?Carboniferous) and coal; irregular planar junction with up to

Thickness
m

up to 1.8

Shingle and chalk scree: clean, fine gravel composed of well-rounded flints, chalk grains and rarer far-travelled erratics, interbedded with beds of interlocking angular blocks of Lower Chalk, which grade up into angular gravel-sized chalk fragments set in a chalk paste; beds dip north-west at about 5°; becoming predominantly gravel at seaward end of section

seen 0.3

The lowest two beds are similar in structure and content to parts of the debris cones which occur along Hunstanton cliffs at the present day. It seems likely that the excavation was made at a point where the old chalk cliff is indented, and the Pleistocene beach is overlain by chalk scree and sludge.

The Hunstanton Park Esker can be traced northwards as an almost continuous ridge of sand and gravel from Ringstead Downs [698 642] to Kimberly Plantation [6960 4145], and from there as isolated patches of sand and gravel at Gedding's Pit [6920 4145], St Mary's Church [6895 4196] and near Warren Farm [6895 4250]. Although now degraded by plantations and ploughing, much of the main ridge has retained its characteristic esker shape, being sinuous and beaded. It is everywhere surrounded by a broad apron of gravelly and sandy head which appears to rest on a veneer of Hunstanton Till mostly less than 1.5 m thick. Small workings for hardcore have, from time to time, revealed crudely bedded gravel composed of the more resistant Hunstanton Till erratics.

Rose (1836) and Whitaker and Jukes Browne (1899) noted several sections at New Hunstanton sea-front [6713 4115 to 6715 4085], prior to the building of the sea-wall, which showed up to 5 m of Hunstanton Till resting on the Hunstanton raised beach: this, in turn, rested on the Carstone.

In Victorian times, sand and gravel was dug for ballast for the railway in several pits [672 402] adjacent to Hunstanton Gasworks. Sections described by Seeley (1866) and Whitaker and Jukes Browne (1899, pp.90–91) showed up to 9 m of cross-bedded sands and shelly gravels largely composed of a suite of Carboniferous, Jurassic, metamorphic and igneous rocks that can be matched with the far-travelled erratics of the Hunstanton Till. The shells, many of which were described as unabraided, include species of the gastropods '*Buccinum, Dentalium, Littorina, Murex, Natica, Purpura* and *Turritella*, and the bivalves *Artemis, Cardium, Corbula, Cyprina, Donax, Mactra, Mytilus, Nucula, Ostrea, Scrobicularia, Tapes* and *Tellina*' all of which occur in The Wash at the present day.

The relationship of the beach deposits to the Hunstanton Till does not seem to have been clearly exposed at this locality. Lamplugh (in Whitaker and Jukes-Browne, 1899, p.91) recorded up to 1 m of 'red-brown clay .... containing scattered stones like a boulder clay' interbedded with the sand and gravel in the highest part of the pit. Jackson (1911) noted that 'Brown Boulder Clay' was irregularly mixed with the sand and gravel. At the time of the present survey, small exposures showed reddish brown clayey till overlying disturbed sand and chalk-rich gravel. All the sections appeared to have been disturbed by cryoturbation and/or solifluction. Parts of a human skeleton were found

near the top of one of the pits in 1897: according to Lamplugh (in Whitaker and Jukes Browne, 1899, p.91) these came from the junction of the highest gravels and overlying sandy head. Keith (quoted in England and Lee, 1991) thought the remains to be Neolithic or earlier.

Heacham to Wolferton

An almost continuous outcrop of Hunstanton Till, probably up to 5 m thick in places, occupies much of the low ground between Hunstanton sea front and Snettisham Scalp. Along much of its length its eastern margin it appears to be a cliff which marked the former limit of The Wash. Small exposures of typical red-brown clayey till with common igneous and metamorphic erratics occur from time to time in the many drainage ditches in the area. These show considerable local lithological variation. Although the 'typical' red-brown clay with small erratics is much the commonest type, mottled brown and grey till, pale brown, chalk-rich till and brown sandy Carstone-rich till is also present. Similarly, the erratic content of the till is locally variable, far-travelled stones being common at some localities and absent at others.

Patches of sand and gravel of similar composition to that of the Hunstanton Park Esker are common in and adjacent to the valley of the Heacham River. One of these [6994 3706] at Church Farm, a kame terrace banked up against a chalk spur, has been worked for aggregate and reveals the internal structure of the deposit. Poorly sorted gravels, cobbles and small boulders, composed largely of flint set in a sandy, chalk-rich matrix derived in part from the Sandringham Sands, Dersingham Beds and Carstone, are irregularly interbedded with beds of sand with common chalk grains. About 99 per cent of the erratics are flint (very predominant) and chalk; the remainder, in order of frequency of occurrence, are oolitic limestone, Carstone and Red Chalk. Igneous and metamorphic rocks are rare. Dips are away from the centre of the mound and are locally disturbed by collapse structures indicative of melting ice.

A sinuous ridge of gravel, possibly a collapsed esker, runs northwards from Round O Wood [691 366] almost to the Heacham River, a fall in height of about 20 m. The gravels have been worked in the wood in pits [6907 3673] up to 3 m deep. Poor exposures showed their grain size and composition to be similar to those of the kame at Church Farm, with igneous and metamorphic rocks again rare.

Flint-rich gravels of similar composition have been extensively worked in pits up to 5 m deep in Gravelpit Belt [6840 3557] near Heacham Bottom Farm.

A line of small, degraded mounds of flint-rich gravel runs southwards from Locke Farm [662 332] to Wolferton [656 290]: they may be the remnants of the terminal moraine of the ice-sheet that deposited the Hunstanton Till. Ditch exposures in one of them [6580 2975] showed the gravels to rest on redbrown till.

An outlier [664 290] of sand and gravel adjacent to Wolferton railway cutting, shown on the map as glacial sand and gravel, may be largely composed of head. Poor exposures in the cutting [6645 2878] showed the following sequence:

	Thickness m
Gravel: ferruginous sandstone and ironstone pebbles set in a grey and white sand matrix	up to 1
Hunstanton Till: red-brown clay with small stones	0.3 to 0.6
Gravel: frost-shattered flints; poorly exposed	c.2
Leziate Beds: brown and white sands	2 seen

This is the most southerly exposure of the Hunstanton Till recorded to date.

Offshore area

The offshore seismic and sampling surveys for the Wash Feasibility Study (Wingfield et al., 1979) showed the till cover to be patchy and to be mostly less than 10 m thick except in the glacially filled valleys (p.143). Short (5 to 35 cm in length) seabed cores were taken from 16 sites in The Wash; most of these proved reddish brown sandy till similar to the Hunstanton Till, but almost all were in weathered material and some might be residual deposits derived from the Chalky-Jurassic till.

Undoubted Hunstanton Till was proved in five partially sampled (bulk, disturbed and U100 samples) boreholes in the south-eastern part of The Wash between Daseley's Sand and Bulldog Sand, offshore from Wolferton.

A borehole [5591 3189] on Daseley's Sand proved 10.5 m of Recent deposits resting on 1.50 m of weathered Hunstanton Till penetrated by peat roots, which rests on Kimmeridge Clay. The till consists of dark red-brown and brownish grey sandy clay with tiny chalk, flint, Triassic and far-travelled erratics. A nearby borehole [5579 3194] proved the Recent deposits to rest on 5.5 m of typical Chalky-Jurassic till.

Other boreholes in the Daseley's Sand area suggest that the Hunstanton Till postdates the Chalky-Jurassic till. A borehole [5729 2916] at the southern end of the sand proved 8.5 m of Recent deposits resting on 2.8 m of typical Hunstanton Till which contains, in addition to far-travelled erratics of coal, Triassic and igneous rocks, streaks of grey Chalky-Jurassic till. This is the most southerly, unambiguous record of the Hunstanton Till to date. Nearby boreholes [5545 2921 and 5698 3055] proved 4.3 m and 6.55 m respectively of Hunstanton Till resting on varved clay and possible interglacial deposits (Gallois, 1979a).

A borehole [6131 2947] near Bulldog Sand proved c.1.1 m of Hunstanton Till overlying c.1.5 m of Chalky-Jurassic till. The junction between the two was not sampled; it is possible that the latter is an erratic within the former. A similarly problematical section was proved in a borehole [6069 2220] drilled in the River Great Ouse, about 7 km south of the nearest well documented occurrence of Hunstanton Till. This showed 7.35 m of deeply weathered dull red-brown and grey clayey till with chalk erratics, passing down into typical grey Chalky Jurassic till. The section is similar to that described from White Hills Wood, Castle Rising where Chalky-Jurassic till is overlain by a weathered brown sandy till (p.148).

Extensive tracts of glacial sand and gravel occur in association with both the Chalky Jurassic and Hunstanton tills in the offshore area. Because they are poorly consolidated, they have been remobilised in part and incorporated into the Recent deposits (see below).

RECENT (FLANDRIAN) DEPOSITS

At the time of the maximum extent of the ice which deposited the Hunstanton Till, about 18 000 years ago, sea level was an estimated 100 to 150 m below OD (Donn et al., 1962). As the ice sheets melted, sea level rose rapidly in the late Pleistocene to about 40 to 50 m below OD at the beginning of the Flandrian (Lambeck, 1991) and, by that time, much of the southern North Sea, the deeper parts of the Wash and the main channel of the River Great Ouse as far inland as Wisbech were reoccupied by the sea. The area now occupied by Fenland[1] became a large, poorly drained basin mostly occupied by freshwater marshes, with brackish marshes in the seaward areas and salt-marshes adjacent to the sea. Throughout much of the present district, these environments gave rise to fresh- and brackish-water peats and salt-marsh clays, with silts and fine-grained sands along the river courses. Marine sands were deposited in the deeper parts of The Wash and in the mouths of the river estuaries.

Sea level probably rose almost continuously in the early part of the Flandrian, until about 6000 BP (before present), with the result that a series of depositional zones was established in which open marine environments passed landward via intertidal silt and mud flats into brackish reed swamps and freshwater fens (Figure 51). Although sea level continued to rise after 6000 BP, periods of relatively rapid rise were punctuated by stillstands and possible slight falls in sea level. The early part of the Flandrian was therefore characterised by the rapid transgression of the marine environments across brackish and freshwater regimes, and the later Flandrian by alternate transgressions and regressions which gave rise to interdigitations of clay and peat (Table 14).

The sources of much of the recent sediment in The Wash, in particular the sand and gravel, were glacial deposits on the floor of The Wash and at outcrop on the adjacent coasts. For much of the Flandrian period The Wash acted as a sediment trap for material moving by longshore drift southwards along the Humberside and Lincolnshire coasts and westwards along the north Norfolk coast. There has been much discussion as to the origin of the finer-gained sediments that have accreted in the intertidal areas in recent years, and the affect that land reclamation and drainage works might have on future accretion and the stability of the coastline. Evans and Collins (1975) concluded that, in the absence of adequate quantitative data, the balance of evidence indicated that the bulk of the fine-grained sediment was also derived from the Quaternary deposits of the adjacent coasts, and that the rivers flowing into The Wash, at least since their entrainment, were only minor contributors. In the 19th and 20th centuries land reclamation appears to have substantially increased the rate of sedimentation in the intertidal areas (Inglis and Kestner, 1958) and has thereby accelerated the rate at which the salt marshes could be reclaimed.

Throughout much of Fenland, the recent deposits can be divided into four stratigraphical units which are, in ascending order, the Lower Peat, Barroway Drove Beds, Nordelph Peat and Terrington Beds (Gallois, 1979a). Only the Nordelph Peat and Terrington Beds crop out in the present district, but all four units have been proved in boreholes. The full sequence is present in much of the eastern part of the district where the higher beds are exposed in the deeper drains. Small areas of Nordelph Peat crop out in the valleys of the Gaywood River, Middleton Stop Drain and River Nar. In the western part of the district, and seawards into the southern part of The Wash, the whole of the Flandrian sequence is represented by marine and estuarine sands and silts.

---

[1] The term Fenland is used here in its broadest sense and includes both the peat areas (the Fens proper) and the reclaimed salt-marsh areas (Marshland).

Presumed depositional environments for the Recent deposits of Fenland are shown in Figure 52. Descriptions of the depositional environments of the Lower Peat and Barroway Drove Beds are given in the Ely (Sheet 173) memoir where they have extensive outcrops (Gallois, 1988).

In the offshore area (Figure 52; 53A), the Flandrian sequences are entirely marine; they have been divided into five lithological groups which can be recognised at outcrop, in boreholes and, to a lesser extent, in seismic sections (Wingfield et al., 1979). These groups are shown on the published geological map as 'clean sand', 'sand with mud laminae', 'mud with sand laminae' 'mussel scalps', and 'saltings'. The last provide a link with the onshore sequence (Figure 53B) because they are the most recently formed Terrington Beds.

In the upland part of the district, the Flandrian Stage is represented only by small strips of alluvial deposits in the floodplains of the larger streams, dry valley deposits in the Ringstead Downs valley, reworked head deposits, and the storm beach gravels and blown sand in the coastal area between Snettisham Scalp and Old Hunstanton.

In the Fenland area, the maximum preserved thickness of the Recent deposits is probably about 35 m in the broad former estuary of the combined Rivers Great Ouse and Nene between Tydd Gote and the modern Nene outfall. In The Wash, where all the Recent sediments are still poorly consolidated and capable of being eroded and redeposited by the strong tidal currents, the thickest proved accumulation is about 30 m beneath the northern part of Thief Sand. It is probably a continuation of the thicker onshore sequences of the former Great Ouse estuary. An isopachyte map for the preserved 'mobile' sediments in The Wash is given in Wingfield et al. (1979).

Away from this estuary, the Recent deposits are mostly between 10 and 20 m thick west of the River Nene, and beneath the Walpoles and Terringtons. They thin eastwards and are less than 10 m thick east of the modern River Great Ouse; thicker sequences occur where they infill the former tidal channels of the larger streams.

In the upland area, none of the Recent deposits probably exceeds 5 m in thickness, and most are less than 3 m.

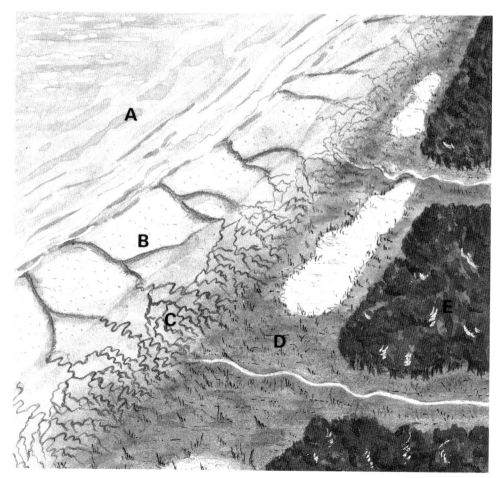

Key to depositional environments and sediment types:

A Subtidal—mostly sand

B Intertidal flat—silt with clay

C Salt marsh—clay with silt

D Peat fen and meres—peat and shell marl

E Deciduous forest—erosional area

Figure 51 Presumed depositional environments of the Recent deposits of the Fenland part of the district.

Table 14 Summary of the Flandrian history of the Fenland part of the King's Lynn district.

Thousands of years BP	Formation in King's Lynn district	Marine transgressions in King's Lynn district	Archaeological period	Pollen zone
0			Modern	
	Period of reclamation	?Medieval flooding ➤	Medieval	
1	– – – – –			VIII
	Terrington Beds	Roman flooding ➤	Roman	
2		Iron age flooding ➤	Iron Age	
3	Nordelph Peat*		Bronze Age	VIIb
4	Barroway Drove Beds		Neolithic	
5		➤		
6	'Lower Peat'			VIIa
7	– – – – –			
	'Lower' Peat in hollows and river channels		Mesolithic	
8				VI
	– – – – –			
9	?Basal sand and gravel			V
10		➤		IV

——————➤   major transgression

* In those parts of the river valleys not reached by the Terrington Beds transgression, the Nordelph Peat continued to form until it was drained and reclaimed during the Medieval period.

FENLAND AREA

Basal sand and gravel

In the period between the retreat of the ice from the northern part of the district about 18 000 years ago and the change to a temperate climate about 10 000 years ago, periglacial climates with high spring runoffs from melting snow gave rise to extensive spreads of solifluction and fluviatile deposits, largely derived from the Sandringham Sands and the poorly consolidated glacial deposits of the district.

Thin, but widespread sheets of poorly sorted clay and gravelly sand were subaerially deposited over much of what is now Fenland and The Wash, with better-sorted, cleaner gravels and sands accumulating in the river valleys. These poorly consolidated deposits were subsequently largely resorted, and in part redeposited, when they became submerged during the successive transgressions in the Flandrian. As a result, the Recent deposits proved in borehole sequences throughout Fenland are invariably underlain by up to 1 m of gravel and sand which have a complex history that depends on their geographical situation. In the deeper river valleys, they consist of well-sorted gravels and sands with many marine shells. Away from the valleys, they mostly consist of poorly sorted, clayey, gravelly sands, often brown and yellow and with root traces, which appear to have been weathered in situ. For convenience, this almost ubiquitous thin deposit is referred to here as the 'Basal sand and gravel' of the Flandrian, for although it was largely formed in the late Pleistocene it has been mostly reworked immediately prior to the deposition of the Recent sediments in many areas.

In the Peterborough district, Horton (1989) described pebbly clays with rootlets that rest on the former land surface and which underlie the Flandrian deposits as the

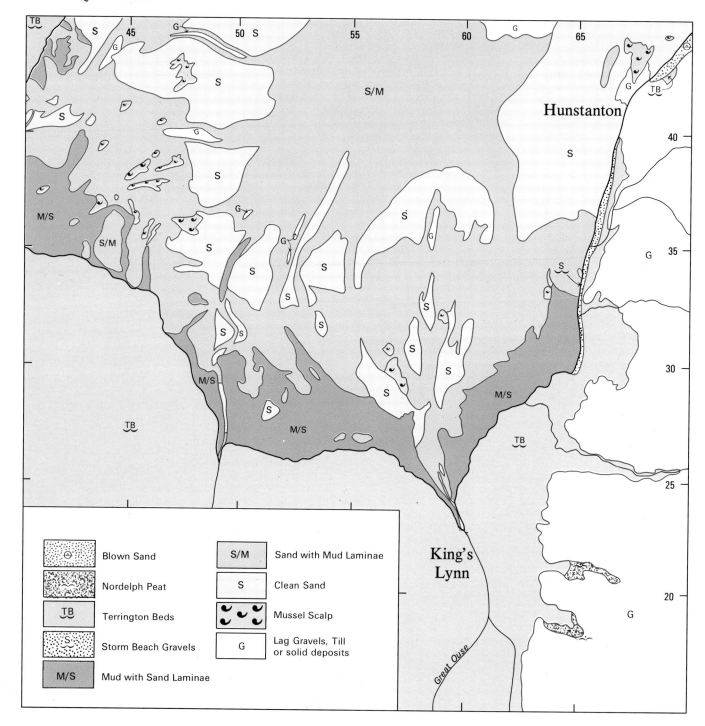

Figure 52 Distribution of the Recent deposits of the district.

'Crowland Bed'. In the Ely district (Gallois, 1988) this type of 'basal gravel' is widespread on the interfluves, and can be traced via better-sorted clayey sandy gravel into the well-sorted shelly sands and gravels which form lag deposits in the major stream and river channels. When The Wash was re-invaded by the sea in the late Pleistocene and early Flandrian, strong tidal currents were active in the main channels and the 'basal gravel' and glacial sand and gravel were locally extensively reworked.

Lower Peat

The Lower Peat is known in the district only from boreholes, being everywhere overstepped by the Barroway Drove Beds; its thickness and age are uncertain. Elsewhere in Fenland, the Lower Peat contains the stumps of deciduous trees rooted into the underlying Upper Jurassic clays and, in places where the peat is several metres thick, a sequence of deciduous and coniferous forests can be traced upwards into raised sphagnum bogs (Gallois, 1988; Horton, 1989). Radiocarbon dating has shown the bulk of the Lower Peat to have formed between about 5400 and 4700 BP with isolated pockets of older peat preserved in the deeper river channels and in hollows which may be between 7000 and 9000 years old (Gallois, 1988). In the Peterborough district, Horton (1989) recorded Lower Peat ranging from about 3400 years to at least 7500 years in age.

Barroway Drove Beds

The Barroway Drove Beds are everywhere overlain by the Nordelph Peat in the district, and are known only from boreholes and the deeper drains close to the eastern edge of Fenland. The marine transgression which deposited these intertidal clays and silts terminated the deposition of the Lower Peat. They consist of soft grey clays and silty clays which were deposited in salt-marshes and shallow-water brackish lagoons cut by a complex network of tidal channels and creeks subsequently infilled with silts and fine-grained sands. Rhizomes of the salt-tolerant reed *Phragmites* occur throughout the clays, and thin mats of rotted, drifted stems of reeds and sedges are common at many levels. Species of bivalves and foraminifers, all indicative of brackish to saline conditions, occur at some levels.

At outcrop in southern and western Fenland, the less compressible silts and fine-grained sands that infill the former creeks within the Barroway Drove Beds form low sinuous ridges or roddons. Their pale-coloured soils contrast with those of the dark grey clays, and this enables the creeks to be mapped in detail on air photographs (e.g. Seale, 1975); their pattern can be seen to be complex. Similar complexity must exist in the Barroway Drove Beds in the present district, but the borehole data are too few for it to be determined.

Nordelph Peat

The Nordelph Peat underlies the Terrington Beds throughout much of the eastern part of the district; it crops out in the downstream parts of the valleys of the River Nar, Middleton Stop Drain and the Gaywood River, just beyond the limit of the Terrington Beds transgression. The Nordelph Peat is exposed in most of the deeper drains on the Terrington Beds outcrop in the King's Lynn, Terrington and Tilney areas, and is exposed from time to time in the intertidal zone (the so-called submerged forest) on the north coast near Old Hunstanton where the salt-marsh deposits of the Terrington Beds have been removed by modern marine erosion. The peat is mostly less than 1m thick in the present district, although up to 3 m has been proved in some of the former river channels.

The Nordelph Peat has a very extensive outcrop and subcrop in Fenland. Radiocarbon dates indicate that its formation began in many areas at about 4000 BP at a time of probable sea-level stillstand, which enabled freshwater marshes to develop on the top surface of the Barroway Drove Beds. In those areas not submerged by the Terrington Beds transgression, in about 2000 BP,

the Nordelph Peat continued to be formed until the fens were drained, mostly in medieval times. In the present district, Nordelph Peat continues to form in a small area in the valley of the Middleton Stop Drain which has been preserved as a natural fen. Where drained, the peat becomes rapidly wasted by oxidation (Plate 18).

The distribution of the Nordelph Peat in its subcrop in the present district is poorly known because of the scarcity of boreholes (Figure 54). In other parts of Fenland, for example in the Peterborough district (Horton, 1989), the Nordelph Peat splits into a lower and upper leaf separated by Barroway Drove Beds, indicating that the rate of sea-level rise during a formation of the Barroway Drove Beds was not constant and that, at times, freshwater fens spread out across the Barroway Drove Beds marshes. A deep excavation at Wiggenhall St Germans, described as containing a 'Lower, Middle and Upper peats' probably exposed the Lower Peat and a lower and upper leaf of the Nordelph Peat.

Adjacent to the eastern edge of Fenland and on the north Norfolk coast, where the Nordelph Peat commonly rests on glacial and older deposits, the trunks and roots of oaks and small trees, some in situ, and woody debris from alder, birch, sallow, buckthorn and pine are common. In more seaward areas, the Nordelph Peat is largely composed of stems and rhizomes of sedges (*Cladium*) and reeds (*Phragmites*). A description of the evolutionary development of the Nordelph Peat and similar peats is given elsewhere (Gallois, 1988, pp.74–78). In southern Fenland, the Nordelph Peat contains thin but extensive lenticular deposits of shell marl which formed in shallow meres within the peat fens. No similar deposit has been recorded in the present district, but a small patch of diatomaceous earth recorded by Mr Young in the Nar Valley probably marks the site of a temporary, small, shallow lake within the Nordelph Peat fen.

Terrington Beds

The Terrington Beds crop out over the whole of the Fenland part of the district, where they form a thin layer, mostly 1 to 2 m thick and everywhere less than 3 m thick, which conceals the older Flandrian deposits. The main Terrington Beds transgression has been estimated, from the age of youngest Nordelph Peat on which it rests, to have been between about 2050 BP and 2600 BP. The transgression caused marine and brackish-water silts and fine-grained sands to be deposited far inland in the major river courses and gave rise to extensive salt-marsh deposits of interlaminated silt and clay which were sufficiently well drained by the time of the Roman occupation for them to have been extensively colonised. A second transgressive pulse in the 3rd century AD caused many of the Roman settlements in southern Fenland to be inundated.

The Terrington Beds have continued to accrete steadily from Roman times until the present day, with the possible exception of a period of transgression and erosion in about the 13th century. The history of the successive

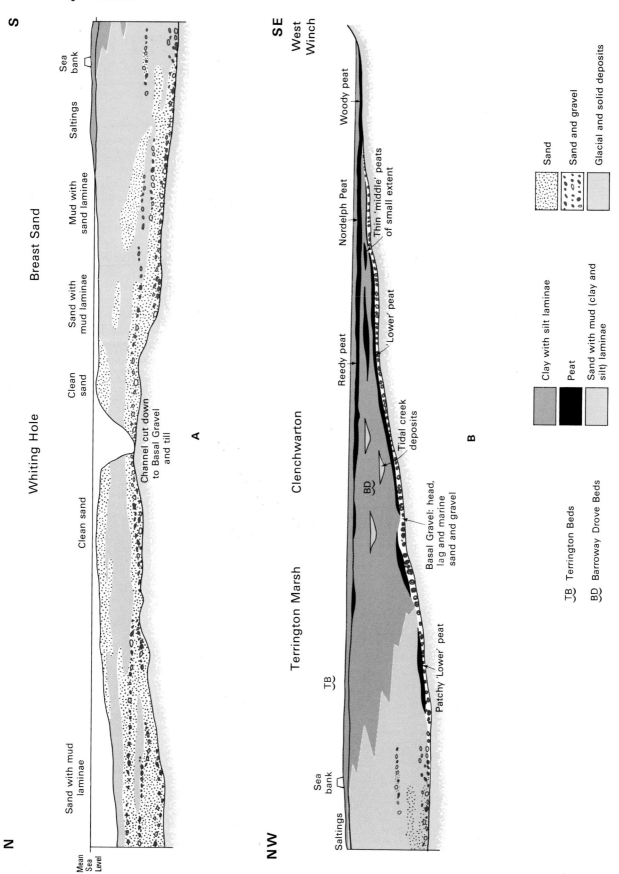

Figure 53 Generalised horizontal sections through the Recent deposits: A=offshore, B=onshore.

Plate 18 Peat shrinkage at Leziate Fen, Ashwicken [702 202].
Over a period of 8 years (1963 to 1971 at the time when the photograph was taken) the ground level fell by
up to 0.5 m due to drainage improvements which led to shrinkage and oxidation of the peat as it dried out.
Hammer is 35 cm long. (A 11807).

reclamations, which converted huge areas of the present district from salt-marsh to arable land, is described in Chapter Seven. Studies of historical charts of The Wash and its coastline (Kestner, 1962) have suggested that the width of the modern salt-marsh, the most recently formed Terrington Beds, stablilises itself at about 1km in the southern Wash. Reclamation of any part of this strip results in rapid accretion until the salt-marsh again reaches stability at about its former width. This suggests that much of the land reclamation in the district, especially that since the 17th century, has been made possible only by the induced sedimentation caused by each successive reclamation.

Details

With the exception of a narrow zone between Snettisham and Wiggenhall St Germans, adjacent to the eastern margin of the fenland, the Recent sequence below the Nordelph Peat is known only from boreholes. Examples chosen to be representative of their areas are given below: additional examples are referred to in Appendix 1.

Lutton to Tydd St Giles

Patches of thin Nordelph Peat are present beneath the Terrington Beds in the area on the landward side of the 'Roman' Bank. In the area close to the bank, the Barroway Drove Beds are represented largely by silt and sand deposited at a time when the area was the combined estuary of the rivers Nene and Great Ouse. The following sequence was proved in a borehole [4101 2011] sited at about 3 m above OD near Poplar House:

	Thickness m	*Depth* m
Terrington Beds: grey and brown silty and sandy clay	2.30	2.30
Nordelph Peat: soft peat	0.15	2.45
Barroway Drove Beds: soft silt and	1.82	4.27
silty, fine-grained sand with some soft clay layers and shells	7.93	12.20

The Tydd St Mary Borehole [4307 1737], sited at about 3 m above OD, proved 32.75 m of Recent deposits including Basal Sand and Gravel, peats, Barroway Drove Beds and Ter-

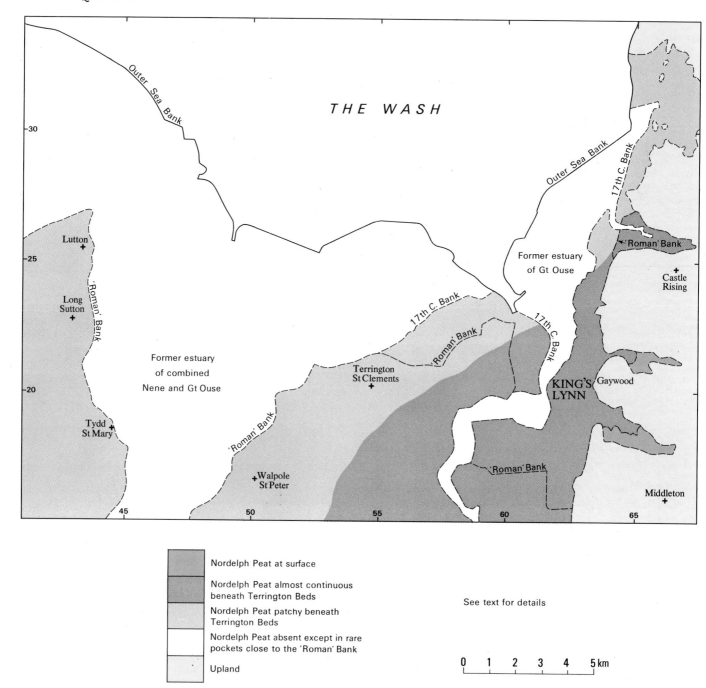

Figure 54   Distribution of the Nordelph Peat.

Legend:
- Nordelph Peat at surface
- Nordelph Peat almost continuous beneath Terrington Beds
- Nordelph Peat patchy beneath Terrington Beds
- Nordelph Peat absent except in rare pockets close to the 'Roman' Bank
- Upland

See text for details

0 1 2 3 4 5 km

rington Beds, but these were not sampled or described in detail.

Former estuary of River Nene

The Nordelph Peat and probably the Lower Peat are absent beneath the large triangular area bounded by the 'Roman' Bank and the present day coastline, except for a few small areas close to the 'Roman' Bank. This large tract of ground includes Lutton and Terrington marshes, and was formerly the estuary for the combined waters of the rivers Nene and Great Ouse. Within it, the Barroway Drove Beds largely consist of silts and sands.

A borehole [4803 2111] at Sutton Bridge proved the following sequence which is typical of the central part of the former estuary:

	Thickness m	Depth m
Soil and subsoil	0.2	0.2
Terrington Beds: brown sandy silt with partings of fine-grained sand and clay	2.8	3.0
Barroway Drove Beds: sand, grey and brown, fine- and medium-grained with a few shells	6.0	9.0

	Thickness m	Depth m
Sand, grey and brown, fine- to coarse-grained with shell fragments and pebbles scattered throughout	5.5	14.5
Clay, grey, sandy and silty interbedded with grey clayey and silty sand with shell debris	2.7	17.2
Sand, grey, silty, gravelly in part and with thin interbeds of sandy clay	1.8	20.0

A large number of site-investigation boreholes drilled for the Wash Feasibility Study on Breast Sand proved thick sandy sequences at the mouth of the former estuary. The following sequence, proved in a borehole [5425 2702] just seaward of the salt marsh at a surface level of 1.9 m above OD, is typical of the area:

	Thickness m	Depth m
RECENT DEPOSITS		
Interlaminated grey-brown silty, sandy clay and fine- to medium-grained, clayey sand	1.50	1.50
Sand, grey-brown, fine- to medium-grained with partings of organic silt	6.50	8.00
Sand, grey-brown, fine- to coarse-grained with scattered shells	4.00	12.00
Sand, grey-brown, silty, fine- to coarse-grained with thin beds and lenses of gravel and partings of clay	10.45	22.45
KIMMERIDGE CLAY	seen 2.75	25.20

Boreholes at the western end of Breast Sand, close to the modern outfall of the River Nene, proved similar sequences but with higher proportions of gravel in their lower parts.

Walpole St Peter to Clenchwarton

The Nordelph Peat is again present, although thin and locally patchy, on the landward side of the 'Roman' Bank between the former Nene estuary and the River Great Ouse. The following sequence was proved in a borehole [5191 1912] at Walpole Cross Keys at 2.9 m above OD.

	Thickness m	Depth m
Made ground: flints and furnace slag	0.70	0.70
Terrington Beds: laminated brown clayey silt resting on grey-brown, clayey, silty, fine-grained sand	1.60	2.30
Nordelph Peat: brown silty clay and peat	0.80	3.10
Barroway Drove Beds: grey-brown silt passing down into fine-grained sand and then into fine- and medium-grained sand with marine shells	10.05	13.15

Several boreholes in the adjacent area proved up to 0.6 m of brown, fibrous Nordelph Peat, but boreholes between Terrington St Clement and near Hay Green [e.g. 5319 1895] mostly proved Terrington Beds (brown silty clays) resting on Barroway Drove Beds (organic-rich, grey silty clays). The following sequence was proved in several closely spaced boreholes [547 201] in Terrington St Clement, all sited at about 3 m above OD.

	Thickness m	Depth m
Soil and subsoil: silty brown clay	0.6	0.6
Terrington Beds: dark brown silt with laminae of clay and fine-grained sand	1.3	1.9

	Thickness m	Depth m
Barroway Drove Beds: grey-brown silt with laminae of sand and organic-rich lenses, passing down at 3.5 m into grey, fine-grained sand with silt laminae	8.6	10.5
Basal Sand and Gravel: gravelly medium- to coarse-grained sand with many shells	6.7	17.2

The junction of the Terrington Beds and Barroway Drove Beds is not clearly defined in the absence of the intervening Nordelph Peat: it is commonly marked in this area by a change from brown silty clays or clayey silts to grey sandy silts and silty sands. The unusual thickness of the Basal Sand and Gravel at this locality, coupled with the large number of shells, suggests that it infills a former tidal channel.

Boreholes near Kenwick Farm [e.g. 5676 1910] proved up to 0.3 m of Nordelph Peat beneath about 2 m of Terrington Beds and resting on organic-rich, predominantly silty Barroway Drove Beds. As elsewhere in the area, the peat is patchy and nearby boreholes [e.g. 5780 1912] proved the Terrington Beds to rest on silty clay Barroway Drove Beds penetrated by peat rootlets.

King's Lynn to the Nar Valley

The Nordelph Peat is present beneath much of King's Lynn. Its thickness is variable over short distances because it is channelled into by the Terrington Beds; it is absent beneath the deeper channels (formerly tidal creeks). The thickest recorded peat in the area is in the Gaywood valley where boreholes [e.g. 6507 2120] have proved up to 3.5 m of woody and reedy peat. Thicknesses of 0.5 to 1.5 m are more common beneath the town.

The following sequence was proved in the Gaywood valley in a site-investigation borehole [6515 2114] drilled for the bypass:

	Thickness m	Depth m
Nordelph Peat: brown peat	2.54	2.54
Barroway Drove Beds: grey silt with sand partings	2.36	4.90
Chalky-Jurassic till	1.65	6.55

The Nordelph Peat, and locally thin representatives of the Barroway Drove Beds and the Basal Sand and Gravel, are exposed from time to time in the deeper drains east of the River Nar between King's Lynn and the Nar Valley. The following sequence was proved in a borehole [6299 1327] in the Nar Valley near Setchey, sited at about 3 m above OD:

	Thickness m	Depth m
Made ground: gravelly sand	0.60	0.60
Terrington Beds: reddish brown clays and silty clays	2.95	3.65
Nordelph Peat: soft fibrous peat	1.06	4.71
Barroway Drove Beds: grey silty clays and silts	1.99	6.70
Lower Peat: fibrous peat	0.69	7.39
Basal Sand and Gravel: gravelly sands	c1.6	c.9.0

OFFSHORE AREA

Clean sand

Clean or relatively clean sand occurs mostly in three types of high-energy depositional environment in The Wash: on the foreshore and its subtidal extension between Heacham and Hunstanton, on the outer margins of the larger banks within The Wash, and in the larger tidal channels. The degree of sorting of the sands is dependent upon the strength and frequency of the wave action, the coarsest and cleanest sands occurring in the more-exposed locations (Plate 19).

The accumulations of clean sand on the banks within The Wash can themselves be divided into three main topographical types. Those which form very large banks, up to 3 km wide and 8 km long, are flat topped over very large areas and are separated from one another by broad shallow channels. They include Roger Sand, Gat Sand, Thief Sand, Seal Sand and Sunk Sand, all of which have offshore margins which slope down into the subtidal area without break in slope. The outer margins of some of these banks, such as the north part of Thief Sand, carry trains of sand waves with approximately parellel crests, mostly 0.5 to 3 m high and with wavelengths from 10 to 50 m, which maintain their regular spacing for lateral distances of several hundred metres. They are mostly concentrated close to low-water mark and are believed to be formed by strong tidal currents. Many of the sand waves have flat tops and carry ripple marks formed by the last ebb tide on their flanks. Very large ripples, up to about 1 m high and with wavelengths of 3 to 5 m, occur between some of the large banks and the larger sand waves, where they form complex patterns produced by the interference of the flood and ebb tides.

Bars of clean sand occur in the Wisbech, Lynn and Old Lynn channels, where the tidal currents are too strong to allow the finer-grained sediments to settle.

Sand with mud laminae

In the intertidal area, sand with mud laminae has an extensive outcrop on the large drying banks in the southern part of The Wash where it is partly protected from wave action by an outer belt of clean sand. In the subtidal area, sand with mud laminae occurs over most of the southern Wash, except in the tidal channels and in shallow water subject to strong wave action, where the bottom currents are too strong to allow mud to settle out. Core samples from the areas of sand with mud laminae mostly contain over 70 per cent clean, medium-grained sand with some coarse shell debris and, in some of the deeper areas where sea bed is more than 12 m below OD, discontinuous layers of gravelly sand and gravel, and scattered pebbles. In the shallow areas, diffuse thin bands of slightly muddy sand are common, together with discrete mud laminae up to 4 mm thick. Erosive channels floored by angular clay clasts are common in the shallower areas.

Mud with sand laminae

Within the intertidal zone there is a gradual landward change from sand with mud laminae to mud with sand laminae. The increase in the proportion of mud gives rise to smooth mudflats crossed by a meandering network of creeks. The creeks are shallow at their seaward end, but become steep sided and very sinuous as they approach the salt-marsh.

Mussel scalps

The common mussel *Mytilus edulis* Linné occurs in communities which cover areas of up to about 1 km^2 in The Wash; these are known locally as scalps. Individual banks may exceed 2 m in thickness. Most of those shown on the published geological map occur in areas of sand with mud laminae, but a significant number occur on sand on the drying banks. There are others at sea bed in the intertidal areas which were proved by coring, but which could not be mapped out accurately. In the scalps shown on the map, an average of about 30 to 40 per cent of each consists of living mussels. Some of the dead parts of the scalps are being activity covered by sand, and the remains of older scalps, now completely enclosed in sand, were recorded in the subtidal area. All these accumulations consist of closely packed mussel shells, with lenses and pockets of sand, and with much interstitial mud which the mussels have filtered from sea water.

Saltings

Between Snettisham Scalp and the western margin of the district near Holbeach St Matthew, the zone between the outer sea bank and high-water mark of medium tides is occupied by saltings (salt-marsh) cut by numerous tidal creeks. Silty clays and clayey silts, carried into the area by spring tides, are trapped there to form interlaminated deposits (Terrington Beds) much of which will eventually be reclaimed by embanking.

UPLAND AREA

Blown sand

Impressive dunes of blown sand, some more than 12 m high, occur on the exposed coastal strip at Hunstanton golf course where they occupy an area up to 450 m wide. The most seaward row of dunes is still active and is receiving sand from the broad expanse of clean sand which fronts it. Inland, the older dunes are now stabilised by vegetation; their size is such that it is probable that they were formed under tundra conditions in the late Pleistocene from extensive spreads of glacial outwash sands which were exposed along the north Norfolk coast at that time.

Alluvium and dry valley and Nailborne deposits

Narrow strips of freshwater alluvium occur in the floodplains of the Babingley River, the River Ingol and the Heacham River, in each case consisting of a maximum of 1 to 2 m of silt and silty clay derived mostly from local Head and glacial deposits. Most sections are floored by reworked gravels and gravelly sands. The dry valley at Ringstead Downs [690 400] (Plate 20) is floored by an al-

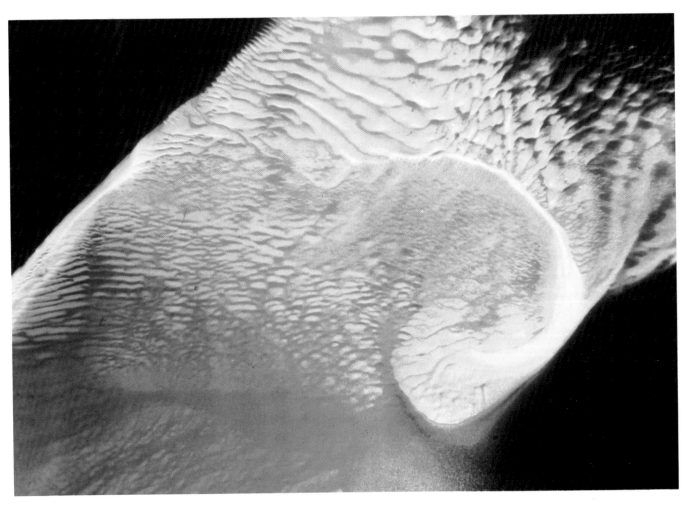

Plate 19   Recurved spit, sand waves and large ripples exposed at low tide, the Hook,
Long Sand.

Large mobile sand banks made up of trains of waves and large ripples of clean sand occur throughout The Wash
in areas of high tidal velocity. Many of these sandbanks are wholly or largely (as here) submerged even at
low spring tides; they commonly have steep sides and enclose areas of deep water, and are a
considerable hazard to shipping. Long Sand, like the other sandbanks, is migrating southwards. On its
north-eastern side (top) large sand waves are asymmetrically shaped by flood tides; on its south western side large
ripples face north-eastwards, reflecting the influence of the last ebb tide. (*Fairey Surveys for Water Resources
Board*).

luvial deposit composed of clay, silt and chalky wash
which has formed at times when the surface run-off has
been sufficiently high, for example due to rapid melting
of snow, for the valley floor to be flooded.

Storm gravel beach deposits

A continuous strip of storm beach gravel, up to 6 m high,
runs for about 11 km southwards from Hunstanton almost to Wolferton. It is composed largely of well-rounded flint and chalk gravel with small numbers of stones derived from the Chalky-Jurassic and Hunstanton tills, and
from the Carstone and Red Chalk outcrops at Hunstanton. This ridge has probably moved steadily eastwards
throughout the Flandrian as sea level rose, and for most
of its length has now reached its eastern limit because it
abuts against higher ground formed by glacial deposits.
Recurved spits at its southern tip and at Snettisham Scalp
indicate that the beach has been active until relatively recently. The encroachment of the salt-marsh around the
southern tip, probably in response to reclamation of the
adjacent salt-marshes to the south, suggests that it is now
inactive.

A narrow strip of gravel, of similar composition but at
a lower height due to degradation, runs from near
Wolferton [646 283] almost to the Babingley River and
has been partially incorporated into the 17th century sea-
bank. This ridge is almost certainly a continuation of the
Hunstanton to Snettisham storm beach, but is separated
from it by the former estuary of the Heacham and Ingol
rivers. Small amounts of gravel and shells in the Terrington Beds, adjacent to the 17th century sea-bank immedi-

Plate 20 Ringstead Downs dry valley, Barrett Ringstead Farm [688 399].
The dry valley runs from Ringstead village to Barrett Ringstead Farm, a distance of about 2 km, and is cut in Lower and Middle Chalk. East of the village, the valley opens out into a series of shallow dry valleys that extend east as far as Brancaster and Docking, and which were the feeders for the late Pleistocene meltwaters that cut the Ringstead Downs valley. The valley forms the southern limit of the Hunstanton Till in this area and was probably initiated as an ice-margin feature. The original gorge-like form of the valley has been modified by cambering (see Plate 17) and late Pleistocene head and alluvial deposits (foreground and centre right). View east. (A 10881).

ately south of the Babingley River [638 268], may indicate the presence of a former ridge of storm beach gravel which ran southwards from the estuary of the Babingley River towards King's Lynn.

Head and head gravel

The head deposits of the district are poorly sorted and poorly stratified materials of local origin which have moved downhill as solifluction or sheet-wash deposits in postglacial times. Where composed largely of gravel they are shown as Head Gravel on the map. There are extensive head deposits adjacent to the upland part of the district, and this reflects both the intensity of the periglacial

cryoturbation and meltwater erosion processes in the late Pleistocene, and the unconsolidated nature of some of the Cretaceous and glacial deposits.

Extensive thick sandy head deposits are present wherever the Leziate Beds outcrop abuts onto low ground, notably between Snettisham and Wolferton, and in the valleys of the Babingley River, Gaywood River and Middleton Stop Drain and their tributaries. Although the bulk of these deposits was laid down in the latest Pleistocene, they are so poorly consolidated that their surface layers are still mobile. At the time of the survey, an unusually heavy rainstorm produced gulleys up to 1 m deep in ploughed fields on the Leziate Beds outcrop on the west side of Roydon Common [around 676 227] which added fans of clean sand to the head on the common.

The sand and gravel associated with the Hunstanton Till also gives rise to extensive sandy head deposits that have mostly been winnowed from the matrix of the gravels. On the higher parts of the Chalk outcrop, east and south east of Hunstanton, patches of sandy head probably contain a wind-blown element of loessic silt and fine-grained sand.

SEVEN
Land use and economic geology

SETTLEMENT

The upland part of the district was probably extensively colonised from very early times, as soon as it was practicable after the retreat of the last Pleistocene ice sheet about 18 000 years ago. Evidence of man's activity prior to the onset of the final ice age is rare, but Palaeolithic flint implements found in gravels at North Wootton, Middleton and Tottenhill are believed to date from the temperate period that immediately preceded the last advance of the ice. Neolithic and Bronze Age material is rare in the district; flint implements have been found at North Runcton and Heacham, and Bronze Age pottery at Heacham. The well-organised flint-mining industry at Grime's Graves, near Thetford, lies only a short distance to the south-east of the present district and it seems reasonable to assume that Neolithic man hunted and fished in The Wash and adjacent coastal areas.

The first real evidence of a settled and complex society comes from the Iron Age, when the district was occupied by a succession of Celtic cultures culminating in the Iceni. Hoards, including magnificent electrum and gold torcs and bracelets, have been found in the Gaywood Valley near King's Lynn, and at Snettisham. These last are the richest array of Celtic treasures yet found in Britain and have been interpreted as the personal jewellery of the local nobility of the lst century BC (Laing, 1979; Stead, 1991). Tumuli of probable Iron Age occur at Castle Rising, Gayton and Middleton.

Romano-British remains are well represented in the area by villas at Grimston and West Newton; a Roman building at Snettisham; pottery debris at Heacham, Holme, Rising Lodge and Snettisham, this last associated with pottery-clay working in the Snettisham Clay; various other remains in the valley of the Heacham River; and iron-bloomery slag at Rising Lodge and Blackborough. Peddar's Way, the Roman road that linked towns in the central part of East Anglia with the north Norfolk coast at Holme next the Sea, runs a few kilometres to the east of the district.

Successive waves of seaborne invaders, including Danes, Vikings and Saxons, settled along the eastern seabord of The Wash, with the result that all but two of the major upland settlements of the district were established before the time of the Domesday Survey. The exceptions are the 'new towns' of King's Lynn and Hunstanton, the former founded in the 11th century as a seaport and the latter in the 19th century as a Victorian holiday resort.

The history of human occupation of the Fenland part of the district is shorter and less well known. The whole of this land area has been reclaimed since Roman times with the result that any pre-Roman remains are now concealed by younger deposits (Terrington Beds). The distribution of known Roman settlements and finds within the Fenland part of district is extremely patchy and, in part, probably reflects the limited research that has been carried out in that area. In contrast, a large number of Roman remains have been revealed by systematic research in the Lincolnshire part of Fenland (see Phillips, 1970 for summary).

Saxon remains, mostly in the form of drainage and embankment works in Fenland, and as pottery and other habitation debris in the upland area, are relatively common in the district. The Fenland area also contains part of what is probably the most impressive construction of the early Medieval period. This is the so-called 'Roman' Bank, a 100 km-long earth bank that formed a continuous barrier to the sea between the Lincolnshire Wolds and the Norfolk uplands. Skertchley (1877, p.13) estimated it to contain about 10.5 million tons of material, mostly silt, and noted that it was considered to be old even at the time of its first written mention in the 12th century. The age of the bank is uncertain and has been the subject of much discussion. It is demonstrably not Roman because it lies many miles seaward of all the Roman drainage works that were inundated by the second Terrington Beds transgression in the 3rd century AD (see p.163). In the present district and in the adjacent Spalding district large man-made mounds of silt abut onto the seaward side of the 'Roman' Bank at many localities. These are the remains of a late Saxon to 15th century salt-making industry which was already well established at the time of the Domesday Survey (see p.184). Almost all these workings postdate the bank, which strongly suggests that the bank itself is late Saxon in age. Evidence of renewed transgression in the 13th century suggests that for a period of several hundred years the 'Roman' Bank was maintained as the outer sea defence; it was eventually superceded by major reclamation works in the 17th century.

The bank, together with its outfall sluices, must have been well maintained for several centuries; even now, large parts of it are intact. It can be traced almost continuously from its entry into the western part of the district at Gedney Dyke [412 260], via Lutton, Long Sutton, Tydd St Mary, Tydd Gote, Wisbech, Walpole St Peter, Walpole St Andrew, Terrington St Clement, Clenchwarton, and along the west bank of the old course of the Great Ouse as far south as Eau Brink [581 160] (Figure 55). To the east of the Great Ouse, its course is not clear because it has been levelled to provide fill for later banks. It is probably the 'North-Sea Bank' that runs from Islington Lodge [580 171] to the River Nar near Whitehouse Farm [617 170].

The course of the bank is not known east of the Nar, but it is assumed that it took the shortest available route

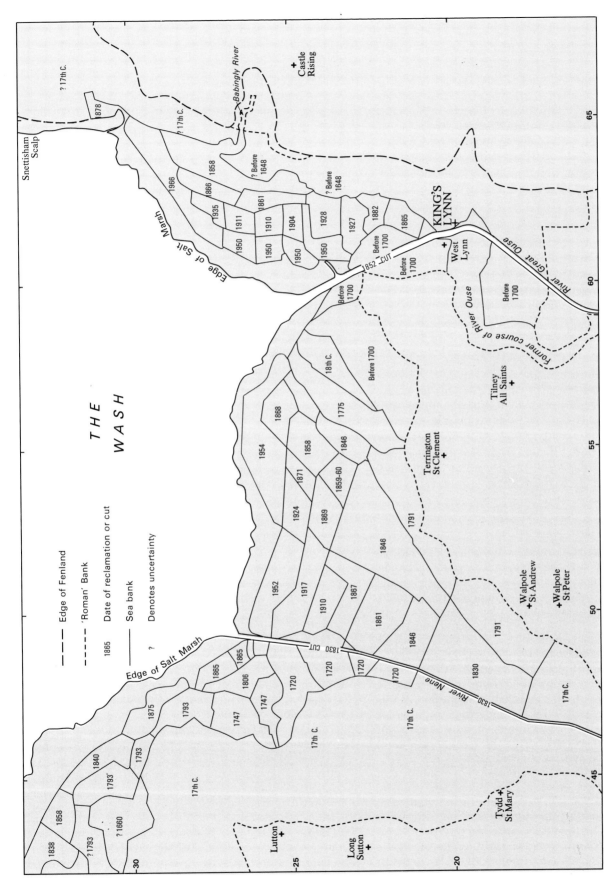

Figure 55   Dates of reclamation of the Terrington Beds.

to the nearby upland and followed the line of an old bank that joins the upland near North Runcton (Figure 54). A short length of bank, which is believed from the field evidence in the present survey to be contemporaneous with the 'Roman' Bank, sealed the estuary of the Babingley River between North Wootton [636 243] and Wolferton [648 267]. Similar short lengths of bank may have enabled the valleys of the Gaywood River and the Middleton Stop Drain to have been reclaimed, but the field evidence in those areas is poor, due to later urban development, and no bank was recorded during the survey.

As with the upland area, all the major settlements of the Fenland part of the district were established by the time of the Domesday Survey, although the population density at that time was very low. Fishing, stock grazing and salt manufacture appear to have been the three most important occupations at that time. The pattern of land use and settlement has changed little since the time of that survey, except in that part of the district seaward of the 'Roman' Bank, where there has been considerable land reclamation (Figure 55). Most of this reclamation appears from the surviving evidence to have taken place since the 17th century, but this may be misleading. Extensive reclamations were probably made between the 9th and the 12th centuries, but were subsequently lost during a period of relatively high sea-level during the 13th century. In addition to the extensive reclamations in the Gedney Drove End and Terrington areas, major works in the 17th to 19th centuries included the entrainment of the rivers Great Ouse and Nene, and the reclamation of their estuaries.

SOILS

There is no modern account of the soils of the King's Lynn district, but the formations that crop out there, the topographical relief and the past and present climates, can be matched in detail with those of adjacent parts of East Anglia where detailed soil surveys have been made in recent years by the Soil Survey of England and Wales. The closest well-documented parallel is the Ely district which contains Jurassic, Cretaceous and Quaternary sequences that are lithologically similar to those of the present district. The following account is largely based, by extrapolation, on the detailed descriptions of the soils and their relationships to the geology given by Mr R S Seale (1975) in the Soil Survey Memoir, and by Mr Seale and the present author in the Geological Survey memoir for the Ely district (Gallois, 1988).

Soils are produced by a combination of processes, of which the two most important in East Anglia are the chemical and mechanical weathering of the geological formations. The most important weathering process at the present time is chemical weathering by groundwater containing dissolved oxygen and carbon dioxide. In the argillaceous formations (the Kimmeridge Clay, parts of the Mintlyn Beds, Snettisham Clay and Gault, and boulder clay where it has a high proportion of these rocks in its matrix) chemical weathering causes a general de-

crease in the crystallinity of the clay minerals and oxidation of iron minerals such as pyrite. As a result, soft grey mudstones are converted to yellow and brown sticky clays. The sandy formations (the Sandringham Sands, Dersingham Beds and Carstone) contain iron minerals such as glauconite and limonite and lesser amounts of siderite and 'chamosite'. These become oxidised to form finely divided hydrated iron compounds that are commonly leached away in solution or as colloids, leaving white, yellow or brown sand residues. The calcareous formations (the Chalk and the more calcareous parts of the Upper Jurassic clays and Gault) are slowly dissolved and any included clastic or mineral matter accumulates as a residue.

The establishment of vegetation intensifies the rate of weathering and produces a pedological soil. Successive generations of plants and animals add humus to the soil and cause a concentration of some minerals and depletion of others. The long-term effect of these changes is to produce a soil profile, a sequence of roughly parallel layers, in which successive changes can be traced from unweathered rock to surface soil.

Variations in soil type are largely related to two factors in the district, the geology and the topographical relief (Table 15). Three main types of site occur within the district:

1. Well-drained sites such as those on the Sandringham Sands, Chalk and glacial sand and gravel where the water table is permanently at some depth below the soil profile.

2. Sites on clays such as the Kimmeridge Clay and Gault where, because of the impermeable nature of the subsoil, perched water tables occur seasonally.

3. Low-lying areas where the water table is within or close to the soil profile for most of the year.

Site-types 1 and 2 are restricted to the upland part of the district, and type 3 to Fenland. An intermediate type of landscape known as 'Skirtland' occurs as poorly drained, low-relief area around the edges of Fenland and in the river valleys.

Four types of soil were recognised in the Ely district (Seale, 1975) and each of which is probably represented in the present district. No quantitative assessment of their relative abundances can be made in the absence of a detailed soil survey, but gley soils are probably the most widely distributed, brown earths are more common than calcareous soils, and organic soils are rare.

Gley soils result from the reducing action of bacteria, micro-organisms and plant roots in poorly drained conditions acting upon ferric and other compounds. Such soils are commonly grey but may contain brown limonitic concentrations in their upper layers if intermittent waterlogging produces alternately reducing and oxidising conditions. They occur on both the upland and Fenland parts of the district.

The Kimmeridge Clay, parts of the Mintlyn Beds, Snettisham Clay and Gault give rise to gley soils (surface-water gley soils) because of their impermeable nature. Downward percolation is restricted with the result that,

Table 15 Simplified classification of the soils of the King's Lynn district, showing the relationship of soil type to geology and landscape

Landscape	Parent material	Soil group	Soil subgroup
Upland	Red Chalk, Chalk and chalky drift	Calcareous soils	Rendzinas
			Brown calcareous soils
	Calcareous parts of Kimmeridge Clay, Gault and boulder clay		Gleyed brown calcareous soils
Fenland	Calareous parts of Terrington Beds		
Upland	Parts of Sandringham Sands and Dersingham Beds; Carstone and sandy and gravelly drifts	Brown earths	Normal brown earths
Skirtland and river valleys	Poorly drained sandy and gravelly solid and drift deposits		Gleyed brown earths
Upland	Most of Kimmeridge Clay and Gault; Snettisham Clay	Gley soils	Surface-water gley soils
Fenland and river valleys	Alluvium, Nar Valley Clay, most of Terrington Beds		Groundwater gley soils
River valleys	Peat	Organic soils	

in winter, a temporary water table is present a little below ground level. In summer, as the clays dry out, lateral drainage occurs through cracks formed in the surface layers. Large areas of the Terrington Beds outcrop also give rise to gley soils (groundwater gley soils and humic gley soils) because of the permanent proximity of the water table to ground level.

Brown Earths are well-drained soils that form on the Roxham Beds, Runcton Beds, Leziate Beds and Carstone, and the more sandy parts of the Mintlyn Beds, Dersingham Beds and noncalcareous drift deposits. Their imperfectly drained equivalents, the gleyed brown earths, form on sandy and gravelly deposits in the skirtlands.

Calcareous soils form on the more calcareous parts of the Kimmeridge Clay and Gault, and on the Chalk and chalky boulder clay where these are relatively well drained and not overlain by noncalcareous superficial deposits. In exceptionally well-drained areas on the steeper chalk slopes, attenuated thin soil profiles (rendzinas) are present in which an organic/mineral-rich surface layer rests on chemically unweathered rock with no intervening weathered layer or with only a poorly developed one. On the lower slopes and in skirtland, where a seasonal water table is present, gleyed rendzinas are formed on the Chalk. On well-drained sites where the calcareous content of the Chalk or boulder clay is mixed with sand and stones from local drift deposits, a more complete soil profile is present and brown calcareous soils are formed; the same rocks give rise to gleyed brown calcareous soils in imperfectly drained or low-lying areas. Parts of the Terrington Beds may also be lo-cally sufficiently calcareous to form a gleyed brown calcareous soil.

Organic soils form on peat; they are confined in the present district to a few small outcrops of Nordelph Peat in the river valleys.

CONSTRUCTION MATERIALS

Building stone

A variety of local rocks, including ironstones, ferruginous sandstones, ferruginous pan, chalk, Red Chalk, flint and drift-derived materials, have been used for building purposes in the upland part of the district. Only one of these, the Carstone of the Hunstanton to Snettisham area, is sufficiently durable to be used on its own for even modest sized buildings. As a result, most of the larger old buildings are built of imported stone or have a framework of imported stone with local materials used to infill the walls. In the Fenland part of the district, including King's Lynn, many quite substantial buildings were also built of locally made brick or have timber frames and brick infillings because of the absence of local stone.

The most common medieval import was Barnack Stone, an oolitic limestone freestone from the Middle Jurassic Lincolnshire Limestone, that was quarried on the banks of the River Welland near Stamford and transported to the district by barge through the Fenland waterways and The Wash. The older parts of St Margaret's Church (c. 1100) and St Nicholas' Chapel (c. 1420), King's Lynn, and the Fenland churches at West Lynn,

Terrington, Long Sutton and Lutton are built largely of this stone. Some idea of the delicacy with which it can be worked can be gained from the carvings and tracery on St Nicholas' Chapel. Later buildings of oolitic limestone, such as the South Gate (1520) and Custom House (1683) at King's Lynn, and Snettisham Church, were built of other Lincolnshire Limestone freestones, such as Clipsham Stone and Ketton Stone, after the Barnack Stone quarries had become exhausted in the 15th century.

Castle Rising Church and the Castle Keep (both c.1140), and the ruined medieval churches at Babingley, Gaywood and Mintlyn all have a framework and ornamental work of Lincolnshire Limestone with wall infillings that reflect the local geology. All are built on the Sandringham Sands and contain ironstone from the Mintlyn Beds, secondary ferruginous pan derived from the same, and flint and rarer quartzite derived from local glacial deposits. In addition, secondary quartzite derived from the local Leziate Beds (see p.77 for details) is abundant at Castle Rising, and iron slag derived from a nearby bloomery of presumed Roman age is common at Mintlyn. By contrast, the wall infillings at Snettisham Church are composed largely of Carstone, Red Chalk and Lower Chalk obtained from local outcrops, together with flint from more distant outcrops of the Middle and Upper Chalk.

Red Chalk, Chalk and flint are common building materials in this north-eastern part of the district: there are especially good examples of garden walls and cottages of these materials in Old Hunstanton. The Red Chalk has not been quarried in the district and all the larger blocks appear to have been collected loose from Hunstanton beach. Some of the smaller material probably came from the Carstone quarries at Snettisham [e.g. 685 349] and from field brash.

Several of the harder beds in the Lower Chalk, notably the Paradoxica Bed, Inoceramus Bed and Totternhoe Stone, can be recognised in local buildings together with less distinctive beds. Chalk was probably obtained for building partly as a byproduct of other chalk workings and partly as fallen blocks from Hunstanton Cliffs. Much of it is susceptible to frost and great care was normally taken to keep it as dry as possible once it had been laid.

Carstone, in the form of loose blocks, has been used as wall infilling in buildings of all ages in the district, but its use as dressed stone does not appear to have become widespread until Victorian times. Jackson (1911, p.62) noted that it made an excellent building stone because it was easily dressed when freshly dug, but hardened on exposure due to oxidation of its contained iron minerals, and was particularly suitable for buildings in a damp or exposed position because it weathered very slowly. Much of the new Victorian resort of Hunstanton and the stations on the railway that linked it to King's Lynn were built of Carstone obtained from Snettisham. Small degraded quarries of unknown age, but probably mostly Victorian, occur in the Carstone near Church Farm, Heacham [6857 3780]; Ken Hill [6731 3481]; Lodge Hill [6697 3396; 6691 3422]; near Hill House Farm, Der-

singham [6985 3061; 6968 3047]; near Sandringham House [6917 2862; 6951 2868] and Appleton House [7095 2711; 7107 2704]. Victorian quarries occur at Snettisham [685 349; 6853 3460; 6875 3470]; Scotch Belt, Sandringham [6832 2827] (reputedly the source of stone for the Feathers Hotel, Dersingham); Carpit Cottages, East Winch [6845 1605]; Blackborough End [6675 1490], and near Park Farm, West Bilney [6965 1417; 6988 1437]. The top 5 m of the Carstone continues to be worked at Snettisham [685 349] in a large shallow pit. The upper 1.5 to 2 m of the workings are in deeply weathered rubbly Carstone that is suitable only for hardcore. The lower beds are more massive and are used as roughly dressed or sawn blocks for the repair of old Carstone buildings.

Ferruginous sandstones in the lower part of the Dersingham Beds have been extensively worked in the district. When weathered, these are sufficiently similar to the Carstone for them to have also been referred to locally as 'Carr' or 'Carstone'. Whereas most of the true Carstone is thickly bedded or massive and provides large blocks, known locally as 'big Carr', the Dersingham Beds sandstones are flaggy and fissile and yield only 'small Carr' (Plate 21). There are also lithological differences. The Carstone is commonly pebbly and medium grained, has a relatively uniform texture and is unfossiliferous, whereas the Dersingham Beds are fine grained and commonly range from densely cemented to highly porous because of the presence of voids formed around rotted fossils.

The basal strata of the Dersingham Beds have been extensively worked in numerous shallow pits in the Sandringham Warren area [e.g. 668 273; 6706 2804; 6629 2846; 6623 2825]. A pit on the Warren [6770 2880] reputedly supplied the stone for the Norwich Gate wall at Sandringham House. All these pits were in deeply weathered material and the usefulness of the sandstone as a building stone must have depended upon the degree of limonitic cementation. At one locality [6734 2839], the sandstones may have been worked underground by means of bell-pits, either as a source of stone or iron ore. Ferruginous sandstones at the same stratigraphical level to those at the Warren have also been worked for building stone at West Newton [6945 2713] and near Rising Lodge [673 222], and as hardcore and inferior stone in the overburden of sandpits at Snettisham and Dersingham.

Superficially similar ferruginous sandstone, comprising large blocks composed of very porous, limonite-cemented sand enclosing fragments of sandstone, bleached flints and rotted modern plant fragments, has been used in some walls and barns on and adjacent to the Mintlyn Beds outcrop. This stone is a ferruginous pan (known elsewhere in southern England as 'chevick' or 'shrave') that forms as lenticular beds or irregular concretionary masses, up to 2 m across and 1 m thick, in the zone of water-table fluctuation in the more poorly drained parts of the Mintlyn Beds outcrop. This pan is commonly sufficiently hard to damage ploughshares, and almost all the material used for building has been dug out primarily to avoid such damage. It may have

Plate 21   Dersingham Beds used as building stone, Sandringham Church.

Flaggy bedded, ferruginous fine-grained sandstones in the lowest part of the Dersingham Beds have been
extensively worked for building purposes in the Sandringham area where they are known as 'small Carr'.
The nave and tower of the church are made of small, roughly dressed blocks of this sandstone, fitted
together without cement, held in place by frames and copings of Jurassic oolitic limestone. (A 10863).

been specifically worked for building purposes in a large
shallow pit [639 157] at Manor Farm, North Runcton. It
is a common, in some cases an almost exclusive constituent of walls and barns in the North Runcton area.

Brick clay etc.

Bricks, and probably tiles, drains and pottery, have
been made from local clays at a large number of localities in the district, all on a small scale and for local use.
Many of the brickpits listed below were probably
worked intermittently for use on the estate on which
they are sited. Bricks were also probably made simply by
building a small temporary clamp on the clay outcrop
whenever the need arose. Such small workings rapidly
become degraded and leave little evidence of their existence; it must be assumed that the smaller workings referred to below are only a fraction of those that formerly existed.

The most extensively worked clay in the district was
the Snettisham Clay. Heacham Brickworks [679 364],
Snettisham Brickworks [6875 3382] and Bawsey Brickworks [685 195] (completely removed by later sand
workings) were in work throughout Victorian times and
into the early part of this century. Brickworks at Ingoldisthorpe Hall [6870 3262]; Brickley Wood, Ingoldisthorpe [6876 3220] and near Life Wood, Dersingham [6872
3138] closed in Victorian times. Small, very degraded
pits of unknown age, some with associated brick rubble,
occur in the Snettisham Clay at Heacham Bottom Farm,
Heacham [6800 3597; 6804 3586]; Ken Hill Wood [6784
3538]; Park Farm [6887 3396] and Park House Farm
[6900 3331], Snettisham; Ingoldisthorpe [6880 3295];
Doddshill, Dersingham [6946 3018]; Folly Lodge, San-

dringham [6760 2798]; Appleton House, West Newton [7050 2723; 7062 2726; 7070 2720], and near Leziate Heath [7024 1962]. Romano-British, late Saxon and Medieval pottery has been found close to the Snettisham Clay outcrop at Park House Farm, Snettisham, and it seems likely that there is a long history of clay working in that area.

Skertchly (1877) noted that large numbers of bricks were made from the Recent clays in Fenland, but that the bricks were commonly of poor quality, being soft, cracked and misshapen. Terrington Beds silty clays, only about 1.5 m thick, have been used for brickmaking at South Lynn [6202 1810] and Night Marsh, Castle Rising [6634 2555]. Shallow excavations in the same beds at West Lynn [6075 1977], Gaywood Park, King's Lynn [6305 1970] and in the reclaimed marsh between Heacham and Hunstanton [6694 3976; 6675 3895] were probably made for brickmaking. The Terrington Beds overlie peat at shallow depth at all these localities and this latter may have been used as a fuel for firing.

The lower, less calcareous beds of the Gault have been used to make bricks and pottery at Pottrow [710 224] and Grimston [7185 2165]. The latter locality is close to the site of a Roman villa.

The Nar Valley Clay was extensively worked for brickmaking in the Nar Valley in Victorian times to produce good quality red, mottled and white bricks (Rose, 1836). Only one of these pits, East Winch Brickyard [706 158], falls within the district. Rose (1835, p.31) recorded a brickpit in the formation in the valley of Middleton Stop Drain, but no evidence was found of it in the present survey.

Whitaker et al. (1893, p.8) noted that bricks were made from the Kimmeridge Clay at West Winch [probably 6305 1515], but the site is now built over.

Extensive shallow workings in chalky boulder clay at Grimston Road, King's Lynn [6485 2245], were probably for brickmaking: some of the large number of marl pits dug in the same material (see below) may also have yielded brick clay.

Aggregate, ballast and building sand

All the sandy and gravelly deposits in the district have at some time been worked as construction materials. In addition, the poorer quality stone from quarries in the Dersingham Beds and Carstone has been used for hardcore, and the dirtier sands from the glass and moulding sand pits in the Leziate Beds have been used for building sand. The better quality gravel resources in the district are Quaternary storm-beach gravels. The older gravels have been largely sterilised by urban development; the younger form the natural sea defence that protects the reclaimed marshes between Snettisham and Hunstanton.

Poorly sorted, chalk-rich gravels have been worked for hardcore or ballast in numerous, mostly small pits in the glacial sand and gravel associated with both the Hunstanton Till and the Chalky-Jurassic till. These include pits in the Hunstanton Park Esker in Hunstanton Park [6980 4029] and at Old Hunstanton [6896 4355]; at

Church Farm, Heacham [7005 3800; 7000 3706]; near Heacham Chalk Pit [6908 3665]; extensive workings in Gravelpit Belt near Heacham Bottom Farm [6832 3556 to 6854 3567]; extensive workings near Cat's Bottom, Babingley [674 268 to 681 272]; several pits at Vincent Hill [6900 2624; 6940 2637; 6960 2625] and a large relatively modern working near Hillington [7035 2535] in the Babingley valley; at Grimston Warren [6794 2133], Waveland Farm [6765 2080] and Leziate [6940 1995; 6932 2000] in the Gaywood valley; at Congham [726 229] and Ashwicken [6980 1875]. Flint-rich gravels are still worked for aggregate, but sporadically and on a small scale, in extensive pits at Blackborough End [676 145] and Foster's End [682 145].

The well-sorted, flint-rich gravels of the storm beach were worked at Snettisham Scalp [648 333 to 649 306] for concreting aggregate for airfield construction during the Second World War. Similar gravels had earlier been worked on a small scale at Wolferton [6465 2852] in a southerly continuation of the storm beach. Flint-rich gravels formed in older, presumed Pleistocene storm beaches have been extensively worked for hardcore, aggregate and embanking materials in pits at North Wootton [6400 2405], South Wootton [6410 2260; 6460 2272 to 6470 2305], Gaywood [6362 2044], Hardwick [6335 1810] and Setchey [634 143; 636 142; 633 144 to 636 141].

In addition to the defunct building-stone quarries listed above, from which waste materials were probably used for hardcore, the quarries in the Dersingham Beds near Rising Lodge [673 222] and those in the Carstone at Blackborough End [674 147] and Park Farm, West Bilney [699 144] have been considerably extended in recent years to provide hardcore. The Dersingham Beds pits at Sandringham Warren [668 273] were reopened to provide sub-base material for the King's Lynn eastern bypass.

Boulder clay was dug at Hunstanton [672 403] for the railway embankments, at Heacham [667 371; 673 374; 673 375] for either the railway or the sea bank, and at Hardwick [636 174] for the embankments of the King's Lynn southern bypass.

OTHER MATERIALS

Glass and foundry sands

The purer parts of the Leziate Beds sands have long been used for glassmaking in the district. Rose (1835, p.177) noted that large quantities of sand were exported from Dersingham to northern England for this purpose. Good quality glass sand is a relatively rare commodity and there are, in consequence, numerous large pits dug for it on the Leziate Beds outcrop. Most of the workings date from the 19th and 20th centuries. The largest of them, at Leziate, is still the basis for one of the most important industries in the district, although glass sand now forms only part of its production. There is no evidence that a large-scale glass-making industry was ever established in the district, probably because of the lack of a suitable fuel. It seems probable, however, that the presence of such large quantities of good quality sand was known from Romano-British times onwards and that it was used to produce glass locally.

In the older workings, the poorer quality sands, those with too much iron, glauconite or clay to be used for glassmaking, were used as building sand or ballast. In the modern workings, much of this material is sorted and upgraded to remove the glauconite and clay, and is then coated with resin to produce foundry moulding sand.

Leziate Beds sands were worked for glassmaking at Snettisham [671 336], Dersingham Common [6872 2971] and Sandpit Cottages, Dersingham [6815 2930] in the 19th century. At this last locality the sand was also worked in an adit that ran eastwards beneath the King's Lynn to Hunstanton road into Jocelyn's Wood [6821 2914]. Sand was taken from the Dersingham pits by barge down Boathouse Creek to The Wash to be transported by coaster to Jarrow-on-Tyne and other northern ports. Coal was commonly delivered on the return journey (hence Coalyard Creek, Wolferton). Sand has been worked in small- and medium-sized pits at Castle Rising [662 247]; Grimston Road [6550 2265] and Sandy Lane [6580 2224], South Wootton; Holding's Hill, King's Lynn [652 205] and Gayton Road, Mintlyn [6633 1993]. Extensive pits [653 243 to 656 236] at Ling Common, North Wootton, were worked during much of the 19th century. These pits, now degraded, are up to 15 m deep, but the remains of the railway by which the sand was taken to a wharf [6320 2480] at North Wootton marshes, until 1861 on the east bank of the estuary of the Great Ouse, are still visible. As at Dersingham, the sand was exported to northern England.

Glass sand has been worked in several large pits on the west side of Roydon Common [6575 2250 to 6788 2188], at Warren Farm [6666 2180] and in a large pit at Grimston Warren [678 217]. This last is a relatively recent working that had a siding which connected it to the King's Lynn to Fakenham railway.

The massive sand pits on either side of the Brow-of-the-Hill road at Leziate are the most extensive workings of their kind in Britain. They were probably begun in the early part of the 19th century as small pits for glass sand. In recent years they have been worked by Hepworth Minerals and Chemicals Ltd (formerly British Industrial Sands Ltd) for glass and foundry moulding sands and they now extend over about km². The full thickness of the Leziate Beds, about 30 m, has been worked in the pits. The sands are mostly clean, well sorted and fine grained, with patchy, irregular, ferruginous staining above the water table, and with the iron-bearing minerals glauconite and pyrite present below (Plate 6). The degree of staining, and hence the iron content, varies considerably over short distances, but the bulk of the sand above the water table is white, grey or pale yellow and relatively free from iron due to leaching by meteoric water.

Sand produced from most deposits in Britain is not suitable for use in the manufacture of colourless glass until it has been beneficiated to certain limits of iron and chromiun content and particle size. The first plant for upgrading sand, which involved more than simple washing, was installed at Leziate in 1935. Prior to this, high-grade silica sand was cheaply available from continental Europe. The plant was of great value during the Second World War when imports from Europe were curtailed. This also provided the initial stimulus to develop beneficiation plants for upgrading lower quality indigenous resources elsewhere in Britain, a trend which has continued.

The chemical requirements of the glass industry vary in detail with the product to be made but, in general, the dry sand feed must contain more than 97 per cent (for coloured containers) to more than 99 per cent (for colourless containers and flat glass) of silica (SiO_2) and less than 0.03 per cent (for colourless glass) to less than 0.25 per cent (for coloured glasses) of iron (Fe_2O_3). However, some variation in these values is acceptable, although overall consistency is of critical importance. The raw sand is upgraded by attrition scrubbing, acid leaching and froth flotation to reduce the iron content and to remove heavy minerals. The bulk of the glass sand produced at Leziate is used to make colourless glass containers. The other major product at Leziate is foundry sands, comprising well-sorted fine- and medium-grained silica-rich sands, which are extensively used in the UK and Europe for making moulds and cores. The finer grades are especially suitable for resin coating, for making small complex castings using the shell-moulding process. Current British production of glass and foundry sands is about 1.9 and 1.5 million tonnes per annum respectively. Leziate is one of the major sources of silica sand in Britain, and an important source of sand for making colourless glass containers.

Nineteenth-century sand pits at Middleton [6667 1680] and Blackborough End [6675 1490; 6726 1439] were probably worked for glass and building sand. Leziate Beds are exposed in the modern pits at Blackborough End [6755 1460], beneath Carstone and glacial sand and gravel, but they have been worked on a small scale and only for building sand.

Thin beds of clean quartzose sand occur at several levels in the Dersingham Beds, notably just below the Snettisham Clay. This horizon has been worked in the floors of the brickpits at Heacham [679 364], Ingoldisthorpe [6871 3220] and Dersingham [6865 3138], where it was presumably mixed with the Snettisham Clay to improve the handling characteristics of the raw bricks. Small pits occur in clean white sands at slightly lower stratigraphical levels in the Dersingham Beds at Drapers Hole, Heacham [6754 3698; 6754 3666], Heacham Parish Pit [6794 3676], Heacham Bottom Farm [6791 3550], Lodge Hill [6682 3408; 6677 3384], Doddshill, Dersingham [6955 3048; 6935 3014] and Folly Lodge, Sandringham [6792 2786]. This last yielded a sharp white sand that was used as an abrasive to clean the Carstone-floored terrace at Sandringham House until the 1930s. The other pits were probably dug for building sand.

There are probably considerable reserves of sand suitable for making glass or foundry moulds and cores at the surface or at relatively shallow depths in the district.

On the geological map, the boundaries of the useful material approximate to those of the Leziate Beds.

Chalk

Chalk has been worked in a large number of pits in the district, almost wholly for agricultural purposes but with some of the harder beds being used for building. Rose (1835, p.136) noted that much of the Lower Chalk of the district was hard enough to be used as a building stone and that some of it had been used, presumably internally, for ornamental architecture. Many of the pits listed below are small and were presumably worked for a single estate. Most of the larger pits are associated with a kiln, and there is evidence, most commonly in the name 'Limekiln Plantation', of the former existence of a kiln. Three pits, at Manor Road, Heacham [682 387], Hillington [723 244] and Gayton [727 197], still work chalk for agricultural lime.

Small degraded chalk pits occur in Hunstanton Park [6934 4148; 6928 4205], at Lodge Farm [6844 4082], Park Road [6743 4059], Lynn Road [6787 3979], near St Edmund's Point [6799 4228], Hunstanton, at Shernborne Hall [7087 3207] and near Church Farm, Heacham [6852 3835; 6913 3778]. Larger pits, mostly of more recent origin and with some exposure, occur at Barrett Ringstead Farm [6892 4003]; Manor Road [682 387] and Lynn Road [688 368], Heacham; Eaton [6962 3583]; Snettisham [6963 3392; 6925 3490; 6983 3321]; Ingoldisthorpe [6925 3210]; Dersingham [7013 3092]; Sandringham [6987 2936]; West Newton [7043 2775; 7062 2762]; Appleton House [7141 2722; 7149 2706]; Flitcham [7245 2663; 7294 2604]; Hillington [723 244], and Gayton [727 197]. Small degraded pits occur at Roydon [7008 2427], Grimston [7252 2123; 7288 2206; 7185 2222] and Well Hall [7220 2015].

Marl

The technique of improving sandy or acid soils by adding marl (calcareous clay) or chalk to them reputedly dates from before the Roman occupation in Britain. There is documentary evidence of marling in Norfolk from the 13th century onwards, and by the 19th century the practice was ubiquitous. In a review of marling in Norfolk, Prince (1960) has estimated that in the 19th century a 40 acre field on the sandy soils of west Norfolk would have required about 4200 tons of marl, creating a pit 10 m deep and 18 m in diameter, every 20 to 30 years.

Chalk was used for improving soils as either lump chalk or burnt lime, where no clay content was required. For the most part, the Chalky-Jurassic till and the Gault supplied the marl in the district. Old marl pits, degraded hollows mostly about 10 m deep and 20 to 60 m across, are abundant on their outcrops and locally reach concentrations of one per field. The presence of a surface layer of cryoturbated flinty sand more than one metre thick over large parts of the till outcrop, combined with the distribution of the pits, suggests that much of the marl may have been used to improve the soils of the till outcrop itself.

Young (1804) noted that oyster-rich bands in the Nar Valley Clay were locally much prized for marling in the Nar Valley.

A small number of marl pits occurs on the Hunstanton Till outcrop in the Heacham area, but there are none on the broad till outcrop between there and Hunstanton, presumably because the chalk content is too low for the till to be of use.

Ironstone

Large amounts of iron-rich slag were recorded at Rising Lodge, Mintlyn and Blackborough End during the survey; similar occurrences were already known from the Ashwicken area. One of the latter, near Holt House Farm [6830 1783], was investigated by Tylecote (1962) who found the slag to be the waste from a number of small Roman shaft furnaces that produced iron by direct smelting (as distinct from the much later blast furnace process) using wood-charcoal as a fuel. The site was dated by geomagnetic methods and pottery to the 2nd century AD. About 350 tons of slag were present and this was estimated to have resulted from the production of less than 70 tons of iron.

The source of iron ore is unknown. Tylecote (1962, p.19) thought it might have come from two marl pits adjacent to the site, but these are both in Chalky-Jurassic till in which ironstone erratics are rare. Some unroasted ore was found at that site in the form of small nodules composed mostly of limonite (72 per cent Fe_2O_3). This type of ironstone is common only in the Snettisham Clay in the district and the ore was therefore presumably obtained from the nearest outcrops at Brow-of-the-Hill and Ashwicken. Limonitic ironstones occur locally in the Leziate Beds and as modern ferruginous pan, but both occur as large irregular masses and are much less pure than the small concretionary nodules found at the site.

The history of the remaining three sites is less well known. That near Rising Lodge [6747 2254] has Roman pottery associated with it and that adjacent to Mintlyn Church [6560 1933] clearly predated the 12th century church because large amounts of the slag are incorporated in its walls. There is no evidence of the age of the Blackborough End slag [6706 1426], other than that it is lithologically similar to that at the other sites and was presumably formed by the same process. The Mintlyn site is located on Mintlyn Beds sands and clays with numerous bands of clay ironstone.

In addition to common weathered (limonitised) clay ironstone at the surface and unweathered siderite mudstone at relatively shallow depths, thick beds of recent ferruginous pan are common in the soil profile. Any of these materials could have been used for iron smelting, although the unweathered siderite would presumably have been best. The Rising Lodge and Blackborough End sites are on till and Leziate Beds respectively. The former is close to sources of sandy ironstone in the Leziate Beds and the Dersingham Beds and the latter to sandy ironstones in the Leziate Beds and Carstone, and to siderite mudstone (weathered and unweathered) and ferruginous pan on the Mintlyn Beds outcrop.

At its most iron-rich, where unweathered in the Hunstanton area, the Carstone contains sufficient 'chamosite' and limonite to be classified as a low-grade ironstone comparable to that of the Jurassic Northampton Sand. Had this been realised when such ores were in common use for iron smelting, it would no doubt have been worked.

Ironstones have formed the basis for two minor local products in the past. Red ochre was reputedly made at Ashwicken, presumably from the ironstones in the Snettisham Clay. Ferruginous (chalybeate) springs issuing from the Mintlyn Beds in the Gaywood area once had a local reputation as a medicine.

Oil shale

The first published record of oil shale[1] in Norfolk was that of William Smith on his geological map of the county (1819) where he noted that the 'Oaktree Clay' [Kimmeridge Clay] was 'part slaty and bituminous as at Kimmeridge in Dorset'. Rose (1835, p.175) subsequently recorded inflammable shales 'that burned like cannel coal' in the Kimmeridge Clay in a brickpit at Southery [probably TL 617 958] (Sheet 159), and oil seepages supposedly derived from the Kimmeridge Clay were noted in the Puny Drain [626 146] at Setchey (Forbes Leslie, 1917a).

It was largely on the evidence of these supposed seepages, and the possible prospect of finding oil shales comparable to those that had been worked at Kimmeridge, that oil-shale exploration began in Norfolk during the First World War. A pilot operation was set up at Setchey by English and Foreign Oil Finance Ltd in about 1916 with W Forbes Leslie as geologist. This work received much publicity from two papers by Forbes Leslie on the occurrence of petroleum in Britain in general (1917b) and on the occurrence in Norfolk in particular (1917a). In the latter account it was stated that at least three 6ft (1.8 m)-thick oil-shale seams were present (pp.16–17) and that two of these could be retorted to yield more than 50 gallons of oil per ton of dry shale. Up to 75 per cent of the oil was described as free-oil filling cavities in the shale (1917b, p.181); yellow sandstones impregnated with bitumen were also said to be present (1917a, p.16).

In 1918 a new company (English Oilfields Ltd) was formed with a share capital of £300 000 to exploit the Norfolk oil shales. Between 1918 and 1919 the thickness and quality of the proved oil-shale seams increased steadily in Forbes Leslie's statements to the English Oilfields shareholders. Yields of 20 to 40 gal/ton were quoted in the 1918 share prospectus, the oil containing 4.5 to 8 per cent sulphur. By September 1919 yields of 50 to 80 gal/ton of 'practically sulphur free' oil had been achieved. By December 1919 an 85 to 95 gal/ton-seam had been discovered and it was envisaged that an exceedingly profitable industrial complex would be built

which, in addition to the oil, would produce cheap Portland cement (from the shale waste and nearby Lower Chalk), high-quality bricks (from the Kimmeridge Clay overburden), electricity (from the waste gas from the retorts) and metalliferous minerals (from unspecified beds beneath the Kimmeridge Clay). In addition, it was claimed that large quantities of free oil had been discovered, together with a widespread 70ft (21.3 m)-thick seam of ozokerite (natural paraffin wax).

The shareholders of English Oilfields approved an increase in the share capital to £1.5 million in 1919 and in the next few years an opencast pit (Plate 22A) and a mine were dug, pilot retorts were operated, a rail link was built to the works and work was begun on three full-scale retorts (Plate 22B) with a view to processing 1000 tons of shale per day. The company acquired a petroleum-exploration licence covering about 1000 sq km of west Norfolk and between 1919 and 1923 drilled more than 50 cored boreholes in the area between the Gaywood River at King's Lynn and the River Wissey at Stoke Ferry.

The full-scale retorts were never completed, the mine and opencast pit were abandoned and there was little activity after about 1923. The company operated on a small scale as a wholesale distributor of oils during the 1920s and 30s and produced oil for use in medicinal soaps under the trade name Icthyol (derivative of the Swiss 'icthyammol', not a contraction of 'icthyosaur oil' as sometimes suggested). However, none of this oil came from Norfolk. The company was wound up in 1966.

Several small syndicates of local landowners were formed to explore the oil shales in their particular parts of west Norfolk at the time when the activity at English Oilfields was at its peak. None of these found seam thicknesses or yields comparable to those reported by Forbes Leslie and all the syndicates were disbanded after only a few boreholes had been drilled.

Few geological details have survived from the English Oilfields boreholes because of the secrecy maintained by the company; the Petroleum (Production) Act of 1918 excluded the necessity to report oil-exploration boreholes and the Mining Industries Act, which compelled companies to provide geological details of their activities to the Geological Survey, was not enacted until 1925. The results of some of the boreholes and the sections proved in the oil-shale mine and opencast works at Setchey, are referred to in publications by their geologist (Forbes Leslie 1917a; 1917b). Unfortunately, these accounts provide no stratigraphically useful information since the thicknesses of the oil-yielding strata are grossly exaggerated, in places by more than 100-fold.

In 1975, the BGS drilled a continuously cored borehole through the full thickness of the Kimmeridge Clay at North Runcton [6404 1624], close to the Kimmeridge Clay outcrop and about 1.5 km down dip from the former workings at Setchey. The borehole proved an unbroken sequence of Beds KC 1 to 47 to be present with groups of thin oil shales concentrated in Beds KC 26 to 46 (Figure 56). A total of 7.17 m of oil shale was proved in over 80 seams ranging from 1 to 47 cm in thickness;

---

1 The term 'oil shale' refers to mudstones that yield appreciable quantities of oil when retorted at 400° to 500°C: they contain little free oil.

Plate 22A   Opencast workings for oil shale in the Kimmeridge Clay, Setchey, c. 1920.

The opencast pit [6268 1452] worked the oil shales (thinly bedded layers in the lower face) in the Wheatleyensis Zone from beneath an overburden of calcareous mudstones in the Hudlestoni Zone and cryoturbated gravelly head (top left).

Plate 22B   Oil-shale retorts, Setchey, c. 1920.

Three retorts were started, but none was completed. The site was linked to the London–King's Lynn railway by its own siding. Construction workers' accommodation can be seen adjacent to the A10. The site was effectively abandoned in 1923.

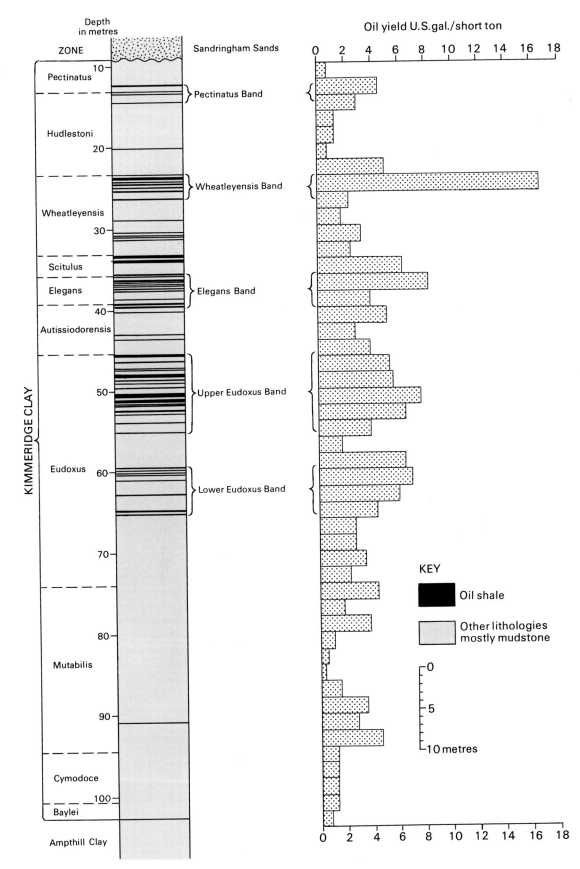

Figure 56
Stratigraphical distribution of oil-shale seams and their yields in the North Runcton Borehole.

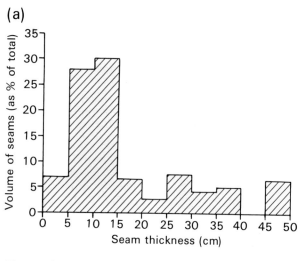

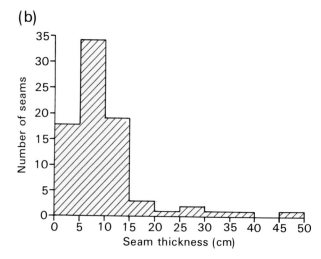

Figure 57   Distribution of oil-shale seams in the North Runcton Borehole by:
(a) volume (b) thickness.

the average seam thickness is 9.0 cm. Eighty two per cent of the seams are less than 20 cm thick and 60 per cent of the total volume of oil shale occurs in seams less than 20 cm thick. The distribution of the thicknesses of the seams and the relationship of seam thickness to volume of oil shale is shown in Figure 57.

The greatest concentrations of oil shales occur in five oil-shale-rich bands, in the lower part of the Eudoxus Zone (Bed KC 29), the upper part of the Eudoxus Zone and the lower part of the Autissiodorensis Zone (Beds KC 32 and 33), the Elegans Zone (Bed KC 36), the upper part of the Wheatleyensis Zone and the basal part of the Hudlestoni Zone (Bed KC 42), and the upper part of Hudlestoni Zone and the lower part of the Pectinatus Zone (Beds KC 45 to 47). Lesser concentrations of oil shale of generally inferior quality occur in the upper part of the Mutabilis Zone (Bed KC 19), the middle part of the Eudoxus Zone (Beds KC 30 and 31) and the Scitulus Zone (Bed KC 37) (Figure 56).

The full sequence was sampled and the potential oil yield was determined for each 2 m-run by retorting at 450°C[1]. Analyses of spot samples showed that individual oil-shale seams have potential yields of up to 40 gal/ton, but only one 2 m-run, that in Bed KC 42, had a potential yield of more than 15 gal/ton (Figure 56). The thinness of the seams, combined with the high (4 to 8 wt per cent) sulphur content of the shale oil, the cost of extraction and retorting, and the environmental problems involved in extraction, make it unlikely that the oil shales of the district will ever be worked.

---

1 Assay methods to determine the potential yield of oil shales measure the weight of oil produced by pyrolysis of a known weight of dry shale. Results are expressed in the laboratory as wt oil/wt shale (per cent), but are usually quoted in US gal. oil/US (2000 lb) ton of shale in publications. For an oil of specific gravity 1.00 (Kimmeridgian shale oils range from 0.95 to 1.05 sp. gr.) 1 wt/wt per cent is approximately 2.4 US gal/US ton.

Salt

The manufacture of salt from salt-impregnated intertidal sediments is not, strictly speaking, a geologically influenced activity, but in the present district it was carried out on such a scale that it produced large amounts of made ground and has, locally, affected the pattern of sedimentation of the Recent deposits.

A Hallam (1960) and H E Hallam (1960) have provided detailed descriptions of the medieval salt industry of the Lincolnshire Fenlands, including the western part of the present district. No account exists for Norfolk, but the field evidence suggests that similar contemporaneous salt industries existed in both counties. The method was to scrape up the surface sediment from the higher part of a silty or sandy tidal flat (the area between high-water-medial and high-water-spring tides), to leach the encrusting salt from the sediment grains to produce a brine concentrate, and to evaporate the brine in large shallow clay pans heated by peat (in Fenland) or by wood. The waste sediment was tipped in irregular heaps, known as saltern mounds, that now occupy several hundred acres, notably at Wolferton, North and South Wootton, West Lynn, Terrington and Gedney Dyke.

The age of the workings has been dated in Lincolnshire by A Hallam (1960) as probably late Saxon to 16th century; the field evidence in the present district supports this. All the saltern mounds, with the exception of those at Terrington, lie on the seaward side of the 'Roman' Bank and on the landward side of the mid 17th century sea-bank. The industry was well established by the time of the Domesday surveys (1066 and 1086). For example, large numbers of salt pans were owned by the Norfolk coastal villages of Terrington, Gaywood and Wootton (Darby, 1971, fig. 32). A large number of inland villages also owned pans and these were presumably sited on the eastern foreshore of the Great Ouse estuary. H E Hallam (1960) has suggested that the industry flourished in Lincolnshire until about the time of the Black Death (1348–49), when the sudden decline in the population combined with renewed flooding in Fenland that made the peat workings inaccessible, caused it to contract rapidly. The industry is believed to have ceased in the 15th century when cheap imported salt became readily available.

Small saltern mounds composed of silt and fine-grained silty sand occur in the reclaimed marsh area near Boathouse Creek, Dersingham [e.g. 6661 3036; 6670 2992; 6680 2984] and between Marsh Farm, Wolferton annd the Babingley River [e.g. 6477 2842; 6486 2813; 6453 2704]. The greatest concentration of saltern mounds in Norfolk occurs along the inner edge of the reclaimed marshland between the Babingley River and North Lynn. Over 70 mounds are visible and others have been landscaped beneath the housing and industrial estates of North Lynn. The mounds vary considerably in size and shape, but are all 2 to 3 m in height. Mounds less than 100 m by 100 m and more than 400 m by 200 m are rare. The largest is at North Lynn, although incomplete due to landscaping it still measures more than 500 m by 400 m; the industry is commemorated in North Lynn by the name Salter's Road. All the mounds in that area, with the exception of a small part of one mound, lie inside the early 17th century estuary bank which in places follows an erratic course to make use of the mounds as embanking material.

Many of the older buildings in West Lynn (formerly known as Old Lynn), including the 13th century church, are built on saltern mounds that lie between the Roman Bank and the former (1604) west bank of the Great Ouse estuary. North Lynn Farm, at that time also on the west bank of the river, is built on one of this group of mounds. The intervening mounds were removed when the new outfall was cut in 1852.

A group of large mounds occurs between Emorsgate and Little London, at Terrington St Clement. These all lie on the landward side of the 'Roman' Bank despite the reference to salt pans being in use in the area at the time of the Domesday surveys. It seems likely, therefore, that the bank shown on the map in that area is medieval in age and that the position of the late Saxon 'Roman' Bank lies to the south of the village, probably along the line of the King's Lynn to Long Sutton road.

Large mounds occur at Gedney Dyke close to the 'Roman' Bank on its seaward side; one mound has been built out over the bank itself. These are the eastern continuation of a group of massive mounds at Holbeach Hurn that must have resulted from one of the biggest concentrations of salt pans in Lincolnshire.

ENGINEERING PROPERTIES

There are few published data for the engineering properties of the Mesozoic and Quaternary rocks of East Anglia, but there is a relatively large amount of unpublished site-investigation data, because numerous structures and earthworks have been founded on these strata. However, much of the unpublished data are unsuitable for analysis because the tested samples lack precise stratigraphical information or have been disturbed during sampling and/or chemically modified after sampling. In some cases it is impossible to determine whether the tested samples were unweathered, weathered, or reworked as solifluted material or incorporated into some other drift deposit, because of the absence of a detailed lithological description. Therefore, the following summary of the engineering properties of the principal formations in the district is based almost wholly on the results of the combined engineering and geological studies carried out for The Wash Feasibility Study (Gallois, 1979b). The properties of samples representative of the range of Ampthill Clay and Kimmeridge Clay lithologies are shown in Table 16, those of representative drift deposits are shown in Table 17.

Soil-test data for the Ampthill Clay and Kimmeridge Clay within the district, and those published for the Gault in southern Fenland (Samuels, 1975), suggest that the engineering properties of all three formations are broadly similar. Where unweathered, they are stiff to hard, fissured, overconsolidated, normally active clays (activity range 0.75 to 1.25) with natural moisture contents mostly in the range 18 to 30 per cent. Undrained shear strengths vary inversely with moisture content from 100 kPa to about 400 kPa. Published data for the strengths of other unweathered Mesozoic clays in England down to depths of about 80 m show similar values and it seems likely that the mass soil properties are greatly influenced by fissures and shear surfaces that were produced by periglacial processes during the Pleistocene. There are few soil-test data for samples from below 80 m, but there is evidence to suggest that mechanical weathering (possibly due to permafrost) is much less below this depth, and that bulk strengths are consequently greater (greater than 1000 kPa), irrespective of stratigraphical horizon. The harder bands in the Upper Jurassic clays, mostly cementstones and oil shales, and the silty mudstones and cemented mudstones in the Gault, are stronger than most of the clays.

The engineering properties of the Sandringham Sands, Dersingham Beds, Roach and Carstone are poorly known. All these formations contain strong calcareously or ferruginously cemented bands, but because of their overall sandy and permeable nature these are mostly weathered to a loose clayey or pebbly sand above the water table. Their surface layers are much affected by cryoturbation and other weathering features.

There are no published engineering data for the Chalk of the district, but its indurated nature, particularly that of the Lower Chalk in the northern part of the district around Hunstanton, suggests that in its unweathered state it is stronger than Chalk at the same stratigraphical level elsewhere in southern England. Within the Chalk, the Paradoxica Bed, Inoceramus Bed and Melbourn Rock form significantly harder bands. As elsewhere, its bulk properties are largely governed by the degree and nature of its fracturing.

Table 16
Engineering
properties of
selected samples of
solid deposits from
the district.

Formation	Borehole	Sample depth m	Natural moisture content %	Bulk density Mg/m³	Atterberg Limits			Unconfined compressive strength kPa
					PL%	LL%	PI%	
	Ongar Hill	72.58	22	2.04	32	74	42	245
	Clenchwarton	14.15*	30	1.86	—	—	—	57
	Tilney All Saints	22.69	27	1.91	28	62	34	175
	Smeeth Lode	12.20*	34	1.83	—	—	—	72
	Walker's Marsh	50.43	21	2.10	30	70	40	325
	Emorsgate	30.17	27	1.97	29	79	50	126
	Emorsgate	37.20	26	1.90	32	70	38	600
	Emorsgate	47.14	23	1.86	31	71	40	220
KIMMERIDGE CLAY	Terrington St John	23.59	28	1.97	30	76	46	118
	Terrington St John	35.14	23	2.08	26	66	40	370
	Terrington St John	43.01	27	2.12	27	63	36	180
	New Common Marsh Farm	22.37*	24	2.03	33	70	37	72
	New Common Marsh Farm	32.79	25	2.03	31	74	43	175
	New Common Marsh Farm	39.38	19	2.09	26	60	34	550
	Ongar Hill	84.97	23	2.08	—	—	—	425
	Gedney Drove End	53.95	22	2.05	36	74	38	415
	Gedney Drove End	62.78	23	2.07	—	—	—	145
	Gedney Drove End	71.82	18	2.14	23	65	42	1010
AMPTHILL CLAY	Gedney Drove End	78.40	21	2.04	20	45	25	425
	Walker's Marsh	58.61	20	2.10	27	67	40	435
	Walker's Marsh	67.80	21	2.09	34	71	37	400
	Spice Chase	42.78	21	2.14	24	59	35	140
	Spice Chase	52.54	24	2.10	29	77	48	190
	Spice Chase	55.65	25	2.02	29	73	44	131
	Spice Chase	65.39	30	2.05	30	71	41	185

* probably affected by weathering

— not measured.

The glacial deposits of the district are laterally and vertically variable in lithology and there are too few test results available for their general properties to be reliably determined. The Chalky-Jurassic till is relatively uniform beneath large parts of Fenland; its natural moisture content in those areas is generally in the range 15 to 25 per cent and its consistency in the range stiff to hard. No comment can be made on the properties of the local glacial sand and gravel, other than to note that it is everywhere of low relative density and composed largely of flint.

The Recent deposits of Fenland are normally consolidated, soft and very soft sediments that are mostly poor foundation materials. In built-up areas such as King's

Lynn and Long Sutton, all the major buildings are built on piled foundations, which pass through the Recent deposits and are founded on the Upper Jurassic clays or on till. There are, in consequence, large numbers of boreholes and foundation pile-holes through the Recent deposits, but little attention has been paid to their engineering properties. They vary considerably in lithology both laterally and vertically, and their properties are difficult to investigate and predict. It was presumed, on the basis of experience elsewhere in Fenland, that any shafts made for The Wash Water Storage Scheme, would have to be frozen where they were dug through the Recent deposits.

Table 17 Engineering properties of selected samples of drift deposits from the district.

Formation	Borehole	Sample depth m	Natural moisture content %	Dry density Mg/m³	Bulk density Mg/m³	Atterberg Limits			Unconfined compressive strength kPa
						PL%	LL%	PI%	
BARROWAY DROVE BEDS	Clenchwarton	2.26	27	1.53	1.95	—	—	—	—
	Clenchwarton	6.50	40	1.36	1.90	—	—	—	14
	Smeeth Lode	3.20	45	1.22	1.77	—	—	—	10
	Wiggenhall	5.53	73	0.91	1.57	—	—	—	24
TILL (BOULDER CLAY)	Ongar Hill	22.3	15	1.91	2.20	—	—	—	206
	Ongar Hill	36.0	16	1.87	2.16	16	30	14	131
	Ongar Hill	43.5	15	1.91	2.19	15	33	18	257
	Ongar Hill	54.52	21	—	2.13	24	54	30	340
	Ongar Hill	62.80	18	—	2.15	19	38	19	190
	Gedney Drove End	29.23	15	—	2.21	18	40	22	470
	Gedney Drove End	41.28	16	—	2.21	16	32	16	325
	Wiggenhall	20.78	14	1.96	2.24	18	40	22	297
	Wiggenhall	21.70	24	1.61	1.99	15	29	14	169
	Walker's Marsh	29.04	12	—	2.31	15	29	14	495
	Walker's Marsh	43.13	15	—	2.06	19	39	20	215
	Spice Chase	32.79	18	—	2.15	22	48	26	220
	Tilney Fen	24.97	18	—	2.18	20	41	21	225

— not measured.

The Barroway Drove Beds are mostly soft, slightly silty or finely sandy clays riddled in their upper part with water-filled peaty root-holes that extend down from the lower surface of the Nordelph Peat. Local lenses of peat and sandy tidal creek deposits occur within the clays. Typical Barroway Drove Beds have natural moisture contents that vary from about 25 to 50 per cent, but they can be over 80 per cent where large numbers of root-holes are present. Dry densities are usually in the range of 1.3 to 1.6 Mg/m³ and are similar to those of the Terrington Beds, but shear strengths are much lower with C_u (undrained cohesion from quick undrained triaxial tests) rarely more than 20 kPa, because of the high organic and moisture contents.

The Nordelph Peat and the 'Lower' Peat vary in texture from woody to fibrous depending on the environment in which they were deposited and their distance from the uplands. Natural moisture contents in the Nordelph Peat are generally high, between 40 and 100 per cent, with values of over 350 per cent occurring in the King's Lynn area (Skempton and Petley, 1970). Dry densities are mostly less than 1.3 Mg/m³. Values for C_u vary with the texture of the peat but are rarely more than 20 kPa where they are fibrous. Soil-test data are rare for the 'Lower' Peat, but its moisture content is generally lower and its shear strength higher than that of the Nordelph Peat as a result of its greater consolidation.

The Terrington Beds are typically soft, dull reddish brown, slightly silty to silty clays with the silt concentrated in laminae. Locally the silt may be replaced by very fine-grained sand. They form a thin surface layer and samples taken for soil testing are commonly contaminated by modern worm and plant action, and have been affected by atmospheric oxidation. Natural moisture contents vary in response to the efficiency of the local artificial drainage system and seasonal conditions, but are generally in the range of 10 to 40 per cent. Strengths vary considerably in response to small amounts of artificially introduced contaminants such as peat or gravel; C_u is usually less than 40 kPa.

No data have been collected during the present survey for the smaller patches of drift deposits such as the Nar Valley Beds, blown sand and storm beach gravels. The first of these comprises lightly overconsolidated clays, sands and peats; the beach deposits are loose sands and gravels.

APPENDIX

Boreholes

The British Geological Survey holds records of over 600 bore-holes in the King's Lynn district and the information obtained from the better documented of these has been used in the preparation of the 1:50 000 geological map. The full records of all the boreholes, together with representative suites of speci-mens in the case of the cored boreholes, are held on open file at the Survey's Keyworth Office. The great majority of the bore-holes in the district were drilled for site-investigation purposes: in general, only those drilled for the Wash Water Storage Feasi-bility Study provide details of the solid geology. Summaries of the geological sections proved in the stratigraphically more use-ful boreholes, most of which are referred to in the text, are given below. The reference number in the BGS filing system is given for each borehole (e.g. TF 52 NW/4 — Symington Farm).

SHEET TF 41

TF 41 NW/15 — Tydd St Mary [4257 1553]
Drilled in 1970 for site investigation.
Ground level 2.74 m above OD.

	Thickness m	Depth m
RECENT		
Terrington Beds	2.74	2.74
Nordelph Peat	0.91	3.65
Barroway Drove Beds	5.49	9.14
Gravelly shelly sand	12.34	21.48
PLEISTOCENE		
Chalky-Jurassic till	2.14	23.62

TF 41 NW/16 — Tydd St Mary [4362 1656]
Drilled in 1970 for site investigation.
Ground level about 3 m above OD.

	Thickness m	Depth m
Made ground		1.21	1.21
RECENT		
Barroway Drove Beds	7.93	9.14
Shelly gravelly sand	10.36	19.50
PLEISTOCENE		
Varved clay	11.93	24.84

TF 41 NW/28 — Tydd St Mary [4307 1737]
Drilled in 1983 for geothermal research.
Ground level about 3 m above OD.
Rock bitted to 48.90 m; cored to 296.66 m.
Stratigraphical classification from Shephard-Thorn and Horton (1985).

	Thickness m	Depth m
QUATERNARY		
Recent deposits	c.32.8	c.32.8
Chalky-Jurassic till	c.10.2	c.43.0
JURASSIC		
Ampthill Clay	c.12.2	55.20
West Walton Beds	15.45	70.65
Oxford Clay		
Upper	19.60	90.25
Middle	22.92	113.17
Lower	13.87	127.04
Kellaways Sand	2.53	129.57
Kellaways Clay	2.09	131.66
Upper Cornbrash	3.09	134.75
Blisworth Clay	1.67	136.42
Blisworth Limestone	2.11	138.53
Upper Estuarine Beds	2.22	140.75
Lower Estuarine Beds	6.02	146.77
Lincolnshire Limestone	1.68	148.45
Lias		
Upper	32.70	181.15
Middle	8.72	189.87
Lower	106.79	296.66

TF 41 NE/2 — Tydd Gote [4508 1776]
Drilled in 1962 for site investigation.
Ground level 4.05 m above OD.

	Thickness m	Depth m
Made ground: sandy silt	3.05	3.05
RECENT		
Silt and shelly sand	19.51	22.56
PLEISTOCENE		
Chalky-Jurassic till	3.34	25.90

TF 41 NE/3 to 14 — South Holland to Walpole Island [4123 1990 to 4745 1736]
Drilled in 1963 for site investigation.
Ground levels mostly about 3 m above OD.

Boreholes 12 to 16 m deep, proving Recent deposits of sand and silty sand with shelly, gravelly sand at base in former estu-ary of River Great Ouse: thick interbeds of peaty clay (Barroway Drove Beds) and thin peat (Nordelph Peat) occur west of Tydd St Mary and thicken westwards.

TF 41 SE/6 — West Walton Highway [4913 1316]
Drilled in 1972 for Wash Feasibility Study.
Ground level 1.58 m above OD.
Rock bitted to 20.73 m; cored to 99.40 m.

	Thickness m	Depth m
RECENT		
Silty clay (Terrington Beds); soft clay with peaty inclusions (Barroway Drove Beds); sand and silt; gravelly sand at base	c.16.3	c.16.3
QUATERNARY		
Interbedded varved clay and Chalky-Jurassic till	c.27.9	44.21
JURASSIC		
Ampthill Clay: Beds 1 to 16 inclusive	24.04	68.25
West Walton Beds: Beds 1 to 16 inclusive	14.06	82.31
Upper Oxford Clay:	17.09	99.40
See Gallois (1979a) for details		

SHEET TF 42

TF 42 NE/1 — Gedney Drove End [4665 2939]
Drilled in 1972 for Wash Feasibility Study.
Ground level 3.72 m above OD.
Rock bitted to 20.13; cored to 100.44 m.

	Thickness m	Depth m
RECENT		
Silt and clay	c.18.0	c.18.0
PLEISTOCENE		
Chalky-Jurassic till	c.25.5	43.50
JURASSIC		
Ampthill Clay: Beds 1 to 29 inclusive	41.80	85.30
West Walton Beds: Beds 1 to 16 inclusive	14.50	99.80
Upper Oxford Clay:	0.64	100.44
See Gallois (1979a) for details		

TF 42 NE/2 — Offshore, near River Nene outfall [4998 2712]
Drilled in 1972 for Wash Feasibility Study.
Ground level (sea bed): 1.80 m above OD.
Bulk, disturbed and U100 samples.

	Thickness m	Depth m
RECENT		
Fine- and medium-grained sand with shelly sand and gravel at base	16.75	16.75
PLEISTOCENE		
Chalky-Jurassic till	7.45	24.20
See Gallois (1979a) for details		

TF 42 SW/1 — Poplar House [4101 2011]
Drilled in 1963 for site investigation.
Ground level about 3 m above OD.

	Thickness m	Depth m
RECENT		
Terrington Beds	2.29	2.29
Nordelph Peat	0.15	2.44
Barroway Drove Beds	1.83	4.27
Sand, shelly in lower part	7.92	12.19

TF 42 SW/4 — Long Sutton [4298 2326]
Drilled in 1897 for water.
Ground level 2.7 m above OD.

	Thickness m	Depth m
RECENT		
Terrington Beds	1.82	1.82
Silt and fine sand	12.20	14.02
Gravel with shells	3.04	17.06
PLEISTOCENE		
Chalky-Jurassic till	66.76	83.82
See Woodward (1904) for details		

TF 42 SW/7 — Sutton Crosses [4232 2120]
Drilled in 1921 for water.
Ground level 2.7 m above OD.

	Thickness m	Depth m
Recent		
Sand and shelly sand	16.8	16.8
PLEISTOCENE		
Chalky-Jurassic till	42.6	59.4
JURASSIC		
West Walton Beds and Oxford Clay	32.3	91.7

TF 42 SE/1 to 4 — Sutton Bridge [4796 2108 to 4804 2108]
Drilled in 1974 for site investigation.
Ground levels about 3m above OD.
Boreholes 10 to 20 m deep, proving 15 m to more than 20 m of Recent deposits, silts and sands on gravelly shelly sands, on Chalky-Jurassic till.

SHEET TF 43

TF 43 SW/1 — Lawyers Farm [?420 330]
Drilled in 1936 for water.
Ground level about 4.2 m above OD.

	Thickness m	Depth m
RECENT		
Mostly sand	18.9	18.9
PLEISTOCENE		
Chalky-Jurassic till	11.6	30.5
JURASSIC		
?West Walton Beds and Oxford Clay	29.2	59.7

SHEET TF 51

TF 51 NW/7 to 32 — A17 Improvement [5130 1996 to 5464 1901]
Drilled in 1975 for site investigation.
Ground levels about 2.5 to 3.0 m above OD.
Proved 6 to 16 m of Recent deposits consisting of Terrington Beds on patchy Nordelph Peat, on silty Barroway Drove Beds, on sands and gravelly shelly sands.

TF 51 NE/1 — Wiggenhall No.1 [5941 1537]
Drilled in 1971 for hydrocarbon exploration.
Ground level (Kelly bush datum) 3.84 m above OD.
Rock bitted to 561.74 m.

	Thickness m	Depth m
QUATERNARY		
Clays and stony clays	13.1	13.1
JURASSIC		
Kimmeridge Clay	60.1	73.2
Ampthill Clay	49.3	109.4
West Walton Beds	11.0	120.4
Oxford Clay	57.0	177.4
Kellaways Beds	3.4	180.8
Middle Jurassic and Upper Cornbrash	17.3	198.1
Lias and ?Rhaetic	146.3	344.4
TRIASSIC		
Sandstones, mudstones and pebble beds	90.2	434.6
?LOWER PALAEOZOIC		
Slates and tuffs	127.14	561.74

TF 51 NE/8 — Tilney All Saints [5791 1779]
Drilled in 1972 for Wash Feasibility Study.
Ground level 2.41 m above OD.
Rock bitted to 9.00 m; cored to 50.15 m.

	Thickness m	Depth m
RECENT		
Clay, silt and fine-grained sand, with thin peat beds; flint gravel at base	c.15.0	c.15.0
JURASSIC		
Kimmeridge Clay: Beds 18 to 30 inclusive	c.35.15	50.15
See Gallois (1979a) for details		

TF 51 SW/2 — Terrington St John [5401 1435]
Drilled in 1972 for Wash Feasibility Study.
Ground level 2.21 m above OD.
Rock bitted to 20.02 m; cored to 45.00 m.

	Thickness m	Depth m
RECENT		
Clay and fine-grained sand with flint gravel at base	c.12.5	c.12.5
JURASSIC		
Kimmeridge Clay: Beds 1 to 18 inclusive	c.31.6	44.10
Ampthill Clay: Beds 41 and 42	0.90	45.00
See Gallois (1979a) for details		

TF 51 SE/28 — Smeeth Lode [5838 1472]
Drilled in 1972 for Wash Feasibility Study.
Ground level 1.07 m above OD.
Bulk, disturbed and U100 samples.

	Thickness m	Depth m
RECENT		
Terrington Beds	1.20	1.20
Nordelph Peat	1.06	2.26
Barroway Drove Beds	5.24	7.50
Channel lag deposit	1.00	8.50
Lower Peat	0.40	8.90
JURASSIC		
Kimmeridge Clay	3.76	12.66
See Gallois (1979a) for details		

SHEET TF 52

TF 52 NW/3 — Walkers Marsh [5268 2530]
Drilled in 1972 for Wash Feasibility Study.
Ground level 2.12 m above OD.
Rock bitted to 14.90 m; cored to 69.18 m.

	Thickness m	Depth m
RECENT		
Clay and sand	c.13.3	c.13.3
PLEISTOCENE		
Chalky-Jurassic till	c.33.4	46.68
JURASSIC		
Kimmeridge Clay: Beds 1 to 10 inclusive	10.22	56.90
Ampthill Clay: Beds 36 to 42 inclusive	12.28	69.18
See Gallois (1979a) for details		

TF 52 NW/4 — Symington Farm [5499 2507]
Drilled in 1972 for Wash Feasibility Study.
Ground level 4.85 m above OD.
Rock bitted to 20.68 m; cored to 75.20 m.

	Thickness m	Depth m
RECENT		
Silt and fine-grained sand	c.19.3	c.19.3
PLEISTOCENE		
Chalky-Jurassic till	c.35.8	55.13
JURASSIC		
Kimmeridge Clay: Beds 1 to 12 inclusive	c.14.1	c.69.2
Ampthill Clay: Beds 40 and 41	c. 6.0	75.20
See Gallois (1979a) for details		

TF 52 NW/5 to 18 — Breast Sand [5114 2881 to 5425 2702]
Drilled in 1972 for Wash Feasibility Study.
Ground levels (sea bed) 1.3m below to 1.9 m above OD.
Boreholes 15 to 26 m deep proving 13 to 26 m of Recent
deposits, mostly sands with gravelly shelly sands in lower part,
on Kimmeridge Clay.

TF 52NW/10 — Offshore, on Breast Sand [5322 2704]
Drilled in 1972 for Wash Feasibility Study.
Ground level (sea bed): 1.30m above OD.
Bulk, disturbed and U100 samples.

	Thickness m	Depth m
RECENT		
Fine-, medium- and coarse-grained sands, shelly in part	16.00	16.00
PLEISTOCENE		
Varved clay	1.50	17.50
JURASSIC		
Kimmeridge Clay	1.45	18.95
See Gallois (1979a) for details		

TF 52 NE/1 — Offshore, near old barrier wall [5729 2916]
Drilled in 1972 for Wash Feasibility Study.
Ground level (sea bed): 1.80 m below OD.
Bulk, disturbed and U100 samples.

	Thickness m	Depth m
RECENT		
Fine- and medium-grained sand and silty sand with scattered shells	8.50	8.50
PLEISTOCENE		
Hunstanton Till	2.80	11.30
Sand and gravel	2.80	14.10
JURASSIC		
Kimmeridge Clay	2.20	16.30
See Gallois (1979a) for details		

TF 52 NE/2 — Offshore, near Hull Sand [5545 2921]
Drilled in 1972 for Wash Feasibility Study.
Ground level (sea bed): 0.40 m below OD.
Bulk, disturbed and U100 samples.

	Thickness m	Depth m
RECENT		
Fine-grained sand and silty sand with scattered shells	13.20	13.20
PLEISTOCENE		
Hunstanton Till	4.30	17.50
Sand and gravel	4.80	22.30
?Till or interglacial deposit	1.70	24.00
JURASSIC		
Kimmeridge Clay	1.00	25.00
See Gallois (1979a) for details		

TF 52 SW/2 — Emorsgate [5321 2006]
Drilled in 1972 for Wash Feasibility Study.
Ground level 2.54 m above OD.
Rock bitted to 23.00 m; cored to 49.52 m.

	Thickness m	Depth m
RECENT		
Clay and fine-grained sand with flint gravel at base	c.23.6	c.23.6
JURASSIC		
Kimmeridge Clay: Beds 6 to 18 inclusive	c.25.9	49.52
See Gallois (1979a) for details		

TF 52 SW/3 — New Common Marsh Farm [5299 2308]
Drilled in 1972 for Wash Feasibility Study.
Ground level 1.57 m above OD.
Rock bitted to 20.77 m; cored to 39.75 m.

	Thickness m	Depth m
RECENT		
Silt and fine-grained sand	c.19.4	c.19.4
JURASSIC		
Kimmeridge Clay: Beds 14 to 19 inclusive	c.20.4	39.75
See Gallois (1979a) for details		

TF 52 SE/1 — Ongar Hill [5767 2444]
Drilled in 1972 for Wash Feasibility Study.
Ground level 3.89 m above OD.
Bulk, disturbed and U100 samples to 52.00 m; cored to 88.20 m.

	Thickness m	Depth m
Topsoil: organic silt	0.40	0.40
RECENT		
Terrington Beds passing down into tidal creek deposits	16.56	16.96
?Lag deposit	4.19	21.15
PLEISTOCENE		
Chalky-Jurassic till	46.97	68.12
JURASSIC		
Kimmeridge Clay: Beds 1 to 13 inclusive	c.13.4	c.81.5
Ampthill Clay: Beds 36 to 40	c.6.7	88.20
See Gallois (1979a) for details		

TF 52 SE/2 — Clenchwarton [5988 2048]
Drilled in 1972 for Wash Feasibility Study.
Ground level 2.11 m above OD.
Bulk, disturbed and U100 samples.

	Thickness m	Depth m
RECENT		
Terrington Beds	1.30	1.30
Nordelph Peat	0.10	1.40
Barroway Drove Beds: peaty clays, and silt and sand	9.80	11.20
?Lag deposit	c.1.8	c.13.0
JURASSIC		
Kimmeridge Clay	c.1.6	14.61
See Gallois (1979a) for details		

TF 52 SE/4 — Balaclava Farm [5549 2339]
Drilled in 1972 for Wash Feasibility Study.
Ground level 3.40 m above OD.
Rock bitted to 30.00 m; cored to 80.36 m.

	Thickness m	Depth m
RECENT		
Silt and fine-grained sand	c.26.2	c.26.2
PLEISTOCENE		
Chalky-Jurassic till	c.35.3	c.61.5
JURASSIC		
Kimmeridge Clay: Beds 1 to 12 inclusive	c. 8.5	69.95
Ampthill Clay: Beds 33 to 42 inclusive	10.41	80.36
See Gallois (1979a) for details		

TF 52 SE/5 — Admirals Farm [5651 2486]
Drilled in 1972 for Wash Feasibility Study.
Ground level 4.14 m above OD.
Rock bitted to 23.71 m; cored to 71.70 m.

	Thickness m	Depth m
RECENT		
Clay, silt and fine-grained sand, with thin peat bed	c.22.0	c.22.0
PLEISTOCENE		
Chalky-Jurassic till	c.32.0	54.00
JURASSIC		
Kimmeridge Clay: Beds 2 to 16 inclusive	17.70	71.70
See Gallois (1979a) for details		

TF 52 SE/6 — Pierrepont Farm [5748 2322]
Drilled in 1972 for Wash Feasibility Study.
Ground level 3.50 m above OD.
Rock bitted to 18.28 m; cored to 83.10 m.

	Thickness m	Depth m
RECENT		
Mostly fine-grained sand	c.20.5	c.20.5
PLEISTOCENE		
Chalky-Jurassic till	c.43.3	63.81
JURASSIC		
Kimmeridge Clay: Beds 1 to 16 inclusive	18.99	82.80
Ampthill Clay: Bed 42	0.30	83.10
See Gallois (1979a) for details		

TF 52 SE/7 — Racecourse Road [5510 2194]
Drilled in 1972 for Wash Feasibility Study.
Ground level 3.49 m above OD.
Rock bitted to 21.08 m; cored to 39.45 m.

	Thickness m	Depth m
RECENT		
Silt and fine-grained sand	c.20.7	c.20.7
JURASSIC		
Kimmeridge Clay: Beds 19 to 28 inclusive	c.18.8	39.45
See Gallois (1979a) for details		

SHEET TF 53

TF 53 SE/1 — Offshore, Daseley's Sand [5591 3189]
Drilled in 1972 for Wash Feasibility Study.
Ground level (sea bed) 1.20 m below OD.
Rock bitted to 13.20 m; cored to 80.02 m.

	Thickness m	Depth m
RECENT		
Fine- and medium-grained sand	10.50	10.50
PLEISTOCENE		
Hunstanton Till	1.50	12.00
JURASSIC		
Kimmeridge Clay: Beds 1 to 30 inclusive	64.50	76.50
Ampthill Clay: Beds 40 to 42 inclusive	3.52	80.02
See Gallois (1979a) for details		

SHEET TF 54

TF 54 NE/1 — Offshore borehole IGS 71/65 [5850 4676]
Drilled in 1971 for research.
Ground level (sea bed) about 32 m below OD.
Rock bitted to 11.50 m; cored to 43.00 m.

	Thickness m	Depth m
RECENT and PLEISTOCENE		
Sand and gravel	11.50	11.50
JURASSIC		
Kimmeridge Clay: Beds 34 to 44 inclusive	31.50	43.00
See Gallois and Cox (1974) for details		

SHEET TF 61

TF 61 NW/237 — North Runcton [6404 1624]
Drilled in 1974 for oil-shale exploration.
Ground level 15.9 m above OD.
Rock bitted to 3.58 m; cored to 106.71 m.

	Thickness m	Depth m
PLEISTOCENE		
Chalky-Jurassic till and gravel	8.87	8.87
JURASSIC		
Roxham Beds	0.78	9.25
Kimmeridge Clay: Beds 1 to 46 inclusive	92.79	102.44
Ampthill Clay: Beds 41 and 42	4.27	106.71
See Gallois (1978) for details		

SHEET TF 62

TF 62 NW/2 — Vinegar Middle [6092 2505]
Drilled in 1972 for Wash Feasibility Study.
Ground level 3.20 m above OD.
Rock bitted to 12.68 m; cored to 84.26 m.

	Thickness m	Depth m
RECENT		
Silt and fine-grained sand	c.10.7	c.10.7
PLEISTOCENE		
Chalky-Jurassic till	c.54.3	64.96
JURASSIC		
Kimmeridge Clay: Beds 12 to 19 inclusive	19.30	84.26
See Gallois (1979a) for details		

TF 62 SW/63 — St Nicholas Chapel, King's Lynn [6171 2048]
Drilled in about 1812 for water.
Ground level about 3 m above OD.

	Thickness m	Depth m
RECENT		
Terrington Beds	2.1	2.1
Nordelph Peat	0.8	2.9
Barroway Drove Beds	2.4	5.3
Lower Peat	0.8	6.1
Shelly silt and clay		
PLEISTOCENE		
Chalky-Jurassic till	9.1	15.2
JURASSIC		
Oxford Clay to Kimmeridge Clay inclusive	192.1	207.3
See Whitaker (1921) for details		

TF 62 SW/96 — North Wootton [6439 2457]
Drilled in 1972 for Wash Feasibility Study.
Ground level 9.26 m above OD.
Rock bitted to 5.38 m; cored to 170.35 m.

	Thickness m	Depth m
RECENT		
Head, stony sand	5.38	5.38
CRETACEOUS		
Sandringham Sands		
Mintlyn Beds	7.52	12.90
JURASSIC		
Roxham and Runcton beds	6.85	19.75
Kimmeridge Clay: Beds 1 to 47 inclusive	99.65	119.40
Ampthill Clay: Beds 1 to 42 inclusive	50.94	170.34
See Gallois (1979a) for details		

TF 62 SW/100 — Gaywood Bridge [6480 2158]
Drilled in about 1919 for oil-shale exploration.
Ground level about m above OD.
Some cores.

	Thickness m	Depth m
RECENT		
Clay and sand	1.2	1.2
PLEISTOCENE		
Chalky-Jurassic till	27.45	28.65
JURASSIC		
Kimmeridge Clay: probably Beds 9 to 34	64.00	92.65

SHEET TF 64

TF 64 SW/1 — Offshore borehole IGS 72/78 [6494 4972]
Drilled in 1972 for research.
Ground level (sea bed) about 18 m below OD.
Rock bitted to 4.00 m; cored to 56.15 m.

	Thickness m	Depth m
RECENT		
Sand with flints	0.20	0.20
CRETACEOUS		
Lower Chalk	7.60	7.80
Red Chalk	3.20	11.00
Carstone	11.20	22.20
Sutterby Marl	1.14	23.34
Skegness Clay	1.26	24.60
Roach	21.70	46.30
Tealby Beds	9.85	56.15
See Gallois and Morter (1979) for details		

TF 64 SW/2 — Offshore borehole IGS 72/77 [6313 4835]
Drilled in 1972 for research.
Ground level (sea bed) about 29 m below OD.
Rock bitted to 6.00 m; cored to 53.20 m.

	Thickness m	Depth m
RECENT		
Cobbles and sand	0.10	0.10
CRETACEOUS		
Roach	9.41	9.51
Tealby Beds	22.59	32.10
Claxby Beds	11.00	43.10
Sandringham Sands		
Mintlyn Beds	5.20	48.30
JURASSIC		
Roxham Beds	1.72	50.02
Kimmeridge Clay	3.18	53.20
See Gallois and Morter (1979) for details		

TF 64 SW/3 — Offshore borehole IGS 71/66 [6117 4475]
Drilled in 1971 for research.
Ground level (sea bed) about 21 m below OD.
Rock bitted to 15.00 m; cored to 45.00 m.

	Thickness m	Depth m
RECENT and PLEISTOCENE		
Mostly sand	c.15.0	c.15.0
CRETACEOUS		
Sandringham Sands		
Mintlyn Beds	c.7.0	22.00
JURASSIC		
Roxham Beds	0.62	22.62
Kimmeridge Clay: Beds 42 to 47 inclusive	22.38	45.00
See Gallois and Cox (1974) for details		

TF 64 SE/11 — BGS Hunstanton [6857 4078]
Drilled in 1970 for research.
Ground level 21.3 m above OD.
Rock bitted to 4.97 m; cored to 128.93 m.

	Thickness m	Depth m
CRETACEOUS		
Middle Chalk	c.1.8	c.1.8
Lower Chalk	c.15.0	16.84
Red Chalk	1.12	17.96
Carstone	18.90	36.86
Roach	16.56	53.42
Dersingham Beds	11.60	65.02
Sandringham Sands		
Leziate Beds	22.00	87.02
Mintlyn Beds and ?Runcton Beds	9.80	96.82
JURASSIC		
Roxham Beds	6.10	102.92
Kimmeridge Clay	26.39	129.31

TF 64 SE/12 — Hunstanton No. 1 [6923 4270]
Drilled in 1969 for hydrocarbon exploration.
Ground level about 10.7 m above OD.
Rock bitted to 862.3 m.

	Thickness m	Depth m
CRETACEOUS		
Chalk and Red Chalk	21.9	21.9
Carstone	18.0	39.9
Dersingham Beds and Roach	18.3	64.0
Sandringham Sands	53.7	117.7
JURASSIC		
Kimmeridge Clay	115.9	233.6
Ampthill Clay	57.6	291.2
West Walton Beds	14.1	305.3
Oxford Clay	30.6	335.9
Kellaways Beds	4.3	340.2
Middle Jurassic and Upper Cornbrash	1.5	341.7
Lias and ?Rhaetic	189.0	530.7
TRIASSIC		
Sandstones and pebble beds	304.8	835.5
PRECAMBRIAN OR LOWER PALAEOZOIC		
Schists and gneisses	26.8	862.3

SHEET TF 71

TF 71 NW/10 — Gayton [7280 1974]
Drilled in 1970 for research.
Ground level about 26 m above OD.
Rock bitted to 4.98 m; cored to 94.71 m.

	Thickness m	Depth m
CRETACEOUS		
Chalk	12.87	12.87
Gault	9.17	22.04
Carstone	8.44	30.48
Snettisham Clay	c. 2.1	c.32.6
Sandringham Sands		
Leziate Beds	c.31.4	63.98
Mintlyn Beds	10.03	74.01
JURASSIC		
Roxham and Runcton beds	5.84	79.85
Kimmeridge Clay	14.86	94.71

REFERENCES

Most of the references listed below are held in the Library of the British Geological Survey at Keyworth, Nottingham. Copies of the references can be purchased subject to the current copyright legislation.

AGER, D V. 1971. The brachiopods of the erratic blocks of Spilsby Sandstone in Norfolk and Suffolk. *Proceedings of the Geologists' Association*, Vol. 82, 393–401.

ALLSOP, J M. 1983. Geophysical appraisal of two gravity minima in the Wash district. *Report of the Institute of Geological Sciences*, No. 83/1, 28–31.

—— 1985. Geophysical investigations into the extent of the Devonian rocks beneath East Anglia. *Proceedings of the Geologists' Association*, Vol. 96, 371–379.

—— 1987. Patterns of late Caledonian intrusive activity in eastern and northern England from geophysics, radiometric dating and basement geology. *Proceedings of the Yorkshire Geological Society*, Vol. 46, 335–53.

—— and JONES, C M. 1981. A pre-Permian palaeogeological map of the east Midlands and East Anglia. *Transactions of the Leicester Literary & Philsophical Society*, Vol. 75, 28–33.

ANDREWS, J. 1983. A faunal correlation of the Hunstanton Red Rock with the contemporaneous Gault Clay, and its implications for the environment of deposition. *Bulletin of the Geological Society of Norfolk*, No. 33, 3–25.

ARCHIAC, A d'. 1847. Rapport sur les fossiles du Tourtia. *Memoires de la Société Géologique de France*, Vol. 2, 291–351.

ARKELL, W J. 1947. The geology of the country around Weymouth, Swanage, Corfe, and Lulworth. *Memoir of the Geological Survey of Great Britain*.

ARTER, G. 1982. Geophysical investigation of the deep geology of the East Midlands. Unpublished PhD thesis, University of Leicester.

ATHERSUCH, J, HORNE, D J, and WHITTAKER, J E. 1985. G S Brady's Pleistocene ostracods from the Brickearth of the Nar Valley, Norfolk, UK. *Journal of Micropalaeontology*, Vol. 4, 153–158.

BADEN-POWELL, D F W, and WEST, R G. 1960. Summer field meeting in East Anglia. *Proceedings of the Geologists' Association*, Vol. 71, 61–80.

BARROIS, C. 1876. Recherches sur le terrain Cretace Superieur de l'Angleterre et de l'Irlande. *Memoires de la Société Géologique du Nord*.

BECKETT, S C. 1977. The Bog, Roos. 42–46 in *INQUA Congress Excursion Guide: Yorkshire and Lincolnshire*. CATT, J A (editor). (Geoabstracts: Norwich)

BIRKELUND, T, CLAUSEN, C K, HANSEN, H N, and HOLM, L. 1983. The *Hectoroceras kochi* Zone (Ryazanian) in the North Sea Central Graben and remarks on the Late Cimmerian Unconformity. *Danmarks Geologiske Undersøgelse, Arbog 1982*, 53–72, Kobenhavn 1983.

BLACK, M. 1972–75. British Lower Cretaceous Coccoliths. 1. Gault Clay. *Palaeontographical Society Monograph*, 1–142.

BOSWELL, P G H. 1927. On the distribution of purple zircon in British sedimentary rocks. *Mineralogical Magazine*, Vol. 21, 310–317.

BOTT, M H P, ROBINSON, J, and KOHNSTAMM, M A. 1978. Granite beneath Market Weighton, east Yorkshire. *Journal of the Geological Society of London*, Vol. 135, 535–543.

BOULTON, G S, JONES, A S, CLAYTON, K M, and KENNING, M J. 1977. A British ice-sheet model and patterns of glacial erosion and deposition in Britain. 231–245 in *British Quaternary studies*. SHOTTON, F W (editor). (Clarendon Press: Oxford.)

BOWEN, D Q. 1989. Amino acid geochronology of landform development in Britain. *28th International Geological Congress, Washington DC*, Vol. 1, 186–187.

BOYLAN, P J. 1967. The Pleistocene mammalia of the Sewerby-Hessle buried cliff, East Yorkshire. *Proceedings of the Yorkshire Geological Society*, Vol. 36, 115–125.

BRIGHTON, A G. 1938. The Mesozoic rocks of Cambridgeshire. 6–15 in *A scientific survey of the Cambridge district*. DARBY, H C (editor). (London: British Association for the Advancement of Science.)

BURNABY, T P. 1962. The palaeoecology of the foraminifera of the Chalk Marl. *Palaeontology*, Vol. 4, 599–608.

CAMERON, T D J, CROSBY, A, BALSON, P S, JEFFERY, D H, LOTT, G K, BULAT, J, and HARRISON, D J. 1992. *United Kingdom offshore regional report: the geology of the southern North Sea*. (London: HMSO for the British Geological Survey.)

CARTER, D J, and HART, M B. 1977. Aspects of mid-Cretaceous stratigraphical micropalaeontology. *Bulletin of the British Museum (Natural History), Geology*, Vol. 29, 1–135.

CASEY, R. 1961a. Geological age of the Sandringham Sands. *Nature, London*, Vol. 190, 1100.

—— 1961b. The stratigraphical palaeontology of the Lower Greensand. *Palaeontology*, Vol. 3, 487–621.

—— 1963. The dawn of the Cretaceous Period in Britain. *Bulletin of the South East Union of Scientific Societies*, Vol. 117, 1–15.

—— 1966. 56–59 and 102–113 in The geology of the country around Canterbury and Folkestone. SMART, J G O, BISSON, G, and WORSSAM, B C. *Memoir of the Geological Survey of Great Britain*, Sheets 289 and 305.

—— 1971. Facies, faunas and tectonics in late Jurassic–early Cretaceous Britain. 153–168 in *Faunal provinces in space and time*. MIDDLEMISS, F A, RAWSON, P F, and NEWALL, G (editors). (Liverpool: Seel House Press.)

—— 1973. The ammonite succession at the Jurassic–Cretaceous boundary in eastern England. 193–226 in *The Boreal Lower Cretaceous*. CASEY, R, and RAWSON, P F (editors). (Liverpool: Seel House Press.)

—— and GALLOIS, R W. 1973. The Sandringham Sands of Norfolk. *Proceedings of the Yorkshire Geological Society*, Vol. 41, 1–22.

—— MESEZHNIKOV, M S, and SHULGINA, N I. 1977. Correlation of the Jurassic–Cretaceous boundary deposits of England, the Russian Platform, the Subarctic Urals and Siberia. *Academia Nauk USSR, Geological Series*, Vol. 7, 14–33. (In Russian.)

CATT, J A. 1991. Late Devensian deposits in eastern England and the adjoining offshore region. 61–68 in *Glacial deposits in Great Britain*. EHLERS, J, GIBBARD, P L, and ROSE, J (editors). (Rotterdam: A A Balkema.)

— and PENNY, L F. 1966. The Pleistocene deposits of Holderness, East Yorkshire. *Proceedings of the Yorkshire Geological Society*, Vol. 35, 375–420.

CHROSTON, P N. 1985. A seismic refraction line across Norfolk. *Geological Magazine*, Vol. 122, 397–401.

— ALLSOP, J M, and CORNWELL, J D. 1987. New seismic refraction evidence on the origin of the Bouguer anomaly low near Hunstanton, Norfolk. *Proceedings of the Yorkshire Geological Society*, Vol. 46, 311–319.

— and SOLA, M A. 1975. The sub-Mesozoic floor in Norfolk. *Bulletin of the Geological Society of Norfolk*, Vol. 27, 3–19.

—— 1982. Deep boreholes, seismic refraction lines and the interpretation of the gravity anomalies in north Norfolk. *Journal of the Geological Society of London*, Vol. 139, 255–264.

CLARKE, B S. 1964. Belemnite orientation in the Hunstanton Red Rock. *Proceedings of the Geologists' Association*, Vol. 75, 345–355.

COPE, J C W. 1967. The palaeontology and stratigraphy of the lower part of the Upper Kimmeridge Clay of Dorset. *Bulletin of the British Museum (Natural History), Geology*, Vol. 15, 3–79.

— 1978. The ammonite faunas and stratigraphy of the upper part of the Upper Kimmeridge Clay of Dorset. *Palaeontology*, Vol. 21, 469–533.

CORNWELL, J D, and WALKER, A S D. 1989. Geophysical investigations. 25-51 in *Metallogenic models and exploration criteria for buried mineral deposits — a multidisciplinary study in eastern England*. PLANT, J A, and JONES, D G (editors). (Keyworth, Nottingham: British Geological Survey; London: The Institution of Mining and Metallurgy.)

COX, B M, and GALLOIS, R W. 1981. The stratigraphy of the Kimmeridge Clay of the Dorset type area and its correlation with some other Kimmeridgian sequences. *Report of the Institute of Geological Sciences*, No. 80/4.

COX, F C, GALLOIS, R W, and WOOD, C J. 1989. The geology of the country around Norwich. *Memoir of the British Geological Survey*, Sheet 161.

— and NICKLESS, E F P. 1972. Some aspects of the glacial history of central Norfolk. *Bulletin of the Geological Survey of Great Britain*, No. 42, 79–98.

CREBER, G T. 1972. Gymnospermous wood from the Kimmeridgian of East Sutherland and from the Sandringham Sands of Norfolk. *Palaeontology*, Vol. 15, 655–661.

DARBY, H C. 1971. *The Domesday geography of eastern England*. (Cambridge: Cambridge University Press.)

DE RANCE, C E. 1868. On the Albian or Gault of Folkestone. *Geological Magazine*, Vol. 5, 163–171.

— 1875. 146 and 434–6 *in* The geology of the Weald. TOPLEY, W. *Memoir of the Geological Survey of Great Britain*.

DONN, W L, FARRAND, W R, and EWING, M. 1962. Pleistocene ice volumes and sea level lowering. *Journal of Geology*, Vol. 70, 206–214.

ENGLAND, A C, and LEE, J A. 1991. Quaternary deposits of the eastern Wash margin. *Bulletin of the Geological Society of Norfolk*, No. 40, 67–99.

ERNST, G, SCHMID, F, and SEIBERTZ, E. 1983. Event stratigraphie im Cenoman und Turon von NW-Deutschland. *Zitteliana*, Vol. 10, 531–554.

EVANS, C J, and ALLSOP, J M. 1987. Some geophysical aspects of the deep geology of eastern England. *Proceedings of the Yorkshire Geological Society*, Vol. 46, 321–324.

EVANS, G. 1965. Intertidal flat sediments and the environments of deposition in The Wash. *Quarterly Journal of the Geological Society, London*, Vol. 121, 209–245.

— and COLLINS, M B. 1975. The transportation and deposition of suspended sediment over the intertidal flats of The Wash. 273–306 in *Nearshore sediment dynamics and sedimentation*. HAILS, J, and CARR, A (editors). (Chichester: J Wiley.)

EVANS, H. 1975. The two till problem in west Norfolk. *Bulletin of the Geological Society of Norfolk*, No. 27, 61–75.

FITTON, W H. 1836. Observations on some of the strata between the Chalk and the Oxford Oolite in the south-east of England. *Transactions of the Geological Society, London*, Vol. 4, 103–389.

FORBES-LESLIE, W. 1917a. The Norfolk oil-shales. *Journal of the Institute of Petroleum Technologists*, Vol. 3, 3–35.

— 1917b. The occurrence of petroleum in England. *Journal of the Institute of Petroleum Technologists*, Vol. 3, 152–190.

FOWLER, A K. 1970. Observations on iron mineral geodes in Norfolk. *Bulletin of the Geological Society of Norfolk.*, No. 18, pp.4–5.

GALE, A S. 1989. Field meeting at Folkestone Warren, 29th November 1987. *Proceedings of the Geologists' Association*, Vol. 100, 73–82.

— and FRIEDRICH, S. 1989. Occurrence of the ammonite genus *Sharpeiceras* in the Lower Cenomanian Chalk Marl of Folkestone. *Proceedings of the Geologists' Association*, Vol. 100, 80–82.

GALLOIS, R W. 1975a. The base of the Carstone at Hunstanton — Part II. *Bulletin of the Geological Society of Norfolk*, No. 27, 21–27.

— 1975b. A borehole section across the Barremian–Aptian boundary (Lower Cretaceous) at Skegness, Lincolnshire. *Proceedings of the Yorkshire Geological Society*, Vol. 41, 499–503

— 1976. Coccolith blooms in the Kimmeridge Clay and the origin of North Sea oil. *Nature, London*, Vol. 259, 473–475.

— 1978. A pilot study of oil shale occurrences in the Kimmeridge Clay. *Report of the Institute of Geological Sciences*, No. 78/13.

— 1979a. Geological investigations for the Wash Water Storage Scheme. *Report of the Institute of Geological Sciences*, No. 78/19, 1–74.

— 1979b. The Pleistocene history of west Norfolk. *Bulletin of the Geological Society of Norfolk*, No. 30, 3–38.

— 1988. The geology of the country around Ely. *Memoir of the British Geological Survey*, Sheet 189.

— and COX, B M. 1974. Stratigraphy of the Upper Kimmeridge Clay of the Wash area. *Bulletin of the Geological Survey of Great Britain*, No. 47, 1–28.

— — 1976. The stratigraphy of the Lower Kimmeridge Clay of eastern England. *Proceedings of the Yorkshire Geological Society*, Vol. 41, 13–26.

— — 1977. The stratigraphy of the Middle and Upper Oxfordian sediments of Fenland. *Proceedings of the Geologists' Association*, Vol. 88, 207–228.

— and MEDD, A W. 1979. Coccolith-rich marker bands in the English Kimmeridge Clay. *Geological Magazine*, Vol. 116, 247–334.

— and MORTER, A A. 1979. The Upper Jurassic and Cretaceous stratigraphy of IGS Boreholes 72/77 and 72/78. *Report of the Institute of Geological Sciences*, No. 78/18, 24–31.

— — 1982. The stratigraphy of the Gault of East Anglia. *Proceedings of the Geologists' Association*, Vol. 93, 351–368.

GIBBARD, P L, WEST, R G, ANDREW, R, and PETTIT, M. 1991. Tottenhill, Norfolk. 131–143 in *Central East Anglia and the Fen Basin field guide*. LEWIS, S G, WHITEMAN, C A, and BRIDGLAND, D R (editors). (London: Quaternary Research Association.)

— — — — 1992. The margin of a Middle Pleistocene ice advance at Tottenhill, Norfolk, England. *Geological Magazine*, Vol. 129, 59–76.

GOLBERT, A V, ZAKHAROV, V A, KLIMOVA, I G, and ROMANOVA, E E. 1972. Main sections of the Berriasian Stage in the Boreal realm within the USSR; Cis-Polar Trans-Urals. 56-71 in *The Jurassic–Cretaceous boundary and the Berriasian Stage in the Boreal realm*. SAKS, V N (editor), HARDIN, H (translator, 1975). (Jerusalem: Israel Programme for Scientific Translation.)

GUNN, J. 1866. Geology of Norfolk. *Geological Magazine*, Vol. 3, 258.

HALLAM, A. 1960. The medieval salt industry in the Lindsey Marshland. *Lincolnshire Architectural and Archaeological Society*, Vol. 8, 76–84.

HALLAM, H E. 1960. Salt-making in the Lincolnshire Fenland during the Middle Ages. *Lincolnshire Architectural and Archaeological Society*, Vol. 8, 85–112.

HANCOCK, J M. 1976. The petrology of the Chalk. *Proceedings of the Geologists' Association*, Vol. 86, 499–535.

— and KAUFFMAN, E G. 1979. The great transgressions of the Late Cretaceous. *Journal of the Geological Society of London*, Vol. 136, 175–186.

HARMER, F W. 1877. *The testimony of the rocks in Norfolk.* (London.)

HARRISON, R K, JEANS, C V, and MERRIMAN, R. 1979. Mesozoic igneous rocks, hydrothermal mineralisation and volcanigenic sediments in Britain and adjacent regions. *Bulletin of the Geological Survey of Great Britain*, No. 70, 57–69.

HATCH, F H, and RASTALL, R H. 1913. *The petrology of the sedimentary rocks.* (London: Allen and Unwin.)

HAWKES, L. 1943. The erratics of the Cambridge Greensand — their nature, provenance and mode of transport. *Quarterly Journal of Geological Society of London*, Vol. 99, 93–104.

HILL, W. 1900. Microscopical structure and mineral ingredients of Gault and Red Chalk. 330–366 in The Cretaceous rocks of Britain. Vol. 1, The Gault and Upper Greensand of England. JUKES-BROWNE, A J. *Memoir of the Geological Survey of Great Britain.*

HORTON, A. 1970. The drift sequence and subglacial topography of parts of the Ouse and Nene basin. *Report of the Institute of Geological Sciences*, No. 70/9.

— 1989. The geology of the country around Peterborough. *Memoir of the British Geological Survey*, Sheet 158.

INGLIS, C C, and KESTNER, F J T. 1958. Changes in The Wash as affected by training walls and reclamation works. *Proceedings of the Institution of Civil Engineers*, Vol. 11, 435–466.

INSTITUTE OF GEOLOGICAL SCIENCES. 1981. 1:250 000 UK and Continental Shelf Series. Bouguer gravity anomaly map. East Anglia, Sheet 52° N 0°W.

— 1982. 1:250 000 UK and Continental Shelf Series. Aeromagnetic anomaly map. East Anglia. Sheet 52°N 0°W.

JACKSON, J F. 1911. *The rocks of Hunstanton and its neighbourhood* (2nd edition). (London: Premier Press.)

JEANS, C V. 1968. The origin of the montmorillonite of the European Chalk with special reference to the Lower Chalk of England. *Clay Minerals*, Vol. 7, 311–330.

— 1973. The Market Weighton structure: tectonics, sedimentation and diagenesis during the Cretaceous. *Proceedings of the Yorkshire Geological Society*, Vol. 39, 409–444.

— 1980. Early submarine lithification in the Red Chalk and Lower Chalk of eastern England: a bacterial control model and its implications. *Proceedings of the Yorkshire Geological Society*, Vol. 43, 81–157.

JEFFERIES, R P S. 1963. The stratigraphy of the *Actinocamax plenus* Subzone (Turonian) in the Anglo-Paris Basin. *Proceedings of the Geologists' Association*, Vol. 74, 1–33.

JUKES-BROWNE, A J. 1887. Geology of east Lincolnshire. *Memoir of the Geological Survey of Great Britain.*

— 1900. The Cretaceous rocks of Britain. The Gault and Upper Greensand of England. *Memoir of the Geological Survey of Great Britain.*

— 1903. The Cretaceous rocks of Britain. Vol. 2, The Lower and Middle Chalk. *Memoir of the Geological Survey of Great Britain.*

— and HILL, W. 1887. On the lower part of the Upper Cretaceous Series in West Suffolk and Norfolk. *Quarterly Journal of the Geological Society of London*, Vol. 43, 544–597.

KAUFFMAN, E G. 1976. British middle Cretaceous inoceramid biostratigraphy. *Annals of the Museum of Natural History, Nice*, Vol. 4, 1–12.

KEEN, D H, JONES, R L, and ROBINSON, J E. 1984. A late Devensian fauna and flora from Kildale, north east Yorkshire. *Proceedings of the Yorkshire Geological Society*, Vol. 44, 385–397.

KEEPING, W. 1883. *The fossils and palaeontological affinities of the Neocomian deposits of Upware and Brickhill.* (Cambridge: Cambridge University Press.)

KELLY, S R A. 1977. *The bivalves of the Spilsby Sandstone Formation and contiguous deposits.* Unpublished PhD thesis, University of London.

— 1984–92. Bivalvia of the Spilsby Sandstone and Sandringham Sands (late Jurassic–early Cretaceous) of eastern England. *Palaeontographical Society Monograph*, 2 Vols., 1–123.

KENNEDY, W J. 1970. Trace fossils in the Chalk environment. 263–282 in Trace fossils. CRIMES, T P, and HARPER, J C (editors). *Special Issue, Geological Journal*, No. 3.

— 1971. Cenomanian ammonites from Southern England. *Special Papers in Palaeontology*, No. 8, 1–133.

KENT, P E. 1947. A deep boring at North Creake, Norfolk. *Geological Magazine*, Vol. 84, 2–18.

— 1962. A borehole to basement rocks at Glinton, near Peterborough, Northants. *Proceedings of the Geological Society of London*, No. 1595, 40–42.

KESTNER, F J T. 1962. The old coastline of The Wash. *Geographical Journal*, Vol. 28, 457–478.

KITCHIN, F L, and PRINGLE, J. 1922. On the overlap of the Upper Gault in England and on the Red Chalk of the eastern counties. *Geological Magazine*, Vol. 59, 194–198.

— — 1932. The stratigraphical relations of the Red Rock at Hunstanton. *Geological Magazine*, Vol. 64, 29–41.

KLIMOVA, I G. 1969. The early Berriasian of western Siberia. *Geologiya i Geofizika*, No.4.

LAING, L. 1979. *Celtic Britain.* (London: Routledge and Keegan Paul.)

LAMBECK, K. 1991. Glacial rebound and sea-level change in the British Isles. *Terra Nova*, Vol. 3, 379–389.

LAMPLUGH, G W. 1888. Report on the buried cliff at Sewerby, near Bridlington. *Proceedings of the Yorkshire Geological Society*, Vol. 9, 381–392.

LARWOOD, G P. 1961. The Lower Cretaceous deposits of Norfolk. 280–292 in The geology of Norfolk. LARWOOD, G P, and FUNNELL, B M (editors). *Transactions of the Norfolk and Norwich Naturalists' Society*, 280–292.

LEE, M K, PHARAOH, T C, and SOPER, N J. 1990. Structural trends in central Britain from images of gravity and aeromagnetic fields. *Journal of the Geological Society of London*, Vol. 147, 241–258.

LE STRANGE, H. 1974. Notes on Hunstanton Red Rock fossils. *Bulletin of the Geological Society of Norfolk*, No. 26, 47–48.

LINSSER, H. 1968. Transformation of magnetometric data into tectonic maps by Digital Template Analysis. *Geophysical Prospecting*, Vol. 16, 179–207.

LORD, A R, and ROBINSON, J E. 1978. Marine Ostracoda from the Quaternary Nar Valley Clay, West Norfolk. *Bulletin of the Geological Society of Norfolk*, No. 30, 113–118.

— HORNE, D J, and ROBINSON, J E. 1988. An introductory guide to the Neogene and Quaternary of East Anglia for ostracod workers. *British Micropalaeontological Society Field Guide*, No. 5. 10pp.

LOWE, J. 1868. The geology of the environs of Norwich. *Report of the British Association*, Vol. 10, 49–73.

MITCHELL, G F, PENNY, L F, SHOTTON, F W, and WEST, R G. 1973. A correlation of Quaternary deposits in the British Isles. *Special Report of the Geological Society of London*, No. 4.

MITLEHNER, A G. 1992. Palaeoenvironments of the Hoxnian Nar Valley Clay, Norfolk, England: evidence from an integrated study of diatoms and ostracods. *Journal of Quaternary Science*, Vol. 7, 335–341.

MORTER, A A. 1975. A Barremian fauna from excavations at Hunstanton beach. *Bulletin of the Geological Society of Norfolk*, No. 27, 29–32.

— and WOOD, C J. 1983. The biostratigraphy of Upper Albian–Lower Cenomanian *Aucellina* in Europe. *Zitteliana*, Vol. 10, 515–529.

MORTIMORE, R N. 1979. The relationship of stratigraphy and tectonofacies to the physical properties of the White Chalk of Sussex. Unpublished PhD thesis: Brighton Polytechnic.

OWEN, E F, RAWSON, P F, and WHITHAM, F. 1968. The Carstone (Lower Cretaceous) of Melton, East Yorkshire, and its brachiopod fauna. *Proceedings of the Yorkshire Geological Society*, Vol. 36, 513–524.

— and THURRELL, R G. 1968. British Neocomian rhynchonelloid brachiopods. *Bulletin of the British Museum (Natural History), Geology*, Vol. 16, No. 3, 101–103.

OWEN, H G. 1971. Middle Albian stratigraphy in the Anglo-Paris Basin. *Bulletin of the British Museum (Natural History) Geology*, Supplement 8, 1–164.

— 1976. The stratigraphy of the Gault and Upper Greensand of the Weald. *Proceedings of the Geologists' Association*, Vol. 86 (for 1975), 475–498.

— 1979. Ammonite zonal stratigraphy in the Albian of North Germany and its setting in the Hoplitinid Faunal

Province. 563–588 in *Aspekte der Kreide Europas*. (Stuttgart: IUGS.)

PEAKE, N B, and HANCOCK, J M. 1961. The Upper Cretaceous of Norfolk. *Transactions of the Norfolk and Norwich Naturalists' Society*, Vol. 19, 293–339.

PENN, I E, COX, B M, and GALLOIS, R W. 1986. Towards precision in stratigraphy: geophysical log correlation of Upper Jurassic (including Callovian) strata of the Eastern England Shelf. *Journal of the Geological Society of London*, Vol. 143, 381–410.

PENNY, L F, COOPE, G R, and CATT, J A. 1969. Age and insect fauna of the Dimlington Silts, east Yorkshire. *Proceedings of the Geologists' Association*, Vol. 224, 65–67.

PERRIN, R M S. 1971. *The clay mineralogy of British sediments.* (London: Mineralogical Society.)

PHARAOH, T C, MERRIMAN, R J, WEBB, P C, and BECKINSALE, R D. 1987. The concealed Caledonides of eastern England: preliminary results of a multidisciplinary study. *Proceedings of the Yorkshire Geological Society*, Vol. 46, 355–369.

PHILLIPS, C W (editor). 1970. *The Fenland in Roman times.* (London: Royal Geographical Society.)

PINCKNEY, G. 1978. *The belemnite genus Acroteuthis in the late Jurassic and early Lower Cretaceous of NW Europe.* Unpublished PhD thesis, University of London.

— and RAWSON, P F. 1974. *Acroteuthis* assemblages in the Upper Jurassic and Lower Cretaceous of northwest Europe. *Newsletters in Stratigraphy*, Vol. 3, 193–204.

PRICE, F G H. 1874. On the Gault of Folkestone. *Quarterly Journal of the Geological Society of London*, Vol. 30, 342–368.

— 1876. On the Lower Greensand and Gault of Folkestone. *Proceedings of the Geologists' Association*, Vol. 4, 135–150.

PRICE, R J. 1977. The stratigraphical zonation of the Albian sediments of north-west Europe, as based on foraminifera. *Proceedings of the Geologist's Association*, Vol. 88, 65–91.

PRINCE, H C. 1960. Pits and ponds in Norfolk. *Erdkunde*, Band XVl, 10–34.

RASTALL, R H. 1919. The mineral composition of the Lower Greensand strata of Eastern England. *Geological Magazine*, Vol. 56, 211–220 and 265–272.

— 1925. On the tectonics of the southern Midlands. *Geological Magazine*, Vol. 62, 193–222.

— 1930. The petrography of the Hunstanton Red Rock. *Geological Magazine*, Vol. 67, 436–458.

ROSE, C B. 1835–36. A sketch of the geology of west Norfolk. *Philosophical Magazine*, Series 3, Vol. 7, 171–82, 274–79, 370–76; Vol. 8, 28–42.

— 1862. On the Cretaceous group in Norfolk. *Proceedings of the Geologists' Association*, Vol. 1, 234–36.

ROSE, J. 1991. Stratigraphic basis of the term 'Wolstonian Glaciation', and the retention of the term 'Wolstonian' as a chronostratigraphic stage name — a discussion. 15–20 in *Central East Anglia and the Fen Basin field guide*. LEWIS, S G, WHITEMAN, C A, and BRIDGLAND, D R (editors). (London: Quaternary Research Association.)

SAMUELS, S G. 1975. Some properties of the Gault Clay from the Ely–Ouse Essex Water tunnel. *Geotechnique*, Vol. 25, 239–264.

SCHWARZACHER, W. 1953. Cross-bedding and grain size in the Lower Cretaceous sands of East Anglia. *Geological Magazine*, Vol. 90, 322–330.

SEALE, R S. 1975. Soils of the Ely district. *Memoir of the Soil Survey of Great Britain.*

SEDGWICK, A. 1846. On the geology of the neighbourhood of Cambridge, including the formations between the Chalk escarpment and the Great Bedford Level. *Report of the British Association* [for 1845], 40.

SEELEY, H G. 1861a. Notice of the Elsworth and other new rocks in the Oxford Clay, and of the Bluntisham Clay above them. *Geologist,* Vol. 4, 460–461.

—— 1861b. Notice on the opinions of the stratigraphical position of the red limestone of Hunstanton. *Annual Magazine of Natural History,* Series 3, Vol. 7, 233–44.

—— 1864a. On the fossils of the Hunstanton Red Rock. *Annual Magazine of Natural History,* Series 3, Vol. 14, 276–80.

—— 1864b. On the Hunstanton Red Rock. *Quarterly Journal of the Geological Society of London,* Vol. 20, 327–32.

—— 1866. Notice of *Torynocrinus* and other new and little-known fossils from the Upper Greensand of Hunstanton, commonly called the Hunstanton Red Rock. *Annual Magazine of Natural History,* Vol. 17, 173–183.

SHEPHARD-THORN, E R, and HORTON, A. 1985. Tydd St Mary Borehole. *Report of the British Geological Survey,* Vol. 16, No. 11, p.2.

SKEMPTON, A W, and PETLEY, D J. 1970. Ignition loss and other properties of peats and clays from Avonmouth, King's Lynn and Cranberry Moss. *Geotechnique,* Vol. 20, 343–356.

SKERTCHLY, S B J. 1877. The geology of Fenland. *Memoir of the Geological Survey of Great Britain.*

SMITH, N J P (compiler). 1985. Map 1: Pre-Permian geology of the United Kingdom (South). 1 to 1M scale. (Keyworth, Nottingham: British Geological Survey.)

SMITH, W. 1815a. *Geological map of England and Wales.* (London: J Cary.)

—— 1815b. *A memoir to the map delineation of the strata of England and Wales, with part of Scotland.* (London: J Cary.)

—— 1819. *Geological map of Norfolk.* (London: J Cary.)

SOLA, M, and CHROSTON, P N. 1978. Gravity measurements in the Wash by the University of East Anglia. *Bulletin of the Geological Society of Norfolk,* No. 30, 105–112.

SPAETH, C. 1971. Untersuchungen am Belemniten des formen kreises um *Neohibolites minimus* (Miller, 1826) aus dem Mittel und Ober Alb Nordwestdeutchlands. *Beihefte Zum Geologischen Jahrbuch,* Vol. 100, 1–127.

—— 1973. *Neohibolites ernsti* and its occurrence in the Upper Albian of northwest Germany and England. 361–368 in The Boreal Lower Cretaceous. CASEY R, and RAWSON, P F (editors). *Special Issue of the Geological Journal,* No. 5. (Liverpool: Seel House Press.)

SPARKS, B W, and WEST, R G. 1965. The relief and drift deposits. 18–40 in *The Cambridge region.* STEERS, J A (editor). (London: British Association for the Advancement of Science.)

SPATH, L F. 1923a. Excursion to Folkestone with notes on the zones of the Gault. *Proceedings of the Geologists' Association,* Vol. 34, 70–76.

—— 1923b. On the ammonite horizons of the Gault and contiguous deposits. *Summary of Progress of the Geological Survey for 1922,* 139–149.

—— 1923–43. A monograph of the ammonoidea of the Gault. *Palaeontographical Society Monograph,* 2 Vols., 1–787.

—— 1924. On the ammonites of the Speeton Clay and the sub-divisions of the Neocomian. *Geological Magazine,* Vol. 61, 73–89.

—— 1926. On the zones of the Cenomanian and the uppermost Albian. *Proceedings of the Geologists' Association,* Vol. 37, 420–32.

—— 1930. On some ammonoidea from the Lower Greensand. *Annual Magazine, Natural History,* Vol. 10, 422.

—— 1947. Additional observations on the invertebrates (chiefly ammonites) of the Jurassic and Cretaceous of East Greenland. 1: The *Hectoroceras fauna* of Southwest Jameson Land. *Meddelelser om Gronland,* Vol. 132, No. 3, 1–69.

STEAD, I M. 1991. The Snettisham Treasure: excavations in 1990. *Antiquity,* Vol. 65, 447–465.

STEVENS, L A. 1960. The interglacial of the Nar Valley, Norfolk. *Quarterly Journal of the Geological Society of London,* Vol. 115, 291–315.

STRAW A. 1960. The limit of the 'last' glaciation in north Norfolk. *Proceedings of the Geologists' Association,* Vol. 71, 379–390.

—— 1979. The geomorphological significance of the Wolstonian glaciation of eastern England. *Transactions of the Institute of British Geographers,* Vol. 4, 540–549.

—— 1983. Pre-Devensian glaciation of Lincolnshire (eastern England) and adjacent areas. *Quaternary Science Reviews,* Vol. 2, 239–260.

STUBBLEFIELD, C J. 1967. Some results of a recent Geological Survey boring in Huntingdonshire. *Proceedings of the Geological Society of London,* No. 1637, 35–44.

SUMBLER, M. 1983. A new look at the type Wolstonian glacial deposits of central England. *Proceedings of the Geologists' Association,* Vol. 94, 23–31.

SWINNERTON, H H. 1935. The rocks below the Red Chalk of Lincolnshire and their cephalopod faunas. *Quarterly Journal of the Geological Society of London,* Vol. 91, 1–46.

—— 1936–55. A monograph of British Lower Cretaceous belemnites. *Palaeontographical Society Monograph,* 1–86.

SYKES, R M, and CALLOMON, J H. 1979. The *Amoeboceras* zonation of the Boreal Upper Oxfordian. *Palaeontology,* Vol. 22, 839–903.

TAYLOR, J H. 1949. The Mesozoic ironstones of England: petrology of the Northampton Sand ironstone formation. *Memoir of the Geological Survey of Great Britain.*

TAYLOR, R. 1823. Geological section of Hunstanton cliff, Norfolk. *Philosophical Magazine,* Vol. 61, 81–83.

TEALL, J J H. 1875. *The Potton and Wicken phosphatic deposits. Sedgwick Prize Essay for 1873.* (Cambridge: Cambridge University Press.)

TORRENS, H S, and CALLOMON, J H. 1968. The Corallian Beds, the Ampthill Clay and the Kimmeridge Clay. 291–299 in *The geology of the East Midlands.* SYLVERSTER-BRADLEY, P C, and FORD, T D (editors). (Leicester: Leicester University Press.)

TRIMMER, J. 1846. On the geology of Norfolk. *Journal of the Royal Agricultural Society,* Vol. 7, 449.

TYLECOTE, R F. 1962. Roman shaft furnaces in Norfolk. *Journal of the Iron and Steel Institute,* Vol. 200, 19–22.

WEST, R G. 1991. On the origin of Grunty Fen and other landforms in southern Fenland, Cambridgeshire. *Geological Magazine,* Vol. 128, 257–262.

—— and DONNER, J J. 1956.   The glaciations of East Anglia and the East Midlands.   *Quarterly Journal of the Geological Society of London*, Vol. 111, 69–91.

WHITAKER, W. 1883.   Excursion to Hunstanton.   *Proceedings of the Geologists' Association*, Vol. 8, 133.

—— 1921.   The water supply of Norfolk from underground sources.   *Memoir of the Geological Survey of Great Britain.*

—— and JUKES-BROWNE, A J. 1899.   The geology of the Borders of The Wash.   *Memoir of the Geological Survey of Great Britain.*

—— SKERTCHLEY, S B J, and JUKES-BROWNE, A J. 1893.   The geology of south western Norfolk and northern Cambridgeshire.   *Memoir of the Geological Survey of Great Britain.*

WHITCOMBE, D N, and MAGUIRE, P K H. 1980.   An analysis of the velocity structure of the Precambrian rocks of Charnwood Forest.   *Geophysical Journal of the Royal Astronomical Society*, Vol. 63, 405–416.

WIEDMANN, J. 1968.   Das Problem stratigraphischer Grenzziehung und die Jura-Kreide Grenze.   *Eclogae Geologia Helvetica*, Vol. 61.

WILKINSON, I P, and MORTER, A A. 1982.   The biostratigraphical zonation of the East Anglian Gault by Ostracoda.   163–175 in *Microfossils from recent and fossil shelf areas.* NEALE, J W, and BRASIER, M M (editors). (Chichester: Ellis Horwood.)

WILLIS, E H. 1961.   Marine transgression sequences in the English Fenlands.   *Annals of the New York Academy of Sciences*, Vol. 95, 368–376.

WILLS, L J. 1978.   A palaeogeological map of the Lower Palaeozoic floor beneath the cover of Upper Devonian, Carboniferous and later formations.   *Memoir of the Geological Society of London*, No. 8.

WILSON, A A, and CORNWELL, J D. 1982.   The IGS borehole at Beckermonds Scar, north Yorkshire.   *Proceedings of the Yorkshire Geological Society*, Vol. 44, 59-88.

WILTSHIRE, T. 1859.   On the Red Chalk of England. Geologist, Vol. 2, 261–278.

—— 1869.   Geological excursion to Hunstanton.   *Geological Magazine*, Vol. 6, 427–30.

WINGFIELD, R T R. 1990.   Glacial incisions indicating Middle and Upper Pleistocene ice limits off Britain.   *Terra Nova*, Vol. 1, 538–548.

—— EVANS, C D R, DEEGAN, S E, and FLOYD, R. 1979.   Geological and geophysical survey of The Wash.   *Report of the Institute of Geological Sciences*, No. 78/18, 1–23.

WOODLAND, A W. 1970. The buried tunnel-valleys of East Anglia.   *Proceedings of the Yorkshire Geological Society*, Vol. 37, 521-578.

WOODWARD, H B. 1895.   The Jurassic rocks of Britain. Vol. 5, The Middle and Upper Oolitic rocks.   *Memoir of the Geological Survey of Great Britain.*

—— 1904.   The water supply of Lincolnshire from underground sources.   *Memoir of the Geological Survey of Great Britain.*

WOODWARD, S. 1833.   *An outline of the geology of Norfolk.* (Norwich.)

WORSSAM, B C, and IVIMEY-COOK, H C. 1971.   The stratigraphy of the Geological Survey borehole at Warlingham, Surrey.   *Bulletin of the Geological Survey of Great Britain*, No. 36, 1–144.

WRAY, D S, and GALE, A S. 1992.   Geochemical correlation of marl bands in Turonian chalks of the Anglo-Paris Basin.   *Geological Society Special Publication*, No. 70, 211–266.

WYMER, J J. 1985.   *The Palaeolithic sites of East Anglia.* (Norwich: Geobooks.)

YOUNG, A. 1804.   *General view of the agriculture of the county of Norfolk.* (London.)

ZIEGLER, B. 1962.   Die Ammoniten-Gattung *Aulacostephanus* im Oberjura (Taxonomie, Stratigraphie, Biologie).   *Palaeontographica*, Vol. 119A, 1–172.

INVENTORY OF CITED FOSSILS

Latinised names only are listed. Species are listed alphabetically regardless of any qualification (e.g. cf., aff., subgenera), and are followed by less specific determinations. Authors of species names are given at the first citation in the text.

GENERAL INDEX

BRITISH GEOLOGICAL SURVEY

Keyworth, Nottingham NG12 5GG
0115-936 3100

Murchison House, West Mains Road, Edinburgh
EH9 3LA   0131-667 1000

London Information Office, Natural History Museum
Earth Galleries, Exhibition Road, London SW7 2DE
0171-589 4090

The full range of Survey publications is available through the Sales Desks at Keyworth and at Murchison House, Edinburgh, and in the BGS London Information Office in the Natural History Museum (Earth Galleries). The adjacent bookshop stocks the more popular books for sale over the counter. Most BGS books and reports can be bought from HMSO and through HMSO agents and retailers. Maps are listed in the BGS Map Catalogue, and can be bought together with books and reports through BGS-approved stockists and agents as well as direct from BGS.

The British Geological Survey carries out the geological survey of Great Britain and Northern Ireland (the latter as an agency service for the government of Northern Ireland), and of the surrounding continental shelf, as well as its basic research projects. It also undertakes programmes of British technical aid in geology in developing countries as arranged by the Overseas Development Administration.

The British Geological Survey is a component body of the Natural Environment Research Council.

HMSO publications are available from:

HMSO Publications Centre
(Mail, fax and telephone orders only)
PO Box 276, London SW8 5DT
Telephone orders 0171-873 9090
General enquiries 0171-873 0011
Queuing system in operation for both numbers
Fax orders 0171-873 8200

HMSO Bookshops
49 High Holborn, London WC1V 6HB
(counter service only)
0171-873 0011   Fax 0171-831 1326
68–69 Bull Street, Birmingham B4 6AD
0121-236 9696   Fax 0121-236 9699
33 Wine Street, Bristol BS1 2BQ
0117-9264306   Fax 0117-9294515
9 Princess Street, Manchester M60 8AS
0161-834 7201   Fax 0161-833 0634
16 Arthur Street, Belfast BT1 4GD
01232-238451   Fax 01232-235401
71 Lothian Road, Edinburgh EH3 9AZ
0131-228 4181   Fax 0131-229 2734

HMSO's Accredited Agents
(see Yellow Pages)

And through good booksellers